Web Mobile-Based Applications for Healthcare Management

Latif Al-Hakim
University of Southern Queensland, Australia

IRM Press
Publisher of innovative scholarly and professional
information technology titles in the cyberage

Hershey • London • Melbourne • Singapore

Acquisition Editor:	Kristin Klinger
Senior Managing Editor:	Jennifer Neidig
Managing Editor:	Sara Reed
Assistant Managing Editor:	Sharon Berger
Development Editor:	Kristin Roth
Copy Editor:	Shanelle Ramelb
Typesetter:	Michael Brehm
Cover Design:	Lisa Tosheff
Printed at:	Integrated Book Technology

Published in the United States of America by
IRM Press (an imprint of Idea Group Inc.)
701 E. Chocolate Avenue, Suite 200
Hershey PA 17033-1240
Tel: 717-533-8845
Fax: 717-533-8661
E-mail: cust@idea-group.com
Web site: http://www.irm-press.com

and in the United Kingdom by
IRM Press (an imprint of Idea Group Inc.)
3 Henrietta Street
Covent Garden
London WC2E 8LU
Tel: 44 20 7240 0856
Fax: 44 20 7379 0609
Web site: http://www.eurospanonline.com

Library of Congress Cataloging-in-Publication Data

Web mobile-based applications for healthcare management / [edited] by Latif Al-Hakim.
     p. ; cm.
Includes bibliographical references and index.
Summary: "This book addresses the difficult task of managing admissions and waiting lists while ensuring quick and convincing response to unanticipated changes of the clinical needs. It tackles the limitations of traditional systems, taking into consideration the dynamic nature of clinical needs, scarce resources, alternative strategies, and customer satisfaction in an environment that imposes unexpected deviation from planned activities"--Provided by publisher.
  ISBN 1-59140-658-7 (hardcover) -- ISBN 1-59140-659-5 (softcover)
  1. Health facilities--Administration--Data processing. 2. Hospitals--Administration--Data processing. 3. Information storage and retrieval systems--Hospitals. 4. Internet in medicine. I. Al-Hakim, Latif, 1946-
  [DNLM: 1. Medical Records Systems, Computerized. 2. Cellular Phone. 3. Computers, Handheld. 4. Internet. WX 173 W364 2007]
  RA971.6.W43 2007
  362.110285--dc22
                              2006033666

British Cataloguing in Publication Data
A Cataloguing in Publication record for this book is available from the British Library.

All work contributed to this book is new, previously-unpublished material. The views expressed in this book are those of the authors, but not necessarily of the publisher.

Web Mobile-Based Applications for Healthcare Management

Table of Contents

Section I:
Web and Mobile Strategies

Section VI:
Conceptual Frameworks

Foreword

Any sufficiently advanced technology is indistinguishable from magic.

– Arthur C. Clarke

It is my honor to write this foreword for Dr. Hakim's book. Changes in international health-care provision coupled with rapid technological advances have resulted in increased opportunities for combining these two sectors.

Health-care institutions around the world have leaned toward both wired and wireless technologies as the backbone of their technology-led strategies. With issues surrounding integration, interoperability, and standards very much at the heart of such strategies, it is inevitable that future health-care IT solutions will have to take note of these and other pressures. If one considers the staggering acceptance and adoption of wireless networks in the home, together with such must-have items as personal digital assistants (PDAs) and the latest mobile telephones—both of which may have the capacity and capability to link to networks remotely and without conventional wires—there is little wonder that such terms as *pervasive and ubiquitous computing* have gained so much favor so quickly.

So is further research required? Yes and no. Xerox's Palo Alto Research Center (PARC) has been working on applications revolving around pervasive computing since the 1980s. New technologies, although of course welcome, can easily coexist alongside established and existing research if a new application can be found for them (such as wireless protocols). Basic (pure) and applied research concepts are therefore of equal importance.

By way of an elementary example, in the United Kingdom, Homerton University Hospital in London recently started a trial of SMS (short-message service) "m-reminders" sent to patients regarding their appointments. A simple-enough concept of course, but where is the benefit? Within the United Kingdom's National Health Service (NHS), missed appointments cost about £300 million per annum (House of Commons Accounts Committee). According to *Health-News*, approximately 30% of people missed their NHS appointments because they simply forgot. A much larger percentage (60%) quoted apathy as their cause for missing appointments: These figures speak for themselves. Simple use of existing technologies can result in vast financial improvements for health-care institutions.

The efficacy of mobile technologies for health has an obvious close relationship with telemedicine and telehealth initiatives. Originally only thought workable for geographically dispersed environments (patients located far away from doctors and hospitals), as social factors came into play (less leisure time, increased pressure of work, and so forth), the application of telemedicine for all is starting to become a reality.

Cutting-edge, almost experimental, research also has a very important role to play. IBM's Almaden Research Center has stated that, by the year 2010, computing will have become so naturalized and accepted within the environment that people will not even realize that they are using computers. The emergence of smart devices with embedded location-centric information allows such technologies to be aware of their context.

With the prevalence of such transmitted information, concerns have been raised over privacy and security, both of which this book considers. It is imperative that technology-based solutions (such as encryption and advanced firewalls) are balanced with the need for accessibility and usability. There is little point in having the most technologically advanced hardware and software solution if most people cannot configure them effectively. The latter half of the book concentrates on applications of technology from the patients' perspective and also illustrates cases from the clinical environment. Behavioral change, which the book tackles, is an important hurdle that needs to be overcome.

Chapter X describes how patients prefer an interface with which they are already familiar. Mobile phones offer SMS, MMS (multimedia messaging service), WAP (wireless application protocol), and HTTP (hypertext transfer protocol) avenues, which can be readily accessed by the patient in order to transmit (and receive) personalized medical information (and, with enough context, clinical knowledge). Another chapter within the "Industrial Applications" section describes an innovative teleradiology project that illustrates the two central constructs of the book: first, the initial idea, which is to transmit in a secure environment X-ray, CT (computed tomography), and MR (magnetic resonance) images for swift diagnosis, and second, how this is to be achieved—by the use of a third-generation PDA-cum-mobile–phone; timely diagnosis can often be vital in such cases.

The book offers both a contemporary overview of opportunities and challenges as well as some insight from the field where applicable. My congratulations go to Dr. Hakim for assembling the chapters in this manner as there is much knowledge and best practice within the book from which both academics and practitioners can learn.

Rajeev K. Bali, PhD
Warwickshire, United Kingdom
June 2006

Rajeev K. Bali is currently a senior lecturer at Coventry University, UK. He is the leader of the Knowledge Management for Healthcare research subgroup, which works under the Biomedical Computing and Engineering Technologies (BIOCORE) Applied Research Group. He is an invited reviewer for several journals, conferences, and organizations. His primary research interests are in health-care knowledge management, clinical governance, e-health, change management, organizational behavior, and medical informatics. He has recently published a text on clinical knowledge management. Dr. Bali has served as an invited reviewer and associate editor for several journals, including Transactions on Information Technology in Biomedicine, *and was the invited guest editor for a special issue in 2005. He is also the associate editor of the* International Journal of Networking and Virtual Organisations.

Preface

Abstract

This Preface addresses the book "Web Mobile-Based Applications for Healthcare Management". It hightlights the importance of Web-based and mobile-based systems for healthcare management and addresses the challenges facing these systems, mainly, accessibility by patients, patient disability and patient's skill and literacy. The book comprises seventeen chapters organised into six sections covering various theoretical and practical issues related to system's strategies, challenges and opportunities, information quality, patient empowerment, applications, and conceptual frameworks. It provides insights and support for healthcare professionals as well as for practitioners concerned with the management of information systems.

Overview

Under current economic conditions, one of the challenges before modern health-care executives is to design systems that reduce costs, improve the quality of health-care delivery, and achieve customer satisfaction in an environment that often gives rise to unexpected deviation from planned activities. Amongst other things, health-care executives are increasingly looking to information technology as an opportunity for developing such systems. Web-based and mobile-based technologies are two examples. These technologies are relatively new. The Web or World Wide Web (known as WWW or W3) began as a network information project at the European Organization for Nuclear Research (CERN), where Tim Berners-Lee, now director of the World Wide Web Consortium (W3C), addressed the need for a computerized collaborative tool helping scientists to communicate and share information (W3C, 2001). The WWW merges the techniques of networked information and hypertext to make an easy, straightforward, but powerful global information system that is accessible to anyone over the Internet via an interface known as browsers. There are three layers of Internet protocols. The first and most important one is the Internet protocol (IP), which defines the datagram or packets (messages, pictures, speech, video, etc.) that carry blocks of data from one node of the network to another.

The recent growth of wireless communication technology (e.g., cellular and digital mobiles) has driven the thinking toward a new version of the IP that accommodates wireless communication. Mobile-based systems already enable Web browsing and e-mail services for mobile users. This feature adds greater capabilities for services requiring person-to-person communication connectivity over mobiles and other wireless devices. The explosion of Web-based and mobile-based technology made real-time online communication between businesses and their customers a reality. These technologies have been adopted in an accelerated fashion in all industries including the health-care industry.

The following statistics can serve as examples. (It is important to bear in mind the time at which these statistics were collected and published as the figures undoubtedly change very rapidly.)

- In December 2001, the National Health Service (NHS) dealt with 5.2 million hits from 171,900 visitors to their Web site. The figures for the same period in 2000 were 2.8 million hits and 24,830 visitors ("The UK's E-Health Services Received a Record Number of Hits Over the Holiday Break," 2002).

- Estimates in 2002 suggest that over 500 million people have access to the World Wide Web, with 50% to 75% of the users having used the World Wide Web at least once to look for health information (Powell & Clarke, 2002).

- The number of Medline searches performed by directly accessing the database at the National Library of Medicine increased from 7 million in 1996 to 120 million in 1997, when free public access was provided; the new searches are attributed primarily to nonphysicians (Sharma, Xu, Wickramasinghe, & Ahmed, 2006).

- A survey conducted in Canada to explore patients' attitude toward health services suggests that Internet users expressed interest in using the Web for several reasons, including learning about their health condition through patient education materials (84%), obtaining information about the status of their clinic appointments (83%), renewing prescriptions (75%), consulting with their health professional about nonurgent matters (75%), and accessing laboratory test results (75%; Rizo, Lupea, Baybourdy, Anderson, Closson, & Jadad, 2005).

- While the traditional media provided the primary source of information on anthrax and bioterrorism during late 2001, a survey in USA finds that 21% of respondents reported searching the Internet for this information. The survey shows that information from health Web sites is trusted slightly more than information obtained from traditional media sources (Kittler, Hobbs, Volk, Kreps, & Bates, 2004).

- A survey conducted in Colorado (USA) finds that the majority of respondents (about 95%) value having access to their medical records and more than half of them support online access (Ross, Todd, Moore, Beaty, Wittevrongel, & Lin, 2005). The study finds the primary determinants of support of Internet-accessible records are not age, race, or education level; rather, they are previous experience with the Internet and patients' expectation of the benefits and drawbacks of reading their medical records.

Dimensions: Accessibility, Timeliness, and Mobility

The World Wide Web Consortium's goal of integrating Web-based with mobile-based technologies is "to make browsing the Web from mobile devices a reality" (W3C, 2006). These technologies, however, do not routinely reduce costs, improve quality, or achieve customer satisfaction unless they create, from the customer's perspective, a value-added service. From the health-care angle, these emerging technologies considerably improve three critical value-added service dimensions in relation to information flow between hospital personnel as well as between hospitals and patients. These dimensions are timeliness, accessibility, and mobility. The third dimension is the result of integrating mobile-based applications with Web-based systems.

Timeliness

Timeliness is a reference to how up to date information is with respect to IS users' needs. It reflects also how fast the information system is updated after the state of the represented hospital system changes. Accurate but out-of-date information may have no value for the decision-making process. Difficulties in updating the information in a timely fashion make the system less valuable. Timeliness also requires real-time information flow between various functions of health-care organizations.

Accessibility

The availability of relevant and complete information when needed is one of the key issues that drive health-care executives to allocate a considerable portion of their budgets to installing advanced information systems. The flow of health-care information is restricted and governed by legislation. This raises the issue of security. Accordingly, the acceptability of a system is directly related to the consistency between accessibility and security issues.

Accessibility and timeliness form key drivers for adopting Web-based systems. The literature provides ample examples. One earlier instance was reported in 1996. The Massachusetts General Hospital in Boston has implemented a Web-based clinical information system that provides physicians in the hospital with access to clinical information for patients they refer to the hospital. The system uses electronic medical records (EMRs) as a middle-layer service (Kittredge, Estey, & Barnett, 1996).

Accessibility and timeliness allow local, state, or even national authorities to receive up-to-date information from all or designated hospitals to guide the authority in emergencies at the national level. A constructive example is the Web-based reporting system implemented by the state of Illinois (USA) launched in 2004. The system provides the state with up-to-the-minute information from more than 200 hospitals in Illinois on the availability of medical resources and beds. This allows it to monitor the status of hospital facilities statewide to "ensure adequate resources are quickly made available for emergency patient care in the event of crisis" (Office of the Governor, 2004). A system with adequate accessibility and timeliness of the flow of information can be used to exchange information at the international level to combat the international spread of SARS (severe acute respiratory syndrome), bird flu, or any other infectious diseases that may pose a serious threat to global health security.

Mobility

Mobility is the ability to move. In mobile-based computing, mobility refers to the capability of a device (e.g., a mobile phone) to handle information access, communication, and business transactions whilst in a state of motion ("Mobility," 2006). Mobility causes mobile-based applications to differ from Web-based applications at two levels (Maamar, 2006).

1. The communication level: Web services are connected with wired channels to the external environment, whereas mobile services (m-services) are connected with wireless channels.

2. The computation-location level: Web services are executed on the service side. M-services are executed on the client side after being transferred from the server side.

At the communication level, the integration of mobile-based and Web-based systems would emulate everything the PC (personal computer) can do while on the move (Olla, Patel, & Atkinson, 2003). The integration allows real-time, anywhere, anytime connectivity to services. It is expected that such integration would save considerable time for hospital personnel and would enhance productivity. Microsoft (2005b) reported the implementation of an integrated system at Moorfields Eye Hospital in London. Effective management of the hospital relies on communication with senior managers while they are away from the office. With the integrated system, "managers can stay in touch with their colleagues from any location, save an hour a day by catching up with e-mail messaging while commuting, and help speed up the decision-making process." Another integrated system, called Mobile-Doctor, implemented at Klinikum der Stadt Hanau hospital (Germany), enables doctors to access information they need on the move. Mobile-Doctor eliminates time-consuming paper documentation and manual processes. It allows doctors to complete critical documentation and immediately submit information needed to claim funds for patient treatments (Microsoft, 2005a). In addition, the integrated system allows medical staff to collect information to research and record patient diagnoses from anywhere in the hospital.

During an emergency, terrorist attack, or war, a doctor may deal with several patients simultaneously; he or she may have no time to prepare a written record in handwriting or via a traditional computerized system. A dictation facility built into the mobile-based application is a good solution. This facility has been used in many hospitals. One constructive example is the innovation referred to as the Battlefield Medical Information System: Tactical (BMIS-T) developed as a result of the first Gulf War in 1991. It is claimed that the solution gives medical providers an all-in-one tool for diagnosing and treating conditions, and capturing patient information (Microsoft, 2004).

Challenges

Web mobile-based technologies seem to be beneficial and promising to both health-care organizations and their patients, but they also face challenges that should be resolved before the technologies become commonplace. The list below is limited only to those related to the patients' perspective of using online and mobile technologies. Interested readers may refer to literature such as Olla et al. (2003) and Sharma et al. (2006) for other related challenges.

Accessibility by Patients

The literature indicates a high rate of deployment of personal computers and mobile phones. Yet, a considerable proportion of society cannot afford computers, constant access to the Internet, or mobile phones. In addition, rural areas in many countries may lack the necessary infrastructure for reliable and high-speed access to information technology (Goslee & Conte, 1998). Furthermore, mobile-based technology may require specific devices that are not yet widely deployed and accordingly, this will limit the full utilization of integrated Web-mobile-based systems.

Patient Disability

Patients with disabilities may face difficulty in accessing the Internet or using mobile phones. Kaye (2000) reports that the rate of having a computer among Americans with disabilities is less than half of their counterparts without disabilities. Kaye also reports that patients with disabilities are less likely to use the Internet and other information technologies.

Patients' Attitude and Perception

There is no doubt that many patients would like face-to-face communication rather than online or wireless communication. Security and privacy issues play an important role in hindering the effective use of the Web and mobile facilities for health-care delivery. Medline health seekers may have the view that Internet companies will collect and share information about their medical status with insurance companies, and fear arises that insurance companies may change their coverage (Sharma et al., 2006).

Patients' Skills and Literacy

Engaging with information technology to promote health and health-care delivery requires specific levels of skill or literacy that allows the users to operate the systems, navigate the Web, retrieve and input information, and grasp the correct contextual meaning of information. At this stage, it is unrealistic to hypothesise that those individuals with access to Web and mobile technologies have the required skill and literacy to use these technologies proactively. Studies show that even in countries with high rates of Internet adoption such as the USA and Canada, over 40% of adults have a basic literacy level below the requirement for optimally participating in the information society (National Assessment of Adult Literacy [NAAL], 2005; Norman & Skinner, 2006; Statistics Canada, 2005).

Book Structure

This book aims to present cutting-edge research on the theory, applications, and challenges facing the implementation of Web-based and mobile-based systems in health-care industries. The book provides insights and support for the following groups of people.

- Professionals and researchers working in the field of Web-based and mobile-based systems information in general and in the sector of health care in particular.
- Practitioners and managers of health-care organizations concerned with the management of Web and mobile systems.

The 17 chapters of the book have been organized into six sections: "Web and Mobile Strategies," "Challenges and Opportunities," "Security, Reliability, and Interpretability," "Patient Empowerment," "Industrial Applications," and "Conceptual Frameworks." The following is a brief description of each section and the chapters included in them.

Section I: Web and Mobile Strategies

The first section of the book comprises three chapters that deal with the Web-based and mobile-based strategies used in the health-care industry. Chapter I, entitled "The Development of a Web Strategy in a Healthcare Organization: A Case History," is by Massimo Memmola, who presents a way of defining a Web strategy by aligning a company's corporate strategy, in which there is an acceptance and awareness of the possibilities that Internet offers by the principal company stakeholders, with a general strategy of utilizing information and communication technology. With the aid of a case study conducted in a private health-care organization, the chapter demonstrates how a health-care organization can develop its own Web strategies. In dealing with the development process of the Web strategy, the chapter takes into consideration the alignment of the Web strategy with the business strategy and the ICT strategy. The chapter also considers the readiness and acceptance of the Internet by inside and outside patients, by outside practitioners, and by the organization's staff.

The application of mobile commerce to health care, namely, m-health, appears to offer a way for healthcare delivery to revolutionize itself. However, little has been written regarding how to achieve excellence in m-health. "A Framework for Delivering m-Health Excellence" by Nilmini Wickramasinghe and Steve Goldberg, the second chapter of this book, serves to address this major void by presenting an integrative framework for achieving m-health, developed through the analysis of longitudinal applied research conducted by industry in conjunction with academe. The chapter illustrates how the mapping of case data to the model enables the attainment of a successful m-health application to ensure the benefits of adopting such a methodology. It stresses that health-care organizations are responding to market challenges by focusing on three key solution strategies, namely, (a) access, which is caring for anyone, anytime, anywhere, (b) quality, which is offering world-class care and establishing integrated information repositories, and (c) value, providing effective and efficient health-care delivery. These three components are interconnected such that they continually impact each other; all are necessary to meet the key challenges facing health-care organizations today. The chapter closes by strongly urging for more research in this area that will further test the developed framework.

The opportunity of innovating business models has basically been linked to continual progress in information and communication technologies. Health care is no exception; information and communication technologies are generally considered the most effective driver for changing organization, improving quality, optimising resources, and so forth, at least in theory. In practice, it is not clear which and how many of these opportunities are really exploited by organizations operating in health care. Chapter III of the book, "Healthcare Organizations and the Internet's Virtual Space: Changes in Action" by Stefano Baraldi and Massimo Memmola, presents the results of a research project aimed at understanding to what degree and how Italian health-care organizations make use of the virtual space made available to them by the Internet. The chapter emphasizes that the Internet has extended the traditional market space by providing new spaces in which economic agents can interact by exchanging information, communicating, distributing different types of products and services, and initiating formal business transactions. It considers four segments of the Internet virtual space: the virtual information space, virtual communication space, virtual distribution space, and virtual transaction space. The chapter explores how Italian health-care organizations actually exploit and use the virtual space offered by the Internet and concludes that the Italian health-care industry does not yet realize and use the full potential of the Internet.

Section II: Challenges and Opportunities

The second section of the book contains three chapters that explore the opportunities and challenges facing the implementation of Web-based and mobile-based systems in relation to three fields: network collaboration, relatively healthy populations, and ubiquitous elder care.

Chapter IV, entitled "An e-Healthcare Mobile Application: A Stakeholders' Analysis" by Niki Panteli, Barbara Pitsillides, Andreas Pitsillides, and George Samaras, presents a longitudinal study on the implementation of an e-health mobile application, know as DITIS, which supports network collaboration for home health care. By adopting the stakeholders' analysis, the study explores the various groups that have directly or indirectly supported the system during its implementation. The system was originally developed with a view of addressing the difficulties of communication and continuity of care between the members of a home health-care multidisciplinary team and between the team and oncologists, often hundreds of kilometres away. DITIS evolved to be much more than that, and even though it was introduced 5 years ago, it is considered a novel application. Despite this, its implementation has been slow, and several challenges, including the system's sustainability, have to be faced. This chapter aims to identify these challenges, and the results of the study point to a diversity of interests and different degrees of support.

New communication technologies have made an impact on several areas of our everyday life, including the areas of health and health promotion. The Internet provides opportunities for personalised interactive health communication at a much larger scale than is possible in face-to-face communication. It has been suggested that only interactive health behaviour-change Web sites that advise, assess, assist, provide anticipatory guidance, and arrange follow-up have the potential to lead to successful behaviour change. Additional factors that may affect the success rate of behaviour-change programs are the reach of and the exposure to such programs. Chapter V, entitled "Behavior Change through ICT Use: Experiences from Relatively Healthy Populations" by Marieke W. Verheijden, elaborates on all of these issues and discusses the following components in more detail: the Internet as a communication channel; the potentials and minimum requirements of Web-based behavior-change programs; the delivery of, reach of, and exposure to Web-based behavior-change programs; and the feasibility and effectiveness of Web-based behavior-change programs. This chapter focuses specifically on relatively healthy populations as opposed to patient populations. The chapter emphasises that conventional mass media generally focus on new discoveries about diseases and their treatment. Much less attention is focused on disease prevention, health behaviour, or early detection. This is in sharp contrast with the general idea that an ounce of prevention is worth a pound of cure, and stresses the need for a focus on applications for relatively healthy populations. This is particularly challenging because (otherwise healthy) patients at elevated risk for cardiovascular disease stated that they were more interested in using Web-based health-promoting programs for information when confronted with a direct medical condition than for prevention purposes. The chapter concludes that the World Wide Web should never fully replace consultations and clinical examinations by health professionals. Depending on factors such as the available resources (time, space, staff, etc.) and the personal preferences of all individuals involved, an ideal mix of intervention approaches may be composed.

A nonnegligible number of elder citizens, who represent a growing fraction of the population in developed countries, has to face a number of daily-life problems stemming from their partial and progressive loss of motor, sensorial, and cognitive skills. This often makes it difficult to live autonomously and, in today's small families, often results in the hospitalization of the people concerned. Chapter VI, entitled "Challenges, Opportunities, and Solutions for Ubiquitous Eldercare" by Paolo Bellavista, Dario Bottazzi, Antonio Corradi, and Rebecca Montanari, overviews the state of the art of solutions for elder assistance, typically at home, and for coordinated-care networking. It argues that wireless sensors and actuators can improve elder life independence, for example, by transforming homes in smart elder-care environments with remote health-status monitoring, remote diagnostics, and facilitated house activities. On the other hand, pervasive wireless computing enables novel opportunities for caregivers, elders, family members, friends, and neighbors to collaborate and coordinate in an impromptu way to provide elder care and social support anytime and anywhere. The chapter points out to the need for advanced context-aware frameworks to properly establish ubiquitous and spontaneous communities of helpers when needed.

Section III: Security, Reliability, and Interoperability

The third section of the book deals mainly with three interrelated dimensions of service quality for Web-based and mobile-based systems, that is, security, reliability, and interoperability.

Wireless technology has broad implications for the health-care environment. Despite its promise, this new technology has raised questions about the security and privacy of sensitive data that are prevalent in health-care organizations. All health-care organizations are governed by legislation and regulations, and the implementation of enterprise applications using new technology is comparatively more difficult than in other industries. Using a configuration-idiographic case-study approach, Chapter VII of the book, "Bringing Secure Wireless Technology to the Bedside: A Case Study of Two Canadian Healthcare Organizations" by Dawn-Marie Turner and Sunil Hazari, investigates security and privacy challenges faced by two Canadian health-care organizations. In addition to interviews with management and staff of the organizations, a walk-through was also conducted to observe and collect first-hand data about the implementation of wireless technology in the clinical environment. In the organizations under examination, it was found that wireless technology is being implemented gradually to augment the wired network. Problems associated with implementing wireless technology in these Canadian organizations are also discussed. Because of different standards in this technology, the two organizations are following different upgrade paths. Based on the data collected, best-practice guidelines for secure wireless access in these organizations are proposed.

The quality and accuracy of online health information is an area of increasing concern for health-care professionals and the general public. Chapter VIII, "Reliability and Evaluation of Health Information Online" by Elmer V. Bernstam and Funda Meric-Bernstam, discusses the problem of how to evaluate online health information. The chapter defines relevant concepts including quality, accuracy, utility, and popularity. It briefly reviews Web search-engine fundamentals and discusses desirable characteristics for quality-assessment tools and the available evidence regarding their effectiveness and usability. The chapter points out that although there is evidence that online information affects health-care decisions, there is little evidence that users are considering quality when accessing health information online. Helping users identify problematic health information online remains an open problem. The chapter emphasizes that the currently available quality-assessment tools cannot be reliably assessed by Web users and may not be effective at identifying problematic health information online. It concludes with advice for health-care consumers as they search for health information. The prudent health seeker will use online resources for education, but review important decisions with their health-care provider. In general, the chapter asserts that government, charity-organization, or academic Web sites are more likely to provide impartial advice compared to company (commercial) Web sites. However, this is a generalization and there may be exceptions.

One of the major challenges in healthcare database integration is the fact that the lack of guidance from central authorities has, in many instances, led to incompatible health-care database systems. Such circumstances have caused problems to arise in the smooth processing of patients between health-service units, even within the same health authority. Due to the lack of uniformity, these systems have very poor interoperability and therefore usability.

There are two potential solutions to the problems of interoperability and automated information processing: redesigning and re-implementing the existing databases or using a database federation. Redesigning and re-implementing existing databases requires large capital investments, and sometimes it is impossible. An alternative solution is to build a database federation in which problems caused by database heterogeneity are remedied by the use of a mediator: metadata. Chapter IX, entitled "Integrating Mobile-Based Systems with Healthcare Databases" by Yu Jiao, Ali R. Hurson, Thomas E. Potok, and Barbara G. Beckerman, discusses the issue of interoperability and focuses on two major challenges in distributed health-care database management: database heterogeneity and user mobility. The chapter designs the prototype of a mobile agent-based data-access system framework that can address these challenges. It applies a thesaurus-based hierarchical database federation to cope with database heterogeneity and utilizes the mobile-agent technology to respond to the complications introduced by user mobility and wireless networks. The functions provided by this system are described in detail, and a performance evaluation is also provided.

Section IV: Patient Empowerment

The new patient-empowerment paradigm promotes the development of novel care approaches in which outpatient monitoring is a basic aspect. Section IV of the book comprises three chapters that deal with issues related to patient empowerment.

Mobile information and communication technologies are advancing rapidly and provide great opportunities for home monitoring applications in particular for outpatients and patients suffering from chronic diseases. Because of the ubiquitous availability of mobile phones, these devices can be considered as patient terminals of choice to provide a telemedical interaction between patients and caregivers. Chapter X, "Utilizing Mobile Phones as Patient Terminal in Managing Chronic Diseases" by Alexander Kollmann, Peter Kastner, and Guenter Schreier, deals with the management of chronic diseases and identifies three phases of monitoring: pretreatment, adoption, and long-term treatment. It emphasizes that the most challenging part is the user interface, that is, to offer the user a method to enter measured data into a system as well as to receive feedback in a comfortable way. The chapter presents a solution for the mobile-phone-based patient terminal developed. The chapter illustrates that mobile data services and transmission protocols like SMS (short-message service), MMS (multimedia messaging service), WAP (wireless application protocol), and HTTP (hypertext transfer protocol) can be used to exchange data and information between patients and their caregivers. These methods have already been evaluated in several clinical trials and feasibility studies, and medical benefits could be demonstrated as well. However, the possibilities for using mobile phones as patient terminals are limited due to small displays, poor resolution, and small buttons for user interaction.

According to the authors, up to now there has not been any method that fulfils all criteria of an ideal patient terminal in terms of high usability, adaptability, flexibility, and low cost. Every method for entering data implies specific advantages and disadvantages. The chapter recommends that when designing a mobile-phone-based home monitoring system, the patient terminal that best fits into a particular monitoring application should be chosen on an individual basis, depending on the requirements, the user group, and the medical demand.

The second chapter in Section IV is Chapter XI, entitled "Considerations for Deploying Web and Mobile Technologies to Support the Building of Patient Self-Efficacy and Self-Management of Chronic Illness" and written by Elizabeth Cummings and Paul Turner. This chapter examines issues relating to the introduction of information and communication technologies that have emerged as part of the planning for the Pathways Home for Respiratory Illness project. The project aims to assist patients with chronic respiratory conditions (chronic obstructive pulmonary disease and cystic fibrosis) to achieve increased levels of self-management and self-efficacy through interactions with case mentors and the deployment of ICTs.

The chapter highlights the fact that in deploying ICTs, it is important to ensure that solutions implemented are based on a detailed understanding of users, their needs, and complex interactions with health professionals, the health system, and their wider environment. Achieving benefits from the introduction of ICTs as part of processes aimed at building sustainable self-efficacy and self-management is very difficult, not least because of a desire to avoid simply replacing patient dependency on health professionals with dependency on technology. More specifically, it also requires sensitivity toward assumptions made about the role, impact, and importance of information per se, given that it is often only one factor among many that influence health attitudes, perceptions, actions, and outcomes. More broadly, the chapter indicates that as ICT-supported patient-focused interventions become more common, there is a need to consider how assessments of benefit in terms of a cohort of patients inform us about individual patients' experience and what this implies for aspects like individualised care or patient empowerment. At this level, there are implications for clinical practice and one-size-fits-all care-delivery practices.

Chapter XII is the third chapter in this section. The chapter is entitled "PDAs as Mobile-Based Health Information Deployment Platforms for Ambulatory Care: Clinician-Centric End-User Considerations" by Jason Sargent, Carole Alcock, Lois Burgess, Joan Cooper, and Damian Ryan. It discusses the broad theme of clinician-centric end-user acceptance toward the adoption of personal digital assistants (PDAs) as mobile-based health information deployment platforms within ambulatory care service settings. Personal digital assistants, ambulatory care, and point of care (POC) are defined and their interrelatedness discussed. Issues, controversies, and problems such as mapping existing workflows, security, and change management are identified, and solutions are suggested for the process of transforming predominantly paper-based ambulatory care systems into electronic point-of-care (ePOC) systems. A current research and development project, the ePOC PDA project, is used as a case study to highlight discussion points. The purpose of this chapter is to illustrate end-user implications and considerations when introducing ePOC systems into ambulatory care service settings and to highlight ways and means of improving future levels of acceptance and support of ePOC systems for clinician end users.

Section V: Industrial Applications

The fifth section of the book features three chapters that deal with three applications of Web-based and mobile-based technologies: image viewing, radio-frequency identification (RFID), and waiting-list rescheduling using GSM (Global System for Mobile Communication) SMS messages.

Teleradiology is the technology of remote medical consultation using x-ray, computed tomographic, or magnetic resonance images. It has been commonly accepted by clinicians for its effectiveness in making diagnoses for patients in critical situations. Because of the huge volumes of data involved in teleradiology, clinicians are not satisfied with the relatively slow data transfer rate. It limits the technology to fixed-line communication between the doctor's home and his or her office. Chapter XIII of the book, entitled "3G Mobile Medical Image Viewing" by Eric T. T. Wong and Carrison K. S. Tong, presents a mobile high-speed wireless medical image-viewing system using 3G (third generation) wireless network, virtual private network, and one-time two-factor authentication (OTTFA) technologies. Using this system, teleradiology can be achieved by using a 3G PDA phone to query, retrieve, and review the patient's record at anytime and anywhere in a secure environment. The chapter emphasises that the system significantly improves the patient-data availability, which is crucial to the timely diagnosis of patients in critical situations.

When dealing with human lives, the need to utilize and apply the latest technology is very important and requires accurate, near-real-time data acquisition and evaluation. At the same time, the delivery of patients' medical data needs to be as fast and as secure as possible. One way to achieve this is to use a wireless framework based on radio-frequency identification. This framework can integrate wireless networks for fast data acquisition and transmission while addressing the privacy issue. Chapter XIV, entitled "Intelligent Agents Framework for RFID Hospitals" by Masoud Mohammadian and Ric Jentzsch, discusses the development of an agent framework in which RFID can be used for patient-data collection. The chapter presents a framework for the knowledge acquisition of patient and doctor profiling in a hospital. The acquisition of profile data is assisted by a profiling agent that is responsible for processing the raw data obtained through RFID and a database of doctors and patients.

"Rescheduling Dental Care with GSM-Based Text Messages" is the title of Chapter XV. This chapter is authored by Reima Suomi and Ari Serkkola. In this chapter, the authors propose a framework for an integrated electronic health platform (e-health). Most of the platform is still at the planning stage, but the first applications are already up and running, among them, dental service appointment rescheduling. In this application, new patients for filling cancelled dental service appointments are searched for from an existing waiting list using GSM SMS messages. The first few months of operation have already shown that the new application, in conjunction with other methods in use, could limit the share of time slots that dentists completely lose through cancellations to under 10% of all cancelled times. The chapter presents and analyses the function of this SMS-message-based dental service appointment reservation system, which is being implemented in Lahti, Finland. The analysis contains a description of the system functions, as well as some assessment of the success from a service provider and customer point of view.

Section VI: Conceptual Frameworks

The significance of aligning IT with corporate strategy s widely recognized, but the lack of an appropriate framework often prevents practitioners from integrating emerging information technologies within organizations' strategies effectively. The sixth and final section of the book features two chapters that address conceptual frameworks for mobile-based and Web-based systems.

Chapter XVI, entitled "Conceptual Framework for Mobile-Based Application in Healthcare" by Matthew W. Guah, addresses the issue of deploying Web services strategically within a mobile-based health-care setting. A framework is developed to match potential benefits of Web services with corporate strategy in four business dimensions: innovation, internal health-care process, patients' pathway, and the management of the health-care institution. The author argues that the strategic benefits of implementing Web services in a health-care organization can only be realized if the Web-services initiatives are planned and implemented within the framework of an IT strategy that is designed to support the business strategy of that health-care organization. The chapter uses case studies to answer several questions relating to wireless and mobile technologies and how they offer vast opportunities to enhance Web services. It also investigates what challenges are faced if this solution is to be delivered successfully in health care. The health-care industry globally, with specific emphasis on the USA and United Kingdom, has been extremely slow in adopting emerging technologies that focus on better practice management and administrative needs. The chapter elaborates certain emerging information technologies that are currently available to aid the smooth process of implementing mobile-based technologies in the health-care industry.

The last chapter of the book, Chapter XVII, is written by Latif Al-Hakim and is entitled "IDEF3-Based Framework for Web-Based Hospital Information System." The chapter presents a framework for a Web-based hospital information system to manage the surgery-management process (SMP). The framework can also be used to manage any other hospital information-system processes. The developed framework challenges the traditional hospital Web strategies with a dual aim: first, to improve customer satisfaction in an environment that often involves unexpected deviation from planned activities, and second, to create a system that is an effective decision-support system for SMP. The chapter identifies factors affecting SMP decisions and employs a descriptive modeling technique known as IDEF3 to map the information flow within and between elements of SMP. IDEF3 is a member of Integrated Definition (IDEF) initiatives developed by US Air Force. The IDEF3 process mapping becomes part of an integrated Web-based system of multiple stages. Each stage has three levels of accessibility. The first level of the Web system is accessible by the public. It allows the public to obtain the necessary information and download the required forms. The second level is accessible by patients and their designated representatives. It allows them to communicate actively with the hospital management and to receive explanations for delays and any other complications. The third level is accessible only by hospital professionals. It allows them to retrieve and update the required information and enables real communication between them during the decision-making process.

References

Goslee, S., & Conte, C. (1998). *Losing ground bit by bit: Low-income communities in the information age.* Washington, DC: Boston Foundation. Retrieved December 6, 2005, from http://www.ncddr.org/cgi-bin/good-bye?url=http://www.benton.org/Library/Low-Income/

Kaye, H. S. (2000). Disability and the digital divide. *Disability Statistics Abstracts, 22,* 1-4.

Kittler, A. F., Hobbs, J., Volk, L. A., Kreps, G. L., & Bates, D. W. (2004). The Internet as a vehicle to communicate health information during a public health emergency: A survey analysis involving the anthrax scare of 2000. *Journal of Internet Research, 6*(1), e8.

Kittredge, R. L., Estey, G. P., & Barnett, G. O. (1996). Implementing a Web-based clinical information system using EMR middle layer services. *Proceedings of AMIA Annual Fall Symposium* (pp. 628-632).

Maarmar, Z. (2006). A mobile application based on software agents and mobile Web services. *Business Process Management Journal, 12*(3), 311-329.

Microsoft. (2004). *U.S. military improves medical care, tactical advantage with wireless point-of-care handheld assistant.* Retrieved June 10, 2006, from http://members. microsoft.com/CustomerEvidence/Search/EvidenceDetails.aspx?EvidenceID=2835 &LanguageID=1

Microsoft (2005a). *Klinikum der Stadt Hanau: Mobile workplace solution takes the strain off hospital administration.* Retrieved June 12, 2005, from http://members.microsoft. com/CustomerEvidence/Search/EvidenceDetails.aspx?EvidenceID=2675&Langua geID=1

Microsoft (2005b). *Moorfields Eye Hospital: London hospital expects to see productivity saving with mobile solution.* Retrieved June 10, 2006, from http://members.microsoft. com/CustomerEvidence/Search/EvidenceDetails.aspx?EvidenceID=3912&Langua geID=1

Mobility. (2006). *Reference.Com.* Retrieved June 12, 2006, from http://www.reference. com/browse/wiki/Mobility

National Assessment of Adult Literacy (NAAL). (2005). *A first look at the literacy of America's adults in the 21st century.* Washington, DC: National Center for Educational Statistics, Institute of Education Sciences, U.S. Department of Education.

Norman, C. D., & Skinner, H. A. (2006). eHealth literacy: Essential skills for consumer health in a networked world. *Journal of Internet Research, 8*(2), e9.

Office of the Governor. (2004, April 26). *Governor announces launch of Web-based hospital resource reporting system.* Retrieved March 20, 2006, from http://www.compressus. com/publicwww4/news/pressreleases/IDPH%20-%20VIGILENT%20Press%20Rel ease%20-%20Apr%2026%2004.pdf

Olla, P., Patel, N., & Atkinson, C. (2003). A case study of MMO2's MADIC: A framework for creating mobile Internet systems. *Internet Research: Electronic Networking Applications and Policy, 13*(4), 311-321.

Powell, J., & Clarke, A. (2002). The WWW of the World Wide Web: Who, what, and why? *Journal of Medical Internet Research, 4*(1), e4.

Rizo, C. A., Lupea, D., Baybourdy, H., Anderson, M., Closson, T., & Jadad, A. R. (2005). What Internet services would patients like from hospitals during an epidemic? Lessons from the SARS outbreak in Toronto. *Journal of Medical Internet Research, 7*(4), e46.

Ross, S. E., Todd, J., Moore, L. A., Beaty, B. L., Wittevrongel, L., & Lin, C. (2005). Expectations of patients and physicians regarding patient-accessible medical records. *Journal of Medical Internet Research, 7*(2), e13.

Sharma, S. K., Xu, H., Wickramasinghe, N., & Ahmed, N. (2006). Electronic healthcare: Issues and challenges. *International Journal of Electronic Healthcare, 2*(2), 50-65.

Statistics Canada. (2005). *Building on our competencies: Canadian results of the International Adult Literacy and Skills Survey, 2003.* Ottawa, Ontario, Canada: Human Resources and Skills Development Canada & Statistics Canada.

The UK's e-health services received a record number of hits over the holiday break. (2002). *KableNet.Com*. Retrieved June 10, 2006, from http://www.kablenet.com/kd.nsf/print-view/A33471B29784EDD480256B37003DC0EF?OpenDocument

World Wide Web Consortium (W3C). (2001). *About the World Wide Web*. Retrieved June 12, 2006, from http://www.w3.org/WWW/

World Wide Web Consortium (W3C). (2006). *Mobile Web initiative*. Retrieved June 10, 2006, from http://www.w3.org/Mobile/

Acknowledgments

The editor is grateful to all those who have assisted him with the completion of this work. In particular, the editor would like to acknowledge his deepest appreciation to many reviewers for their time and effort. Comments and amendments suggested by them were incorporated into the manuscripts during the development process and significantly enhanced the quality of the work.

Thanks are due to a number of staff at Idea Group Inc. who have provided time, coordination, support, and encouragement at all stages as the project has progressed. The editor is indebted in particular to Dr. Mehdi Khosrow-Pour and Jan Travers who provided needed support and coordination. Appreciation also goes to Kristin Roth, Michelle Potter, Sharon Berger, and Jamie Snavely who gave their time willingly to describe many issues related to the preparation of this work and shared their experiences with me.

Additional acknowledgments are owed to Professor Eberhard Bischoff of the University of Wales Swansea (UK), who provided assistance and help that allowed the editor to complete this project during his academic visit at the university. Special thanks go to the staff of the University of Southern Queensland for all of their assistance in seeing this work completed, in particular, Dr. Heather Maguire, Dr. Geoff Cockfield, and Ms. Jenna Schott.

Finally, the editor would like to thank his family, particularly his wife Aliah, daughter Saba, and his two sons Samer and Ahmed, for their patience during the course of editing the book.

List of Reviewers

Among 39 chapter proposals initially received, only 26 of them had been accepted for possible inclusion in this book. Each chapter of the book has been blind reviewed by at least two reviewers. Some chapters had been reviewed more than once before final acceptance. The number of accepted chapters is 17. Below is the list of reviewers.

Rajeev Bali	Coventry University, United Kingdom
Paolo Bellavista	University of Bologna, Italy
Elmer Bernstam	University of Texas, USA
Eberhard Bischoff	University of Wales Swansea, United Kingdom
Elizabeth Cummings	University of Tasmania, Australia
Matthew W. Guah	University of Warwick, United Kingdom
Sunil Hazari	University of West Georgia, USA
Yu Jiao	Oak Ridge National Laboratory, USA
Alexander Kollmann	ARC Seibersdorf Research GmbH, Austria
Heather Maguire	University of Southern Queensland, Australia
Massimo Memmola	Catholic University, Italy
Masoud Mohammadian	University of Canberra, Australia
Niki Panteli	University of Bath, United Kingdom
Jason Sargent	University of Wollongong, Australia
Reima Suomi	Turku School of Economics and Business Administration, Finland
Barbara Roberts	University of Southern Queensland, Australia
Mark Toleman	University of Southern Queensland, Australia
Paul Turner	University of Tasmania, Australia
Marieke Verheijden	TNO Quality of Life, Netherlands
Nilmini Wickramasinghe	Illinois Institute of Technology, USA
Eric Wong	The Hong Kong Polytechnic University, Hong Kong

Section I

Web and Mobile Strategies

Chapter I

The Development of a Web Strategy in a Healthcare Organization:
A Case History

Massimo Memmola, Catholic University, Italy

Abstract

Many writers have described the advantages that the Internet can bring to a health-care organization in terms of consistent improvements in efficiency and efficacy, the reduction of access time to services, and an improved awareness of these. Bankruptcy costs and devastating failures of investments in technology would have us believe that the go-to Internet has taken place, at least in the healthcare field, with a certain improvisation and without a thorough knowledge of the full potential that the Internet offers. This chapter presents a way to define a Web strategy by aligning a company's corporate strategy, in which there is an acceptance and awareness of the possibilities that the Internet offers by the principal company stakeholders, with a general strategy of utilization of information and communication technology.

Introduction

In recent years, there has been a veritable literary frenzy on the theme of healthcare and the Internet. In these works, however, there has not been adequate attention paid to the role that potential users of the site (patients, general practitioners, healthcare personnel, students, private doctors, other healthcare organizations, etc.) could and should play in the process of defining the Internet strategy.

To overlook these aspects while planning the information content and services of a Web site would have the same consequences as failing to do a stakeholder analysis while planning corporate strategy. This, in fact, allows the identification and analysis of the importance of people, groups, or institutions that can influence, positively or negatively, corporate activity, consequently determining the success or failure of a strategy. The objective, in the final analysis, is to identify their expectations, needs, and requirements so as to ensure correct alignment with the corporate strategic policy. This process is much more important when you are about to change the logistics of the production or provision of a product or service.

To put it more simply, you cannot hope to develop a high-profile Web strategy when patients do not have access to the Internet or lack technological skills, when the doctors of the organization are not interested and do not want to collaborate in providing the various content for services to be offered by the site, or when the general practitioners (GPs) do not regard the Internet as a tool that can improve their working conditions or results.

Objectives and Research Methodology

This study had the aim of determining the best methodological approach to take for a healthcare organization to determine and develop its own Web strategy. In particular, our intention was to look for useful knowledge that would allow us to understand the following:

- How does one determine the best strategy for the "colonization" of the virtual space made available by the Internet?

- How does one use the Internet to the best of its potential to provide information, offer services, make transactions, and interact (in real time) with patients, doctors, and whoever has an interest in any way in the organization?

- What will the impact be on corporate strategy and clinical, organizational, and management processes?

- How does one establish a presence on the Internet and how long will it take?
- Which performance measurement should be applied and how does one evaluate the benefits and success of the strategies chosen?

This research is part of a more wide research project (Baraldi & Memmola, 2006), started in 2002 and still ongoing, aimed at determining how much and in what way healthcare organizations can exploit the potential of the Internet to create value in the eyes of the principal stakeholders. Analysis took place outside the organization, evaluating, through the Web site, the final product of the Web strategy of each healthcare structure.

However, to obtain a complete picture of the subject under investigation, it was necessary to go inside the organization to share with those responsible for it the preparation, determination, and eventual implementation of the Internet strategy, critical aspects and development times, and the expectations and needs of the people involved directly or indirectly.

The experimental research (Yin, 2003) was undertaken thanks to the cooperation of a relatively small private healthcare organization (150 beds and about 300 employees).

The logistics that inspired the setting up of the project is briefly described in Figure 1, which shows the route to determining a Web strategy through the appropriate alignment of the following:

- The corporate strategy of the organization (business strategy).
- The general strategy of utilization of information and communication technology supporting corporate processes (ICT strategy).
- The degree of acceptance and awareness of the possibilities of the Internet by The personnel (not only medical, but also technical and administrative).
- The degree of acceptance and awareness of the possibilities of the Internet by patients, GPs, and other corporate stakeholders (outside acceptance of the Internet).

The final two points of this list particularly required the carrying out of a survey that was different for each type of corporate stakeholder: patients, general practitioners, and personnel (healthcare, technical, and administrative).

This method made available a mass of information particularly relevant for the understanding of the best way a healthcare organization should develop an Internet presence (Web strategy). The colonization of the virtual space the Internet provides can be achieved, therefore, through an effective evaluation of the following:

Figure 1. Research methodology

- Corporate-strategy orientation.
- Information gained from the survey aimed at evaluating inside and outside acceptance of the Internet.
- The impact on clinical, organizational, and managerial processes.
- Costs and benefits associated with the contents once activated.
- Performance measure mechanisms through the determination of the KPA (key performance area) and the KPI (key performance indicator), which enable the evaluation of the success of the accomplished initiative.

The model presented in these pages leads to a rationalization of the complex process through which an organization determines the informative or service content of its own Web site.

To explain the type of strategic approach used by a healthcare organization to determine the contents of its Web site and to evaluate how and how much the potentiality of the Internet is being exploited, we have used Angehrn's model (Angehrn, 1997), known by the acronym ICDT (Information, Communication, Distribution, Transaction).

Figure 2. The ICDT model (Angehrn, 1997)

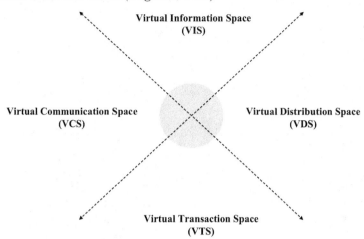

[Angehrn, 1997]

The model (Figure 2) gets its name from the segmentation of the Internet's virtual space into four main sectors: the virtual information space (VIS), virtual communication space (VCS), virtual distribution space (VDS), and virtual transaction space (VTS). This segmentation emphasizes that "the Internet has extended the traditional market space by providing new spaces in which economic agents can interact by exchanging information, communicating, distributing different types of products and services and initiating formal business transactions" (Angehrn, 1997, p. 362).

Angehrn's model has allowed us to build a map of the Internet presence of healthcare organizations aimed at providing a clear and immediate perspective of the Web strategy pursued by the enterprise (Figure 3).

The Web site is ideally segmented into a series of minimum units of analysis (MUAs). An MUA represents an area inside the site in which the contents (information or services) are homogeneous and represent in this way a precise area of occupation of the virtual space offered by the Web. This might not involve just one page of the site but might be spread over several pages, or the same page may be shared by other MUAs. In the VIS area, the MUAs that can be activated are mainly informative as, for example, those that provide general information about the organization: its history, clinical specialties, waiting times, and any research being carried out. In the VDS area, on the other hand, an MUA is normally dedicated to providing a service such as online consulting, educating, or personalizing programs for health monitoring. Each MUA is then placed in its own virtual area of competence (VIS, VDS, VCS, VTS); the greater the level of technological sophistication (LOS) from

Figure 3. Creation of the map of the Internet presence of a healthcare organization

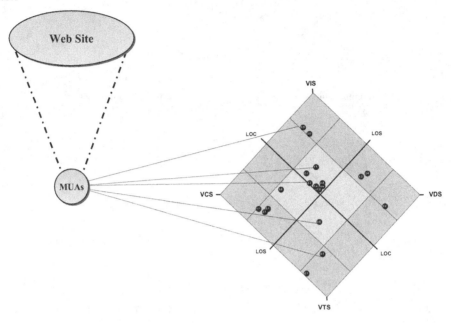

the outside, the greater the level of personalization of contents (LOC), as evaluated along the orthogonal axis of the figure.

Main Findings

Research at MD Hospital, a small private healthcare organization (150 beds and about 300 employees) in the north of Italy, began in the last 4 months of 2004 and was completed in the first 2 months of 2005. Top management had just completed the process of determining a strategy for 2005 to 2007, in which great emphasis had been placed on the need to develop a suitable presence on the Internet that could help communication and support, in some way, new corporate stances. However, the strategic approach to the Internet was decidedly low profile (Figure 4a), with a limited quantity of content of an informative nature, determined at the time of registration of the domain some years earlier.

After we shared with top management the methodological approach by which the development of the case under study would proceed, the first problem we met in-

Figure 4. Evolution of the Internet positioning of MD Hospital due to alignment with the business strategy

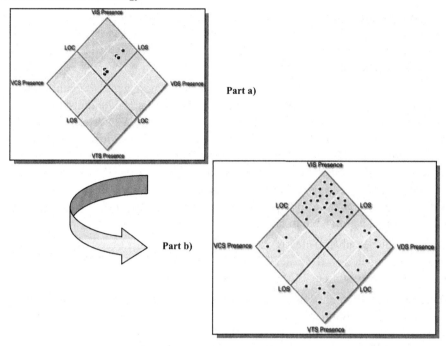

Part a)

Part b)

volved the composition of the group of work and, in particular, the assignation of the ownership of the process of determining the Web strategy.

Aware that the Internet can bring about a really dramatic change in the genetic code of any organization, including those of healthcare, and as a direct consequence of the way business is carried out (or rather the way treatment services are provided), we felt that active involvement by the top management (both head office and the medical and administrative managers) was critical. Moreover, if they are the principal actors of corporate strategy, why ask others (the information officer or a consultancy firm) to be responsible for and to coordinate a process that should teach one how to use the potential of the Internet to pursue this strategy?

Furthermore, enthusiastic sponsorship by senior levels of the organization makes the employees realize, when they are eventually involved in the project directly or indirectly, the importance of what is happening. Active participation by the staff is a factor we believed to be of the utmost importance. To this end, starting with the preliminary phases of the project, meetings were held with all the doctors and the nursing and administrative staff in which the objectives and expected results of the project were explained, in particular, the level and type of their involvement.

Aligning Web Strategy with Business Strategy

It is fundamental for any strategy guide to explain and detail strategic objectives as best as possible, and to express them clearly in writing and share them with those who have responsibility in the organization and on who depends, in the final analysis, the attainment or not of these objectives.

Recent literature on balanced scorecard systems and, more generally, multidimensional performance measurement stress the necessity to translate the strategy into action, but it has only highlighted this necessity (Kaplan & Norton, 2004). The moment an organization decides to determine its Internet position, the removal of any ambiguity regarding its future direction becomes, in the opinion of this writer, imperative for the success of the project. If, in fact, you accept the assumption that the Internet is a lever available to top management to pursue and determine corporate strategy, how can you hope to use this tool efficiently and effectively if it is not clear for what you have to use it?

It is also indispensable to clearly identify and then share the objectives of the Web strategy, or rather, how the potential that the Internet offers should be used to support the corporate strategy of the organization.

Therefore, the first phase of the project involved identifying, given the aim expressed in the corporate strategic plan, which contents, information, or services to activate in each of the areas in which the Internet's virtual space occupies (VIS, VDS, VCS, VTS).

To this end, Tables 1, 2, 3, and 4 show the results obtained relating to the most relevant content. (For the VIS area, content relating purely to the organization such as a general description of the organization, its history, or how to get to it have been omitted.)

Each table shows, for each area, the contents proposed for activation and the strategic aims that can be supported in a complementary or alternative way. For example, activation of a service that informs patients in real time about waiting times for the various types of treatment helps to facilitate access to services and greater transparency of levels of efficiency of the organization, and it creates value for the patient. In the VDS area, activation of online consultation helps to make patient-doctor relationships more loyal through greater interaction, branding, and so forth.

At the end of this phase, the map of Internet presence was drawn up in light of the information emerging from the corporate strategic plan (Figure 4b).

Table 1. Alignment of business strategy and Web strategy in the VIS area

SITE CONTENTS	STRATEGIC AIMS
Clinical Specialities	• Evaluation of inside expertise • Branding
Waiting Times	• Greater attractiveness of MD's offering • Facility of creation of new contacts
Scientific Information	• Support for international development strategy • Transparency of levels of efficiency of organization
MD/Tour Virtual Architecture	• Creation of value for patients • Facilitation of access to MD's services • Creation of value for specialist doctors and for GPs
Training/Congresses	• Greater participation and loyalty of GPs • Increase of specialized contacts • Improved image of MD as an advanced organization • Improved image as specialized organization from a scientific point of view

Table 2. Alignment of business strategy and Web strategy in the VDS area

SITE CONTENTS	STRATEGIC AIMS
Online Consultation	
Education	• Creation of value for patients • Participation and loyalty of GPs and patients
Health-care Glossary	• Branding
Personal Health Scorecard	• Greater interaction • Improved image of MD as specialized organization
Drug Information	• Continuous patient support during treatment
First Aid & Self-care Guide	• Evaluation of internal expertise • Greater awareness of MD's offering
Health Lifestyle Planners	

Aligning the Web Strategy with Inside and Outside Acceptance of the Internet

Aligning the Web strategy with the business strategy is far from enough. Literature unanimously agrees on the need to subject any creation process of the strategy to confrontation with the environment in which the organization works. Environmental analysis is, in fact, the corporate process that gives top management the information necessary to clearly understand the principal trends that condition the environment

Table 3. Alignment of business strategy and Web strategy in the VCS area

SITE CONTENTS	STRATEGIC AIMS
Virtual Medical Rooms	• Development and sharing of medical/scientific knowledge • Sharing of best practices
Forum	• Facilitated interaction between doctors/specialists and GPs • Loyalty of GPs • Creation of value for patients
Chat Line	• Participation and loyalty of GPs and patients • Branding • Greater interaction • Improved image of MD as advanced organization

Table 4. Alignment of business strategy and Web strategy in the VTS area

SITE CONTENTS	STRATEGIC AIMS
Online Appointment Booking	• Creation of value for patients
Payments For Treatment	• Participation and loyalty of GPs and patients
Distance Refertation	• Greater interaction
Complaints And Suggestions	• Improved image of MD as advanced organization and closer to users' needs
Customer Satisfaction	• Facilitated access to organization
Useful Forms	• Improved efficiency in undertaking of operative mechanisms

(the main economic-social factors, analysis of competitors and expectations of consumers, technological advances, regulations, etc.) and consequently formulate strategic alternatives (Piggot, 2002).

With a substantially analogous logic, the next step should be development of the Web strategy. Presence on the Internet determined from information emerging from a corporate strategic plan must be confronted with the level of knowledge of the Internet inside and outside the organization; in other words, it is necessary to check the validity of the choices made by confronting expectations, needs, and the level of the technological skills of the employees, patients, and any other stakeholders deemed relevant in this process.

To state the obvious, it would be perfectly useless to create a high-profile strategy if most patients do not have access to the Internet or do not intend to use the Internet to educate themselves or out of interest during their treatment. Similarly, one would not set up a consultation service online if the doctors refused, for various reasons, to collaborate in providing the service.

Aligning the Web Strategy with Outside Acceptance of the Internet: The Patients

Evaluation of the level of acceptance and knowledge of the Internet by patients was undertaken using a questionnaire in the last 2 months of 2004.

The research involved outpatients, too, as all patients have the same questions. There were, however, different means of medication. Inpatients received medication on the basis of direct interviews in the ward by the office of internal relations on several occasions. Outpatients received medication after completing a questionnaire at reception.

Each questionnaire was in two parts: the first to determine the patient's profile regarding his or her level of acceptance and awareness of the Internet, and the second to obtain the necessary information to check the feasibility of the Web strategy pre-selected for him or her. This procedure was the same for the questionnaires completed by the other stakeholders investigated (general practitioners and personnel).

The relatively high redemption rate, with an average value of about 60%, resulted in the composition of a sample of analysis deemed to be statistically significant with respect to the average population of patients who had access to the organization in the 2 months the study lasted. Particularly, the questionnaire was completed by 52% of outpatients (729 out of a total number of 1,402) and by 67% of inpatients (1,112 out of a total of 1,663).

The sample studied comprised a quota of outpatients of 40% (729 out of a total of 1,841 questionnaires received) and a larger group of inpatients, which was 60% (1,112 questionnaires out of a total of 1,841). In the sample investigated, there was a high incidence of patients over the age of 60 (about 26%) and a larger more relevant cluster of patients aged between 30 and 60 (about 57%). Regarding the educational profile, patients having an average to high background (high-school diploma or degree) made up 43% of the sample; the remaining patients had a middle-school diploma (26%), had an elementary-school education (25%), or did not answer the question (about 6%).

As expected, the younger and more educated patients were more technology savvy: 81% of patients under 30, 65% of those between 30 and 60 (Figure 5a), and around 8 out of 10 patients with a high-school diploma or degree (Figure 5b) stated that they used a personal computer (PC) and regularly used the Internet.

The use of technology literally collapses in the "weaker" bands both for age (only 2 patients out of 10 over 60 years old said that they used a PC or the Internet) and

Figure 5. Internet use by patients of MD Hospital

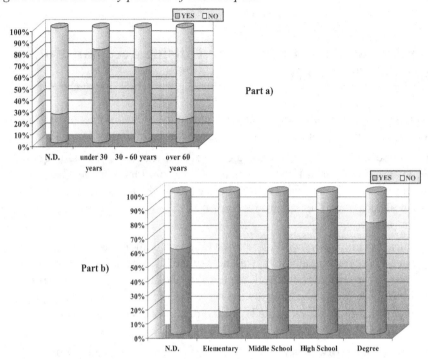

education (with a percentage that varies from about 15% for those with elementary-school education to 46% for those with a middle-school diploma). It should be pointed out that these groups, as already mentioned, made up over 50% of the sample.

Focusing attention on only the more educated patients (those with a high-school diploma or degree), the Internet was mainly used to look for information or for electronic post services.

The Internet was used to a lesser degree to look for information about healthcare, probably due to a lack of awareness of the role that the Internet can play in choosing a medical specialist or healthcare organization in which to be treated, which is still the case for Italian patients. There was a particularly low level of interest in e-commerce services and online interaction (forums, chat lines). The former, it can be reasonably assumed, lies in the risks involved in making online transactions. In the second case, the result was probably influenced by the rather advanced age of the analysis sample (the use of this type of service is widespread among younger Internet users) and on the particularly widespread use of other communication technologies such as the mobile phone.

The analysis revealed a very low level of use of electronic post services to interact with medical specialists, with medical doctors in general, or with healthcare organizations in general.

In the first section of the questionnaire, we wanted to investigate the efficacy of the previous Web strategy of the organization, and obtained, inevitably, rather disappointing results: Few patients stated that they knew of the Web site of the MD Hospital or had visited it (only about 17% of patients who made up the sample); when they had, the main reason was to look for information about the organization.

The second section of the questionnaire was, instead, more practical, regarding the approach to the new Internet presence of the organization. To this end, the tables shown in Figures 6, 7, 8, and 9 report the results of the investigation not only for patients but also for the other stakeholders taken into consideration. A green square indicates a high level of interest in the content, a yellow triangle is low interest, and a red circle is limited or even zero interest.

For the contents of the VIS area, patients said that they were very interested or quite interested in getting information about access times for various treatments (in 63% of cases) and, for individual clinical specialties, a list of services available (58%) and the members of the medical team (51%). Patients were less interested in being able to contact the medical specialists by e-mail (34%) or to read their curricula vitae (CVs; 32%).

Figure 6. Report on the evaluation of inside and outside acceptance of the Internet: VIS area

VIS Area Contents Report
(Virtual Information Space)

Minimum Unit of Analysis	Patients	Staff Commitment			GPs
		Activation	Development	Maintenance	
VIS Area Contents (*Virtual Information Space*)					
Clinical Specialties					
+ List of clinical services	■	■	■	■	■
+ Medical Team	■	■	■	■	■
+ Curriculum Vitae	△	■	△	△	■
+ Contact medical staff by e-mail	△	N. A.	N. A.	N. A.	■
+ Publications	N. A.	■	△	■	■
+ Research	N. A.	■	△	■	■
Training/Congresses	N. A.	■	△	■	■
Projects and business initiatives	N. A.	■	△	△	N. A.
Annual Report	N. A.	■	△	△	N. A.
Waiting Times	■	■	△	△	■
MD/Tour Virtual Architecture	■	■	△	△	N. A.

(N. A. = Not asked)

For the content of services able to be activated in the new portal of MD Hospital, patients showed considerable interest in the VDS area, stating they were very interested or quite interested in being able to do the following (Figures 7 and 8):

- Book appointments online (63%).
- Have distance refertation of their diagnostic investigations (59%).
- Use the educational services for pathologies (58%).

There was slightly less interest in the following:

- Programs for monitoring or improving patients' state of health (53%).
- Online consultation (57%).
- How to use the principal drugs (52%).

Figure 7. Report on the evaluation of the inside and outside acceptance of the Internet: VDS area

VDS Area Contents Report
(Virtual Distribution Space)

Minimum Unit of Analysis	Patients	Staff Commitment			GPs
		Activation	Development	Maintenance	
VDS Area Contents (*Virtual Distribution Space*)					
Online consultation	☐	☐	△	△	☐
Educational	☐	☐	●	●	N. A.
Personal Health Scorecard	☐	△	●	●	N. A.
Health Lifestyle Planners	☐	△	●	●	N. A.
Telemedicine	N. A.	☐	●	●	N. A.
Scientific Information	N. A.	△	△	△	△
Drug Information	☐	☐	●	●	N. A.

(N. A. = Not asked)

Figure 8. Report on the evaluation of inside and outside acceptance of the Internet: VTS area

VTS Area Contents Report
(Virtual Transaction Space)

Minimum Unit of Analysis	Patients	Staff Commitment			GPs
		Activation	Development	Maintenance	
VTS Area Contents (*Virtual Transaction Space*)					
Booking/Paying appointments on line	□	□	△	△	△
Distance refertation	□	□	●	●	□
Look for a doctor	□	□	●	●	N. A.
Complaints and suggestions	□	□	●	●	N. A.
Training courses	N. A.	□	N. A.	N. A.	□

(N. A. = Not asked)

In contrast, there was much less interest in the interactive online services in the VCS area such as the forum and the chat line (18% of responses) as is shown in Figure 9.

Aligning the Web Strategy with Outside Acceptance of the Internet: General Practitioners

Using the Web to increase the involvement of GPs is one of the principal aims of the strategic plan of MD Hospital.

This is no surprise. The general practitioner plays an important interface role between the National Health Service and patients, directing them and in many cases guiding them in their choice of specialist and/or the organization in which they might resolve their health problems.

The questionnaire was given to doctors during training courses organized at MD Hospital and sent by e-mail. Intentionally, the sample was prevalently composed of people who work inside the catchment area of the hospital. Even if it is not statistically significant (43 completed questionnaires were received), it allows interesting evaluations about the attitudes of the GPs to the use of ICT technologies in general, and to the Internet content of the future portal of MD Hospital in particular.

The GPs who constituted the sample of investigation demonstrated a good level of confidence about telecommunication technologies: PC use, Internet use, and use of electronic post services were particularly high (the number of doctors who claimed to be frequent or quite frequent users of PCs and the Internet was about 70%). The Internet was used principally for professional updating and markedly less for looking for information about healthcare organizations. It was very rarely used to communicate with other professionals or take distance courses, and almost never used to communicate by electronic post with patients.

Regarding the VIS area of the site of MD Hospital, the GPs showed a general interest in the activation of content that offers information about activities carried out by the clinical specialties of the organization, such as the list of services available, the composition of the medical team, publications, research activity, and so forth. Here, the number of positive responses (those "very interested" or "quite interested") reached 95%.

For the more advanced content (Figures 7, 8, and 9), the GPs were particularly interested in online consultations (80% positive responses) in the VDS area, participating in forums involving medical personnel of the organization (70%) in the VCS area, and remote access to the results of diagnostic examinations (85%) in the VTS area. Regarding this area in particular, medical personnel held that it was useful not only to be able to find out through the site about the services offered by the organization (in 90% of cases), but also to be able to carry out online enrollment (85%), to download support materials (85%), and to use distance learning courses (72%). It was seen as slightly less of a priority to be able to make online payments to enroll for the courses (65%).

There was less interest in booking appointments online (54%) in the VTS area, and the availability of content of a scientific nature (51%) and the virtual library in the VDS area of the site.

Aligning the Web Strategy with Inside Acceptance of the Internet: The Members of Staff

The investigation about the degree of acceptance and awareness of the Internet in the organization was carried out in December 2004 and involved all 305 members of the staff. Apart from the average redemption rate, which was significantly high (204 questionnaires were completed, 67% of the total), it should be pointed out that the sample involved three main clusters:

Figure 9. Report on the evaluation of inside and outside acceptance of the Internet: VCS area

VCS Area Contents Report
(Virtual Communication Space)

Minimum Unit of Analysis	Patients	Staff Commitment			GPs
		Activation	Development	Maintenance	
VCS Area Contents *(Virtual Communication Space)*					
Forum	△	△	●	●	☐
Chat-line	△	●	●	●	N. A.
Virtual Medical Rooms	N. A.	△	△	△	N. A.

(N. A. = Not asked)

- **Nursing staff and radiology technicians:** representing just under half of the sample (48%) with a redemption rate of 81% (98 of 121 questionnaires were completed).

- **Medical personnel:** representing just under a third of the sample (29%) with a redemption rate of 58% (out of a total of 102 doctors, 59 questionnaires were received).

- **Administrative and technical staff:** representing just under a quarter of the sample (23%) with a redemption rate of 57% (47 out of 82 questionnaires were received).

Focusing immediately on the key messages for the development of a Web strategy, it can be affirmed that, generally, the informative contents in the VIS present few problems: All employees at different levels and in different ways were in favor of activating the content proposed (Figure 6).

It should be pointed out, however, that with this general alignment, there was not always such a high commitment regarding the involvement of individuals in the planning or development of the content, or, though to a lesser degree, the successive maintenance phase. In fact, the percentage of staff who stated they were "very willing" or "quite willing" to be involved in the planning of the information content or the site's services did not exceed 30% of the total, while, for maintenance, the percentage rose to 50%. Such a low level of inclination to participate is even more pronounced when we take a more detailed look at the results regarding the medical-personnel cluster, especially concerning activation of the more complex service content (online consultations or personal health scorecards in the VDS area, for example), which demand a more intense cooperation: Here, the percentage of staff who claimed to be very or quite willing to participate fell to 5%.

More specifically, the content that raised most interest were those indicating waiting times for services; here, the willingness of the personnel (medical, nursing, and administrative) to be involved in the planning (30%) and successive maintenance (47%) of the site was good. There was overall positive support for involvement of the medical personnel (at least from the general average of results) in the content involving medical practice such as the indication of the clinical specialties, the list of services available, and the composition of the various teams and their curricula vitae.

The area concerning the VDS was particularly problematic because here the Internet not only redefines the doctor-patient relationship, but profoundly impacts the operative mechanisms that involve the medical and nursing staff (Figure 7).

For this reason, this phase of the project was particularly delicate not only from a technological point of view and for the estimation of costs, but also for the reengineering of clinical processes and related performance-evaluation mechanisms. Here the personnel of MD Hospital were principally focused on three main areas:

- Telemedicine
- Online consultation
- Education

Analyzing the profile of the strategic benefits that may result from the activation of such service content, it must be stressed that, according to the personnel in general, but particularly the medical staff, there is no resulting reduction of costs: On average, only 33% of those interviewed (48% for telemedicine) claimed that the activation of these contents could bring financial benefit. The value added would provide the following:

- **An improvement of the organization's image:** according to 83% of telemedicine staff, 86% of those in educational services, and 56% of those in online consultation.
- **Greater innovation:** 81% of telemedicine staff, 82% of those in educational services, and 85% in online consultation.
- **Patient satisfaction:** 85% for online consultation, 84% for educational services, and 80% for telemedicine.

- **Improvement in quality:** to a relatively lower degree than for the previous variables with 74% of personnel agreeing for telemedicine and educational services, and 62% for online consultation.

Distance programs for the management and improvement of patients' health seem, on the other hand, to have met with less approval: The percentage of personnel who believe that the activation of theses contents would result in benefits in terms of reduction of costs, improvement of quality and image, greater innovation, and patient satisfaction is around 15%.

For the VTS, the situation was generally the same as for the VDS area. In this case, the aspects most involved concern management (Figure 8).

In this area, focus shifts toward managerial concerns. Once more, however, there is scarce commitment by the medical staff and the personnel in general, who do not want to be or cannot be involved in the planning and development phase and the successive maintenance of the various service content.

The aspects that attracted most attention were online booking of appointments and distance refertation. Again, it can be affirmed that the personnel of MD Hospital hold that activation of these contents generates strategic benefits not so much for the reduction of costs (only 58% of those interviewed said that they very much or quite agreed), but rather results in the improvement of the following:

- **The organization's image:** 90% of those interviewed believe that activation of online booking brings benefits here, 83% believe the same for online refertation.

- **Innovation:** the percentage here reaches 87% both for online bookings and remote refertation.

- **Customer satisfaction:** 89% for online booking of appointments, 86% for remote refertation.

- **Overall quality of services offered to patients:** 76% for the booking of appointments online, 81% for remote refertation.

In the VTS area, the activation of content that might enable the patient to quickly and easily search for a doctor in the organization in order to book an appointment or make suggestions or complaints may generally be agreed to merit activation, but it is very far from being seen as bringing strategic benefits. Personnel who very much or quite agree that activation of these contents results in cost savings, improvement of the organization's image, or patient satisfaction were never more than 30% of those questioned.

In this final area in which the Internet's virtual space can be occupied (VCS), there was a very low level of commitment by the personnel of MD Hospital. There was little interest in the interaction offered by the Internet (forum and chat line): Only 43% claimed to be very much" or quite in favor of activating these contents. Even fewer were those who were willing to collaborate in the planning of these contents and successive maintenance (on average, 20%). There is an amber light too for Virtual Medical Surgery, an area of virtual space on the site of MD Hospital where doctors and paramedical personnel could exchange material, papers, opinions, and guidelines on treatment protocols, creating in some cases a real, valid system of management of the organization's knowledge. The results of the investigation carried out (percentages that are around 50% for those who claim to be in favor of activation and successive involvement in the development and maintenance phases) suggest there should be further verification of the functioning and usefulness of such content.

Alignment of the Web Strategy with Inside and Outside Acceptance of the Internet: Summing Up

Taking an overall look at the results obtained in the three areas of analysis (patients, GPs, staff) leads to some observations about the plausibility of the Web strategy concerning the variables investigated.

Let us look at the points that emerge on one particularly crucial aspect: online consultation. Corporate strategy, consequently the alignment between corporate strategy and Web strategy, indicates that the activation of these contents is necessary to support corporate branding and to initiate a process of the creation of loyalty, for both general practitioners and patients, through greater interaction, which is part of a general policy to create value for the patient. The testing of this decision with respect to the other corporate stakeholders and aligning the Web strategy with the level of acceptance and knowledge of the Internet brought completely unpredictable results.

Patients and doctors, in fact, were in favor of and particularly interested in the activation of these contents, supporting and confirming, in a sense, the choices of senior management. The internal personnel (though here we really refer to the doctors employed by the organization), while recognizing the importance of activating the contents in pursuance of corporate aims, declared little interest or rather little inclination to collaborate in the development and planning phase or in the successive maintenance phase of the service itself.

The problem, clearly, is no small one. The approach selected allowed us, however, to see this possible cause of failure of the strategy right from the beginning, and to intervene immediately in the planning stage, in which decisions taken can be changed with little consequence since the correlated investments have not yet been made.

The reason for the obstructive behavior of the medical personnel at MD Hospital was not because of lack of knowledge (we do not feel ready or we do not want to use this type of technology), but more simply due to the consequences that activation of this service in virtual space would have by reducing real time for direct contact with patients. The workload for the medical personnel to deal with 150 online consultations a year has been estimated as equivalent to 40 working days.

This inevitably has consequences: The activation of a high-profile Web strategy requires a redefinition of corporate organizational processes, and a rethinking of operative mechanisms for determination and assignation for each operative unit and each manager of the objectives and correlated systems of evaluation of performance. All of this in anything but virtual reality is indisputably real.

Alignment of the Web Strategy with the ICT Strategy

Developing an Internet strategy means confronting the enabling technology that makes materially possible all that has been discussed on the preceding pages.

It is not one of the aims of this chapter to make a technical evaluation of what and how much technology should be acquired and installed to support the choices made in terms of strategies for the colonization of the virtual space of the Internet. We would, however, like to make some observations regarding the entity and particularly the composition of the investment involved.

For the former, it should be pointed out that in the literature (Coile, 2002; Robert & Racine, 2001) there is major disagreement between those who see the Internet as a technology that offers substantially good value and those who hold instead that an effective Web presence requires a considerable monetary investment. We believe it is necessary to put these views into context.

Investment in the development of a high-profile Internet strategy, as for any other investment, should of course be measured according to the size and the general financial situation of the organization. That is not all: If the investment is the right one, then it will bring its own rewards, producing new value over time. This, however, is not a peculiarity of the Internet.

Unlike other technologies more directly involved in providing treatment services (computerized axial tomography [CAT] or nuclear magnetic resonance [NMR], for example), in the case of the Internet, it is considerably more difficult to evaluate the return on an investment and, paradoxically, the entity of the investment. During this project, it was practically impossible to estimate, in a way that would not be subjective or arbitrary, the strategic benefits in terms of branding, corporate image, the creation of patient loyalty, and so forth—benefits that, moreover, are only potential.

As well as the quantification of investment, two further factors of complexity must be considered.

First, activation of a Net presence should involve not only the application of closely correlated technology (Web server, development software, software for content management, hosting, Web infrastructure), but also a revisiting of existing technology that is not immediately compatible (from simple adjustment of individual work positions to a revisiting of equipment and software for distance refertation).

The human factor, therefore, is an essential component of investment; it has already been emphasized that the provision of virtual contents that are much used services (online consultation, education, programs for monitoring and improving patients' state of health) brings an additional workload (real) for the medical personnel. This is, however, something that involves all the site's contents and, to a considerable degree, all the personnel in the organization. For example, updating informative contents about the organization (responsibility of the external-relations office) or dealing with the booking of appointments or online complaints is more work for the people taking the bookings and the public-relations office.

Figure 10 shows, with a purely monetary criterion and not economic competence, the entity of investments necessary in the 3 years involving the Web strategy of MD Hospital. In particular, with the term *external resources*, we have indicated

Figure 10. The investment plan for the development of the Web strategy of MD Hospital

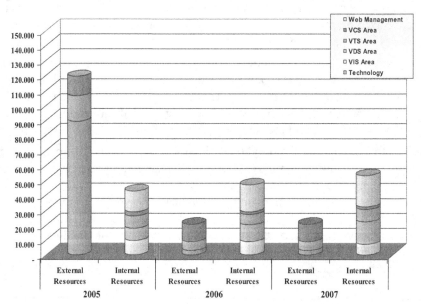

the acquisition costs of the technologies and of outside consultation regarding the site in general, and activation of the contents in the individual areas in which the Web space was organized. With the term *internal resources*, we have instead indicated both the costs involved in activating an office responsible for managing the site (Web management) and the costs generated by the increased workload of the organization.

It must be underlined that this component progressively increases in importance as traffic to the Web site increases, and there is a consequent greater use of the Web services.

The Development Process of the Web Strategy

Planning the Web strategy systematically and rationally is not enough: It is necessary to watch over the activation process, too, with the utmost attention to prevent the project failing because of critical situations generated during this phase. As with the planning of the corporate strategy, here it is necessary to decide who is going to do what by when and in which order of priority, determining at the same time the mechanisms to evaluate how much progress has been made, or rather the performance accomplished compared with the aims of the project.

While the first phase of the project in MD Hospital involved a work group composed mainly of top management and a limited number of other staff, in the development phase we tried to get as widespread participation as possible by all the personnel. Naturally, it was the results that emerged from the evaluation of the level of acceptance of the Internet inside the organization that revealed how particularly important motivation and the quality of this collaboration were. It was decided to present the results of the planning phase in full, using this as an opportunity to confront possible problematic situations, correlated solutions, and, in general, prospects of Internet use in support of corporate strategic objectives.

Development of the strategy was entrusted to work teams for the various clinical and technical-administrative operative units. Each team was normally composed of a manager and at least two other members belonging to the same operative unit. Each team was responsible for the development of one or more items of content regarding information or services.

The external-relations office, for example, was entrusted to produce content regarding the organization in general (its profile, history, and news; how to access services; etc.). Content relating to the various clinical specialties (services provided, research activity, composition of teams, CVs, etc.) were the responsibility of the teams of the relative operative units. Organization of the more complex content concerning services (online consultation, education, programs for monitoring and improving patients' health) was entrusted to teams in which most members worked in different operative units. Each team had deadlines according to the order of priority assigned on activation of the various contents.

Coordination was entrusted to the office of external relations, while verification of the coherence of the work regarding indications emerging in the planning phase was the responsibility, as it could only be, of the top management.

A Web marketing plan was established with the aim of informing corporate stake-holders about the times and scope of the initiative to pursue the effectiveness of the Web presence. Since Italian healthcare organizations cannot advertise their activities, action undertaken here involved the following:

*   Getting listed on the principal search engines.

*   The development of collaboration with portals relating to healthcare themes in general.

*   The development of collaboration with portals relating to specific patholo-gies.

*   The preparation of promotional materials and a newsletter in both paper and electronic forms.

Figure 11. Measurement of performance with respect to company stakeholders

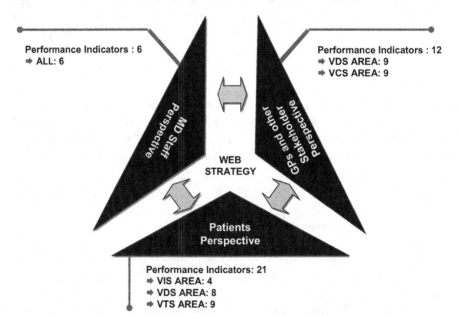

Monitoring and evaluation of the results obtained regarding predefined objectives is an essential condition of any strategic plan so that even the best orchestrated are not just academic exercises. The indications that emerge from the evaluation of performances inevitably constitute the starting point for the formulation of successive strategies.

For this we decided to develop a system of indicators (Figure 11) with three principal perspectives, one for each of the stakeholders held to be relevant in the planning phase of the Web strategy (patients, GPs, staff).

In general, for each perspective, there is at least one KPA corresponding to the areas of the Web space in which contents are predisposed to satisfy the precise need for information of the stakeholder.

Consequently, the patients' perspective is in three KPAs: the VIS, VDS, and VTS; there are no indicators for the VCS area simply because results from the investigation on the patients led to a definition of a Web strategy that does not foresee the availability of informative or service content for patients in this area (Table 5).

The same can be said for the GPs' perspective, which has only two performance areas, VDS and VTS, because it is particularly in these two areas that content specifically aimed at satisfying the needs of this type of stakeholder have been developed and

Table 5. The KPI system for perspective of patients

WEB SPACE	PROGRESS	KPI	CALCULATION	OWNER	TARGET			ACTUAL 2005
					2005	2006	2007	
VIS	1	Traffic to clinical specialties	Number of accesses to clinical specialties	External Relations	7,200	8,800	12,000	9,435
	2	Index of delocalization of MD	Number of non-local out-patients and inpatients/total number of patients	Health Management	not active in 2005			
	3	Traffic to international area	Number accesses to international area	Health Management/ General Management	90	110	150	103
	4	Index of internationalization of MD	Number of international patients/total number of patients	Health Management / General Management	not active in 2005			

Table 5. continued

VDS	5	Traffic to educational material	Number of accesses to educational material	Scientific Coordinator	3,600	4,400	6,000	4,570
	6	Index of branding on educational material	Number of requests for further information/number of accesses to educational material	Scientific Coordinator	20%	25%	30%	17%
	7	Traffic to online consultations	Number of accesses to online consultation	External Relations	450	550	750	2,300
	8	Index of use of online consultation (patients)	Number of consultations requested by patients	Scientific Coordinator	81	99	135	120
	9	Index of efficacy of online consultation (patients)	Number of consultations with positive outcomes/ number of consultations requested	Scientific Coordinator	100%	100%	100%	100%
	10	Traffic to medical glossary	Number accesses to medical glossary	Scientific Coordinator	3,600	4,400	6,000	3,121
	11	Traffic to drug information	Number of accesses To drug information	Scientific Coordinator	3,600	4,400	6,000	3,041
	12	Traffic to online appointment booking	Number of accesses to online booking of appointments	External Relations	3,600	4,400	6,000	5,431
VTS	13	Index of use of online booking	Number of online bookings made	External Relations	2,520	3,520	5,400	2,410
	14	Index of frequency of online booking compared to traditional booking methods	Number of online bookings/total number of bookings	External Relations	**not active in 2005**			
	15	Index of efficacy of online booking service	Number of online bookings carried out/ number of online bookings	Bookings	90%	95%	100%	100%
	16	Traffic to remote refertation	Number of accesses to remote refertation	External Relations	3,600	4,400	6,000	4,971
	17	Index of use of remote refertation service	Number of reports requested online	Laboratory	2,880	3,740	5,400	2,831
	18	Index of exploitation of remote refertation service compared to traditional methods	Number of online reports/ total number of reports	Laboratory	**not active in 2005**			

Table 5. continued

19	Traffic to online complaints and suggestions	Number of accesses to on-line complaints and suggestions	External Rela-tions	270	330	450	342
20	Index of use of online complaints and suggestions service	Number of online complaints	External Rela-tions	216	`281	405	171
21	Index of exploitation of online complaints service compared to traditional methods	Number of online complaints/total number of complaints	External Rela-tions	**not active in 2005**			

which therefore need to be monitored particularly carefully to verify the success (or failure) of the strategy adopted (Table 6).

For the personal perspective of the staff, a different approach was taken. The doctors, nurses, laboratory staff, and technical-administrative personnel make up the structure that has to ensure the correct functioning of the portal through continual updating of the informative content and provision of the real services available to patients, more than any other provider of interest to the organization. The Web strategy of a healthcare organization can be carried out only if the personnel (or at least those in a position of responsibility) undertake to provide the information and services that constitute the essence of the Internet presence of the organization. In no performance area were anomalies found, but a system of indicators was established to check how and to what degree (how efficiently and how effectively) the staff worked to attain these predefined objectives (Table 7).

In each area, one or more indicators of performance were activated to monitor as thoroughly and efficiently as possible the capacity of the strategy to fulfill the expectations of potential users of the site, determining, in the final analysis, the success or not of the project.

Regarding the perspective of patients, for example, in the VDS area, three indicators were activated that enable the monitoring of the success of the services in terms of the following.

- Traffic (number of visitors to the unit minimum of reference analysis).
- Use of the contents of the service (number of consultation requests made compared to the number of visitors recorded).
- Effectiveness of the service (number of consultations concluded satisfactorily compared with the number of consultations recorded).

Table 6. The KPI system for the GPs' perspective

WEB SPACE	PROGRESS	KPI	CALCULATION	OWNER	TARGET			ACTUAL 2005
					2005	2006	2007	
VTS	1	Traffic to training area	Number accesses to training area	External Relations	1,200	1,300	1,400	2,100
	2	Index of use of training area by GPs	Number of GPs enrolled through site /number of GPs participating in training courses	Scientific Coordinator	70%	80%	90%	59%
	3	Index of use of download function of training area	Number of downloads /number GPs participating in training courses	Scientific Coordinator	70%	80%	90%	49%
VDS	4	Traffic to online consultation	Number of accesses to online consultation	External Relations	450	550	750	2,300
	5	Index of exploitation of online consultation (GPs)	Number of consultations requested/number of accesses to online consultation	Scientific Coordinator	100	150	200	57
	6	Index of efficacy of online consultation (GPs)	Number of consultations with positive outcomes/ number of consultations requested	Scientific Coordinator	100%	100%	100%	100%
	7	Traffic to virtual library	Number of accesses to virtual library	External Relations	1,300	1,500	1,700	1,950
	8	Index of total use of virtual library	Number of user IDs requested	Scientific Coordinator	250	500	800	321
	9	Index of use of virtual library for research purposes	Number of research requests/number of user IDs authorized	Scientific Coordinator	30%	40%	50%	40%
	10	Index of users' activity in virtual library	Number of subjects of discussion proposed/number of user IDs authorized	Scientific Coordinator	10%	10%	10%	12%
	11	Index of use of virtual library for bibliographical research	Number personal bibliographical searches /number of user IDs authorized	Scientific Coordinator	2%	5%	10%	1%
	12	Index of use of virtual library for requests for documentation	Number of documentations completed/number of user IDs authorized	Scientific Coordinator	1%	3%	5%	1%

Table 7. The KPI system for the personnel perspective

WEB SPACE	PROGRESS	KPI	CALCULATION	OWNER	TARGET			ACTUAL 2005
					2005	2006	2007	
ALL	1	Index of sharing Web-strategy compared to medical personnel	Number of doctors who request user IDs for remote management of Web pages/total doctors at MD	External Relations	40%	50%	70%	37%
	2	Index of sharing Web-strategy compared to technical personnel	Number of administrative technicians who request user IDs for remote management of Web pages/total administrative technicians	External Relations	60%	70%	80%	64%
	3	Index of sharing Web-strategy compared to nursing personnel	Number of nurses who request user IDs for remote management of Web pages/total nurses	External Relations	60%	70%	80%	81%
	4	Index of productivity of MD personnel	Average number of pages published by authorized MD users	External Relations	3	5	7	4.3
	5	On-time response by doctors to online requests for consultation	Number of late online consultations (over 48 hours)/number of online consultations requested	External Relations	10%	5%	1%	0%
	6	Index of commitment to Web strategy	Interlocutors participating in coordination meetings on Web strategy/number of interlocutors at Web-strategy meetings	External Relations	90%	90%	90%	78%

A particularly critical aspect in the definition of a strategy and the planning of performance-measurement mechanisms is the identification of objective values at the budget level, in other words, the partial results or milestones that have to be attained year by year to attain the final strategic objective. This is even more critical for decisions regarding the Web strategy in situations such as that of MD Hospital, for which there are no historical references (the preceding Internet position was vastly different) or benchmark values (for example, total traffic recorded for the site).

We decided, therefore, to identify an objective value at a general level (numbers of accesses to the home page of the site recorded) and to successively derive objectives for each KPI. Not having available, as stated above, any parameters of reference, we took our indicators from the propensities of Internet use provided by the investigation of the patients. We estimated, very conservatively, that of the around 36,000 patients (inpatients and outpatients) who visit the organization in one year, 25% (around 9,000 patients) would have used the new portal of MD Hospital at least once to look for information about the organization or to access the services offered. We deliberately tried to simplify as much as possible the algorithm of quantification of the objective, without taking into account, for example, the possibility that the same patient may access the site several times in the course of a year (or a month or even a week). In fact, the requirement of the top management of MD Hospital was not so much to have an objective calculated on the basis of a rigid and rigorous scientific approach, but rather to have a threshold value from which investment could be justified and, secondly, some elements of evaluation of performance that could be completed and refined successively. With this in mind, objectives were fixed for 2006 (11,000 accesses to the home page with an increase of 25% compared to 2005) and 2007 (15,000 within a year, an increase of 35%).

Subsequently, the objective values for each KPI in the different areas of performance relating to the perspective of patients and perspective of GPs were derived in the same way. For the patients' perspective, it was decided to not activate those indicators fundamentally aimed at verifying the weight of the Internet route compared to more traditional means of contact with the patient, such as, for example, the online booking of appointments or the number of reports available on the Web out of the total number of reports produced by the laboratory. The reason for this decision is the management's wish to have available, probably at the end of the first year, further knowledge that can be used to define objectives while in greater cognition of the facts.

An obviously different approach needed to be taken to define the objectives for the KP of the personnel area. In this case, a process of negotiation was started with the various managers of the medical, nursing, and technical-administrative staff similar to that for the preparation of the corporate budget and the budgets of each area of responsibility (departments rather than individual operative units). The degree of innovation of the processes of analysis, together with a consequent lack of experience and a substantial absence of indications in reference literature, has led us to the following:

- **A definition of the objectives at an overall corporate level:** There has not, therefore, been an articulation and consequent negotiation of such objectives for the two distinct levels of accountability typical of healthcare organizations (the department on the first level, the individual operative unit on the second level).

- **A putting off of the definition of the incentive mechanisms tied to performance until 2006:** That is when top management and other corporate decision makers will have a clearer picture of how and to what degree the Web strategy can support the attainment of corporate strategic objectives.

Finally, analysis of the final results of 2005 (final columns in Tables 5, 6, and 7) indicates that in general the predefined objectives have been largely achieved due to the cautious approach that characterized the preceding phase of the program and the novelty effect of the content, information, and services offered by the new portal.

What emerges, valid for the three perspectives of performance identified, is that it is necessary to change direction by reviewing the objectives and adapting them in the light of what has emerged during the first year. In a way, the need for this change of direction had already been felt in the same phase of planning the 2005 objectives because of the problems previously identified.

In the details regarding the perspective of patients, there emerge some interesting facts, particularly regarding those contents for which results were not in line with expectations, such as the traffic recorded for drug information or the glossary of medical terms, or the fact that the number of online appointment bookings was slightly lower than expected despite the high volume of traffic recorded in this area. These involve, according to the evidence, different situations. In the first two cases, it is likely that the informative contents offered aroused little interest in the visitors to the portal (they must therefore be made more interesting or more accessible) or there was an error in evaluating the objectives due to the lack of experience that has been previously mentioned. For online bookings, the potential that seemed to be due to the great interest aroused by the content (witnessed by the number of accesses) was unfounded and did not translate into an effective use of the service. It should be investigated whether this is due to an overcomplicated booking procedure (and work is already under way to rectify this) or to a lack of trust or to fear of technology by the patients.

The same considerations can be made for the GPs' perspective for training courses offered by and for the organization.

Finally, the results obtained for the personnel perspective have been judged as good overall by the corporate management.

Limitations and Further Research Areas

The project was undertaken in MD Hospital and was of an exploratory nature, with the aim of evaluating in the field and determining a methodological approach resulting from research ongoing for almost 4 years.

It is evident that the validation of models and observations presented in these pages can only occur following further confrontation with clinical evidence both from inside, in the case of research, and outside through a survey to investigate how and in what ways to determine an Internet strategy.

It should further be remembered that in this work, attention was focused exclusively on the public use of the Internet, but in reality, Web technology may be used to set up a private Web presence (intranet) aimed at the employees and internal resources of the organization rather than a public presence with limited access (extranet) aimed at suppliers, customers, and other authorized stakeholders. A complete and exhaustive picture of the phenomena investigated would necessarily involve a widening of the hypotheses and models of research in these areas of analysis.

Conclusion

It's much more difficult than I thought!

This statement, made by one of the directors of the organization in which the research reported here took place, very accurately reflects the typical view held in organizations regarding approaches to and ways of determining a Web strategy—the strategy that determines the Net presence of the organization. The problem lies not in the choice of the site's graphics or which images and text must be inserted, but rather how this presence in the virtual world, which generates consequences in the real world, can be best exploited to pursue corporate strategic objectives.

If we accept this assumption, organization of the informative content and services that a healthcare organization can or should offer on its own site must never be the result of improvisation or ignorance of the potential of the Internet, and neither should it be left in the hands of a consultancy or suppliers of technology except for purely operational aspects.

It is vital, instead, to prepare a proper strategic plan for the colonization of Internet space using a system that is in some ways similar to that for determining corporate strategy.

There is an important moment in planning when it is vital to check the alignment and coherence of the Web strategy with respect not only to the business strategy, but also to the level of awareness and acceptance of the Internet shown by the principal corporate stakeholders (patients, doctors, personnel, etc.) and the general use of other technologies (ICT strategy). The activation phase of the project has many critical elements that involve personnel of the organization, both clinical and technical-administrative, in identifying objectives and the assignation of connected responsibilities, the determination of a system that allows effective monitoring of performances, and the supply of necessary feedback for modifying and improving the strategies in the successive cycle.

References

AA.VV. (2000a). *Winning the loyalty of the e-health consumer.* Deloitte & Touche in, Given R. et al. (2000). Winning the loyalty of the e-health consumer, Deloitte Research. Available at http://www.dc.com/research.

AA.VV. (2000b). HealthCast 2010: Smaller world, bigger expectations. PricewaterhouseCoopers in, J. Rodgers et al. (2000). HealthCast 2010: Smaller world, bigger expectations, PricewaterhouseCoopers.

AA.VV. (2001a). *Understanding the NHS market for eHealth.* Deloitte & Touche in, Given R., et al. (2001a). Understanding the NHS market for ehealth, Deloitte Research. Available at www.dc.com/research.

AA.VV. (2001b). *Strategy and e-health.* Deloitte & Touche in, Given R. et al. (2001b). Strategy and e-health, Deloitte Research. Available at http://www.pwcglobal.com/healthcare.

AA.VV. (2002). Stanford-Makovsky Web credibility study 2002: Investigating what makes Web sites credible today. Retrieved from http://www.webcredibility.org in, Fogg, B.J. et al. (2002). Stanford-Makovsky web credible torday. A research report by the Stanford Persuasive Technology Lab & Makovsky & Co. Standford University. Available at http://www.webcredibility.org

Alemi, F. (2000). Management matters: Technology succeeds when management innovates.*Frontiers of Health Service Management, 17*(1), 17-30.

Angehrn, A. (1997). Designing mature Internet business strategies: The ICDT model. *European Management Journal, 15*(4), 361-369.

Assinform. (2004a). *Il rapporto Assinform sul mercato dell'IT nelle regioni Italiane.* Retrieved October 2004, from http://www.assinform.it

Assinform. (2004b). *Rapporto Assinform sull'ICT in Italia: 2004.* Retrieved October 2004 from http://www.assinform.it

Baraldi, S., & Memmola, M. (2006). How healthcare organisations actually use the Internet's virtual space: A field study. *International Journal of Healthcare Technology and Management, 7*(3/4), 187-207.

Buttignon, F. (2001). Strategia e valore nella Net-economy. *Il Sole 24 Ore, Milan, Italy.*

Coile, R. C. (2002). *The paperless hospital: Healthcare in a digital age.* Chicago: Health Administration Press.

De Luca, J. M., & Enmark, R. (2000). E-health: The changing model of healthcare. *Frontiers of Health Service Management, 17*(1), 3-15.

Elango, B. (2000). Do you have an Internet strategy? *Information Strategy: The Executive's Journal,* (pp. 32-38).

Fattah, H. (2000). Failing health. *Media Week, 10*(17), 100-103.

Flory, J. (1999). Healthcare communications approaches for an online world. *Marketing Health Services,* (pp. 25-30).

Fox, S., & Rainie, L. (2002). *The on line healthcare revolution: How the Web helps the Americans take better care of themselves* (Executive summary of research). Retrieved 2002, from http://www.pewinter.org

Glaser, J. P. (2000). Management response to the e-health revolution. *Frontiers of Health Service Management, 17*(1), 45-50.

Glaser, J. P. (2002). *The strategic application of information technology in healthcare organizations.* Jossey-Bass.

Goldsmith, J. (2000). How will the Internet change our health system? *Health Affairs, 19(1), 148-157.*

Goldstein, D. E. (2000). *E-healthcare: Harness the power of the Internet e-commerce & e-care.* Aspen Publication.

Kaplan, R. S., & Norton, D. P. (2004). *Strategy maps.* Boston: Harvard Business School Press.

Malcolm, C. (2001). Making a healthcare Web site a sound investment. *Healthcare Financial Management,* (pp. 74-79).

Malec, B., & Friday, A. (2001). *The Internet and the physician productivity: UK & USA perspectives* Working paper, EHMA Congress. Granada: EHMA Congress.

Minard, B. (2001). CIO longevity: IT project selection and initiation in the healthcare industry. *IT Healthcare Strategist, 3*(12), 1-12.

Minard, B. (2002). CIO longevity and IT project leadership. *IT Healthcare Strategist, 4*(1), 3-7.

Nicholson, N. (1999). *The Internet and healthcare.* Chicago: Health Administration Press.

Piggot, C. S. (2002). *Business planning for healthcare management.* McGraw Hill.

Porter, M. E. (2001). Strategy and the Internet. *Harvard Business Review, 79*(3), 63-78.

Robert, M., & Racine, B. (2001). *e-Strategy.* McGraw-Hill: NY.

Shapiro, C., & Varian, H. R. (1999). *Information rules.* Harvard Business School Press: Boston.

Solovy, A., & Serb, C. (1999). Healthcare's 100 most wired. *Hospitals and Health Network, 73*(2), 43-51.

Streeter, A. (n.d.). *Local doctors take to the Internet.* Retrieved 2000, from http://www.news-press.com

Vittori, R. (2004). Web strategy. *Franco Angeli Editore, Milan, Italy.*

Yin, R. K. (2003). *Applications of case study research* - Second Edition. Sage Publication, Thousands Oaks, California, USA.

Chapter II

A Framework for Delivering m-Health Excellence

Nilmini Wickramasinghe, Illinois Institute of Technology, USA

Steve Goldberg, INET International Inc., Canada

Abstract

Medical science has made revolutionary changes in the past decades. Contemporaneously, however, healthcare has made incremental changes at best. The growing discrepancy between the revolutionary changes in medicine and the minimal changes in healthcare processes is leading to inefficient and ineffective healthcare delivery, and is one, if not the significant, contributor to the exponentially increasing costs plaguing healthcare globally. Healthcare organizations can respond to these challenges by focusing on three key solution strategies, namely, (a) access, as in caring for anyone, anytime, anywhere, (b) quality, delivered by offering world-class care and establishing integrated information repositories, and (c) value, which is created by providing effective and efficient healthcare delivery. These three components are interconnected such that they continually impact on the other and are all necessary to meet the key challenges facing healthcare organizations today.

The application of mobile commerce to healthcare, namely, m-health, appears to offer a way for healthcare delivery to revolutionize itself. However, little if anything has been written regarding how to achieve excellence in m-health. This chapter serves to address this major void by presenting an integrative framework for achieving m-health, developed through the analysis of longitudinal applied research conducted by INET in conjunction with academe. After presenting this framework and discussing its key inputs, we then illustrate how the mapping of case data to the model enables the attainment of a successful m-health application to ensue and the benefits of adopting such a methodology.

Introduction

Currently the healthcare industry in the United States as well as globally is contending with relentless pressures to lower costs while maintaining and increasing the quality of service in a challenging environment (Blair, 2004; European Institute of Medicine, 2003; European Union [EU] Health and Consumer Protection, 2005; Frost & Sullivan, 2004; Kulkarni & Nathanson, 2005; Lacroix, 1999; Lee, Albright, Alkasab, Damassa, Wang, & Eaton, 2003; National Center for Health Statistics, 2002; National Coalition on Healthcare , 2004; Organization for Economic Co-operation and Development [OECD], 2004; Pallarito, 1996; *Plunkett's Healthcare Industry Almanac*, 2005; Russo, 2000; Wickramasinghe & Silvers, 2003; World Health Organization [WHO], 2000, 2004). It is useful to think of the major challenges facing today's healthcare organizations in terms of the categories of demographics, technology, and finance. Demographic challenges are reflected by longer life expectancy and an aging population, technology challenges include incorporating advances that keep people younger and healthier, and finance challenges are exacerbated by the escalating costs of treating everyone with the latest technologies. Healthcare organizations can respond to these challenges by focusing on three key solution strategies, namely, (a) access, as in caring for anyone, anytime, anywhere, (b) quality, delivered by offering world-class care and establishing integrated information repositories, and (c) value, which is created by providing effective and efficient healthcare delivery. These three components are interconnected such that they continually impact on the other and are all necessary to meet the key challenges facing healthcare organizations today.

In short, the healthcare industry is finding itself in a state of turbulence and flux (European Institute of Medicine, 2003; National Coalition on Healthcare , 2004; Pallarito, 1996; Wickramasinghe & Mills, 2001; World Health Organization, 2000, 2004). Such an environment is definitely well suited for a paradigm shift with respect to healthcare delivery (von Lubitz & Wickramasinghe, in press). Many experts within the healthcare field agree that m-health appears to offer solutions for healthcare delivery and management that serve to maximize the value proposition for healthcare. However, to date, little if anything has been written regarding how to achieve excellence in m-health. This chapter serves to address this apparent void.

Integrative Model for m-Health

Successful m-health projects require a consideration of many components. Figure 1 provides an integrative model that serves to capture all these key factors that we have identified through our research. What is highlighted by this figure is the need to address so many aspects in order to achieve m-health excellence, as we now expand upon.

Figure 1. A mobile e-health project delivery model

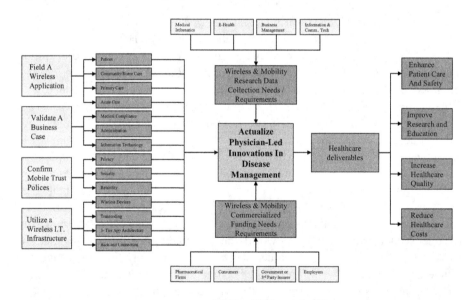

* Information and Communication Technology   © Goldberg & Wickramasinghe, 2004

Figure 2. Web of healthcare players (Adapted from Wickramasinghe et al., in press)

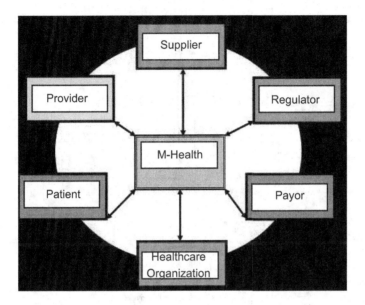

Web of Players

The first consideration is the people component. Any healthcare initiative, be it wired or wireless, must first and foremost be aware of all key actors that are involved in the delivery of healthcare. It is useful to think of these various actors as a web of players because they interact at different levels and degrees depending on the specific action or procedure. Figure 2 depicts the web of players that must be considered for healthcare in general and m-health in particular. From this figure, it is possible to see that m-health requires input and coordination between and within suppliers, payers, healthcare organizations, providers, regulators, and the patient if excellence is to truly ensue. Furthermore, all these players are represented in Figure 1, where their specific roles regarding any m-health project are also noted. For example, patients, providers, and healthcare organizations are amongst the primary inputs while payers and regulators as well as various suppliers are modifiers to the process of actualizing the physician-led solution.

Figure 3. IT architecture (Adapted from Wickramasinghe et al., in press)

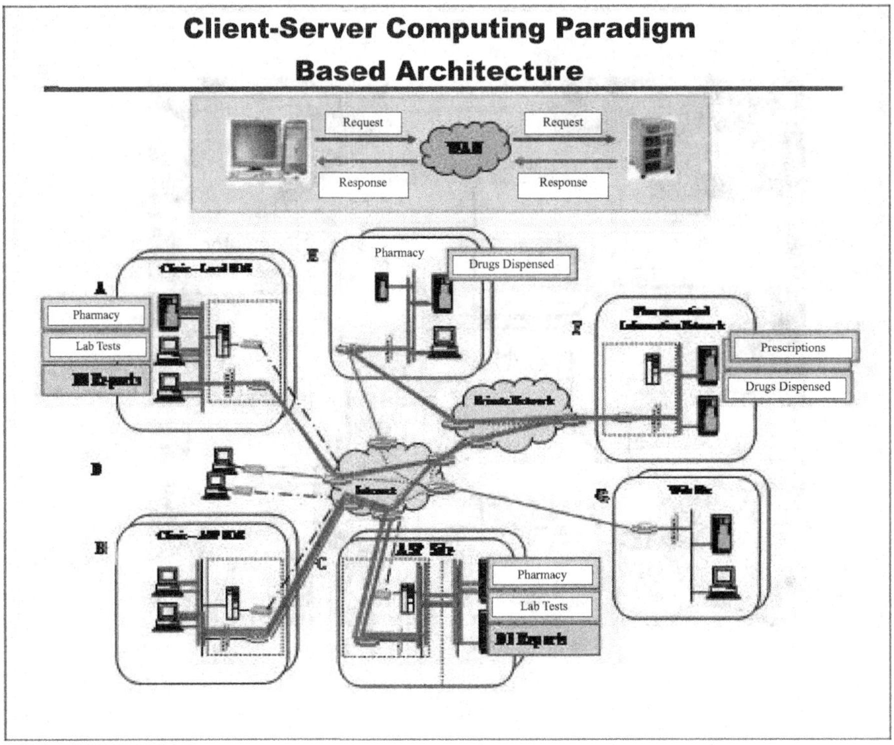

IT Architecture

The next important consideration is concerned with the existing IT infrastructure and architecture. Any m-health solution must leverage off existing IT architecture, wired and wireless, at the respective locations of the web of key players. Typically, in today's techno-centric world, this involves understanding the client-server computing paradigm as depicted in Figure 3. To support such a client-server architecture, special attention must be paid to the ICT infrastructure. The ICT infrastructure includes phone lines, fiber trunks and submarine cables, T1, T3 and OC-xx, ISDN (integrated services digital network), DSL (digital subscriber line), and other high-speed services used by businesses as well as satellites, earth stations, and teleports. A sound technical infrastructure is an essential ingredient to the undertaking of e-health and m-health initiatives. Such infrastructures should also include telecommunications, electricity, access to computers, the number of Internet hosts, the number of ISPs (Internet service providers), and available bandwidth and broadband access. Such IT infrastructure and architecture components form a key input into the designing and development of any m-health project as can be seen in Figure 1.

Figure 4. A standardized mobile Internet (wireless) environment

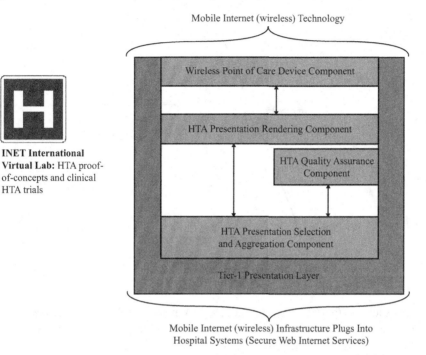

IT Architecture and Standard Mobile Environment

By adopting a mobile or wireless healthcare delivery solution, it is possible to achieve rapid healthcare delivery improvements that impact both the costs and the quality of healthcare delivery. This is achieved by using an e-business acceleration project that provides hospitals with a way to achieve desired results within a standardized mobile Internet (wireless) environment. Integral to such an accelerated project is the ability to build on the existing infrastructure of the hospital. This then leads to what we call the three-tier Web-based architecture (Figure 4).

In such an environment, Tier 1 is essentially the presentation layer, which contains the Web browser. However, no patient data is stored within this layer, thereby ensuring compliance with international security standards and policies like HIPAA (Health Insurance Portability and Accountability Act). Tier 2 provides the business logic including, but not limited to, lab, radiology, and clinical transcription applications; messaging of HL7 (Health Level 7), XML (extensible markup language), DICOM (digital imaging and communication in medicine), and other data protocols; and interface engines to a hospital information systems (HIS), lab information system (LIS), and radiology information systems (RIS), as well as external messaging

Figure 5. HIPAA triangle (Adapted from Fadlalla & Wickramasinghe, 2004)

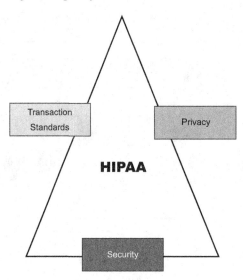

systems such as Smart Systems for Health (an Ontario Healthcare IT infrastructure project). Finally, there is the Tier 3 architecture, which consists of the back-end databases like Oracle or Sybase.

Security and Regulatory Conformance

The third primary input in Figure 1 is concerned with security, privacy, and reliability issues. Given the nature of healthcare data, adherence to adequate standards in this regard is imperative to the ultimate success of m-health delivery. In the United States, security, privacy, and standards for electronic submissions and the exchange of healthcare information are covered by HIPAA (Fadlalla & Wickramasinghe, 2004; *Health Insurance Portability and Accountability Act*, 2001; Moore & Wesson, 2002). It is useful to conceptualize this as a HIPAA triangle (Figure 5) that highlights the fundamental elements of the HIPAA regulation, namely, security, transaction standards, and privacy.

Security

According to HIPAA, a number of security criteria must be met by all electronic healthcare transactions. Some of these criteria directly affect how healthcare systems can be accessed as well as how the key players may interact with these systems. Table 1 details some extracts of the HIPAA security requirements (readers inter-

Table 1. HIPAA security requirements (Adapted from Fadlalla & Wickramasinghe, 2004)

Extracts from HIPAA Concerning Security Requirements
• Establishment of trust partnership agreements with all business partners
• Formal mechanisms for accessing electronic health records
• Procedures and policies to control access of information
• Maintaining records of authorizing access to the system
• Assuring that system users receive security-awareness training, and the training procedures are periodically reviewed and updated
• Maintaining security configuration including complete documentation of security plans and procedures, security incident reporting procedures, and incident recovery procedures
• Communication and network control including maintaining message integrity, authenticity, and privacy. The encryption of messages is also advocated for the open network transmission portion of the message.
• Data authentication to ensure that data is not altered or destroyed in an unauthorized manner

ested in the complete HIPPA security requirements are referred to *HIPAA Security Requirement Matrix*, 2002). Essentially, these security criteria fall into three main categories, namely, administrative, physical, and technical. Table 1 summarizes the major issues and levers under each of these categories, as well as identifying which are required and which are optional.

Transaction Standards

The standards for electronic health-information transactions cover transactions including claims, enrollment, eligibility, payment, and coordination of benefits. Succinctly stated, the aspect of HIPAA referring to transaction standards can be thought of in terms of practice standards and technical standards, as can be seen in Table 2.

Table 2. Practice and technical standards

STANDARD SETS
Practice Standards
Healthcare Common Procedure Coding System (HCPCS) This standard contains the Level II alphanumeric HCPCS procedure and modifier codes, and their long and short descriptions. These codes, which are established by CMS's Alpha-Numeric Editorial Panel, primarily represent items, supplies, and non-physician services not covered by the American Medical Association's CPT-4 codes. This standard does not contain the American Medical Association's CPT-4 codes.
ICD-9 Diagnosis Codes International Classification of Diseases, ninth revision, Clinical Modification ICD-9-CM
ICD-9 Procedure Codes International Classification of Diseases, ninth revision, Clinical Modification ICD-9-CM
Technical Standards
Technical Standards Adoption of electronic data interchange (EDI) using healthcare industry implementation guidelines and other standards such as XML and X12

Privacy

The final element of the HIPAA triangle deals with ensuring the privacy of health-care information. Specifically, the Federal Register (Vol. 67, No. 157) details all the rules that must be adhered to with respect to privacy. The purpose of these rules is to maintain strong protection for the privacy of individually identifiable health information, addressing the unintended negative effects of the privacy requirements on healthcare quality or access to healthcare, and relieving unintended administrative burdens created by the privacy requirements. Thus, theses privacy requirements cover uses and disclosures of treatment and payment information, and create national standards to protect individuals' medical records and other personal health information. Specifically, they achieve the following:

- Give patients more control over their health information.
- Set boundaries on the use and release of health records.

- Establish appropriate safeguards that healthcare providers and others must achieve to protect the privacy of health information.

- Hold violators accountable with civil and criminal penalties that can be imposed if they violate patients' privacy rights.

- Strike a balance when public responsibility requires the disclosure of some forms of data, for example, to protect public health.

For patients, this means being able to make informed choices when seeking care and reimbursement for care based on how personal health information may be used. Specifically, privacy requirements help patients, as follows:

- Enable patients to find out how their information may be used and what disclosures of their information have been made.

- Generally limit the release of information to the minimum reasonably needed for the purpose of the disclosure.

- Give patients the right to examine and obtain a copy of their own health records and request corrections.

For the average healthcare provider or health plan, the privacy regulations require activities such as the following:

- Providing information to patients about their privacy rights and how their information can be used.

- Adopting clear privacy procedures for its practice, hospital, or plan.

- Training employees so that they understand the privacy procedures.

- Designating an individual to be responsible for seeing that the privacy procedures are adopted and followed.

- Securing patient records containing individually identifiable health information so that they are not readily available to those who do not need them.

Mapping Case Study to Model

During the past 6 years, INET has used an e-business acceleration project to increase ICT project successes. Today, INET is repurposing the e-business acceleration project into a mobile e-health project to apply, enhance, and validate the mobile e-health project-delivery model. Such a model provides a robust structure and enables the expeditious realization of the particular m-health initiative without compromising success. INET's data provide the perfect opportunity to examine the components of our model (Figure 1) as they are both rich and longitudinal in nature. In mapping the data and specific business case, we have drawn upon many well-recognized qualitative techniques including conducting both structured and unstructured interviews, in-depth archival analysis, and numerous site visits. (Goldberg, et al., 2002a, 2002b, 2002c, 2002d, 2002e; and Wickramasinghe & Goldberg, 2004, capture and substantiate the findings discussed, while Kavale, 1996; Boyatzis, 1998; and Eisenhardt, 1989, detail the importance and richness of the methodologies we have adopted in presenting the following findings.)

E-Business Acceleration Project Background

Understanding the need for an e-business acceleration project starts with the core practice in ICT project delivery, the system development life cycle (SDLC). Typically, these are 1- to 5-year cycles that focus on reengineering large and complex business processes. When engaging in these projects, major changes occur in the way people work, the way they are compensated, and the way they engage others in the delivery of goods or services. Once an organization engages in an SDLC project, it quickly formalizes a change-management team to prevent potential disruptions to delivery due to process, organizational, and technological change. Unfortunately, even with this attention to rigor and consideration of the people issues, it is well documented that many of these SDLC projects fail. For instance, a major part of the SDLC is software; these projects have a high failure rate as presented in Figure 6.

Figure 6. Software Project Statistics 1995: Software project delivery success (Source: The Standish Group, 1995)

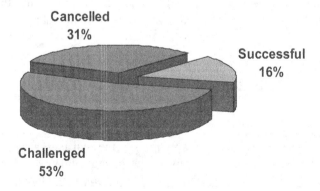

Table 3. Project-impairment factors

Project-Impairment Factors		% of the Responses
1.	Incomplete Requirements	13.1%
2.	Lack of User Involvement	12.4%
3.	Lack of Resources	10.6%
4.	Unrealistic Expectations	9.9%
5.	Lack of Executive Support	9.3%
6.	Changing Requirements & Specifications	8.7%
7.	Lack of Planning	8.1%
8.	No Longer Needed	7.5%
9.	Lack of IT Management	6.2%
10.	Technology Illiteracy	4.3%
11.	Other	9.9%

A closer look at the reasons behind these failures can be found in papers such as *The CHAOS Report* (Standish Group International Inc., 1994). The scope and approach of this landmark survey provides expert comments on IT project failures. It was conducted among 365 IT managers from companies of various sizes and in various economic sectors.

Opinions about why projects are impaired and ultimately cancelled rank incomplete requirements and lack of user involvement at the top of the list. Please refer to Table 3 for a list of project-impairment factors.

To increase project success, INET created an e-business acceleration project to narrow an SDLC project scope from meeting hundreds or even thousands of requirements to meeting just a few high-impact requirements. As a result, INET projects accomplish the following:

• Engage users very early in the project to identify, prioritize, and select the right set of requirements the first time and apply Internet and wireless technology to maximize user involvement.

- Minimize the need for resources in technology, processes, and people by developing an ICT application that can be developed, tested, and quality assured within days.

- Demonstrate results early with a pilot project to set realistic expectations, achieve executive sponsorship faster, and prevent changes to requirements once in field.

- Release low-cost, simple-to-use, pervasive, and commercialized ICT solutions to make planning much easier and significantly reduce technology educational costs and time cycles.

The purpose is to reengineer a large and complex delivery process in small manageable chunks, in a much shorter time cycle, with minimum impact on the way people work. Once a couple of projects are successfully accepted by the user community, many INET projects can happen concurrently to scale results and accelerate SDLC achievements. This is presented in Figure 7, "Refocusing the SDLC."

The INET e-business acceleration project's success has been documented in healthcare (Wickramasinghe & Goldberg, 2004). It meets the need to enhance healthcare delivery, under a medical model, in making small incremental changes and scaling success with international peer review and acceptance. This becomes even more important in m-health initiatives because many are still skeptical about the adoption of m-health in facilitating healthcare delivery, thereby making failures or inferior

Figure 7. Refocusing the SDLC

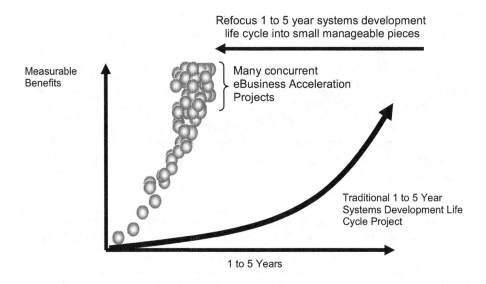

results due to poor implementation a setback in trying to encourage embracement of this new paradigm, as well as naturally being costly and unacceptable.

A New ICT Infrastructure to Support E-Business Acceleration Projects

INET's challenge was finding an ICT infrastructure to support the delivery of an accelerated e-business project. After 5 years of investigation and practical experience, INET has defined this new ICT infrastructure as follows:

1. Simple and low-cost technology: Internet applications, wireless technology, cell phones, and/or personal digital assistants (PDAs).
2. Next-generation ICT professional: ICT players who have made the transition to a new culture of meeting the highly responsive and evolutionary needs of end users.
3. Demonstrable processes to accelerate ICT project delivery: A rigorous project-delivery process that reduces time cycles by 75% with a 50% reduction in project costs.

For an in-depth look at INET findings, let us begin with a definition of an ICT infrastructure. It can be defined as a set of technologies, such as computer hardware, devices, printers, applications, software, wired and wireless networks, and other technology components. The infrastructure is supported by ICT players who research, develop, deliver, service, and sell technologies to enhance information-systems operations. These players also deliver new systems using a rigorous systems-development life-cycle methodology. In healthcare, an ICT infrastructure can be referred to as e-health and typically contains configurations similar to that depicted in Figure 3.

What is different today? In the past, many people resisted the use of technology to exchange and communicate information. It is well documented that people will not use technology if it is too complex or too costly. What is new today is the commercialization and acceptance of wireless technology. Wireless technology eliminates the costs and complexity in communicating and exchanging information at the point of need, regardless of location or time. This is evident by the widespread use of PDAs and cell phones along with wireless data networks. Today, over 33% of physicians use PDAs, and by 2005, it is predicted to grow to 50%.[1] In 2003, PDA usage among medical residents was 85%.[2] Over 172 million American consumers use cell phones[3] with the potential to access healthcare ICT systems anywhere, anytime.

Even with wireless technology removing the cost and usability barriers to enhance communications and the exchange information, INET quickly discovered a new barrier when trying to engage the ICT industry in an e-business acceleration project. This is manifesting itself as user frustration with the responsiveness of the ICT industry to meet their demands. INET believes this is the result of a project-engagement gap between the ICT 1- to 5-year system-development cycle and users' demand for immediate and quick results. The e-business acceleration project was designed to bridge the gap between ICT and the user. From INET's experience, ICT needs a cultural change to meet the highly responsive and evolutionary needs of an e-business acceleration project. The INET experience clearly showed there was no consensus on an ICT infrastructure (people, process, and technology) to support both benefits of wireless technology and rapid project-delivery achievements. INET spent 4 years to research and develop a road map to help the ICT industry make this transition. To release a solution and build consensus, INET founded and chaired the first wireless IT committee for the Information Technology Association of Canada (ITAC). After the inaugural year, the group delivered the Wireless IT Infrastructure Procurement Guidelines in Healthcare.[4]

At the same time of the ITAC wireless IT committee, INET started to gather evidence on an ICT infrastructure that can support the rapid and concurrent delivery of INET e-business acceleration projects. After a surveillance of Internet usage, INET selected healthcare market research in 2003. This industry is quickly shifting traditional quantitative research methods, such as face-to-face or telephone interviewing, to online surveys using the Internet. The use of the Internet has shortened field time cycles by as much as 75% with the added benefit of reducing costs upward to 50% when compared with telephone studies. To confirm these findings, INET completed over 20 market-research engagements involving the United States, United Kingdom, Canada, Japan, Denmark, Italy, Spain, Australia, Germany, and France. To deliver these e-business acceleration projects, INET assembled and used an ICT infrastructure with the following key attributes:

- The heart of the INET ICT infrastructure is form-generation technology components to create custom online surveys within 3 days. This is an instrument with the programming capability to deliver custom surveys (forms) on demand; field multiple projects at the same time; integrate with many online panel methods; support, at a minimum, all standard survey-question types including single and multiple selections, grid and three-dimensional grid, open text or verbatim, and numeric entry; and securely integrate video and audio clips and other embedded technology objects.

- The INET hosting environment employs a fully redundant technical platform, hosted in a state-of-the-art data center, and capable of handling over 50,000 surveys per hour. INET also assures these platforms are secure and when required, can conduct an HIPAA-compliant audit.

- The technology-hosting servers have redundant connectivity to the Internet and are managed on a 7-day/24-hour basis.

- When required, it can connect to wireless networks and devices using transcoding technology components. A transcoding component sits on top of the three-tier application architecture to redirect applications for use by a smaller screen, synchronizes data, and supports many types of wireless devices from laptops to cell phones (Figure 5).

Using the INET ICT infrastructure, an INET e-business acceleration project for market researchers begins with a questionnaire written in an MS Word document; it is then programmed and hosted for the Internet. When required, the survey is translated into another language to field in a non-English-speaking country. Then INET fields the study with an online panel of physicians; for instance, INET sourced almost 200,000 physicians worldwide to participate in online research. An online panel process begins by e-mailing an invitation to the targeted respondent to participate in an Internet survey. With the use of an anonymous electronic ID, privacy is upheld and enhanced. There is no personal-information capture in the online survey. Once a panel member clicks on a unique URL (uniform resource locator) in the e-mail, he or she is immediately connected to a secure Internet site to complete the survey. Typically, an online panel's response rate is 23% and upward to 40%. The total time to program and test a survey is 3 days. The timing in the field is approximately 5 to 7 days regardless if the survey is done by 100 or 2,000 respondents, and delivery of the data file happens within 24 hours.

In summary, INET reuses the lessons learned from delivering market-research data-collection projects and using wireless technology to show how an ICT infrastructure can support INET e-business acceleration projects. In this way, continuous improvement is achieved and the future state is always improved through analysis and diagnosis of the current state outcomes and implementation of appropriate prescriptive measures. The next step is to gain better acceptance of an e-business acceleration project in healthcare. This will occur as the industry embraces e-health and wireless possibilities.

Achieving ICT Project Success in Healthcare

To achieve ICT project success in healthcare, INET is mapping the e-business acceleration project to the mobile e-health project delivery model, and then vetting the model with a use-case scenario. INET sees this model, combined with use-case scenarios and a peer-review process, as a rigorous mechanism in the delivery of local and international mobile e-health projects.

For the past 5 years, the primary sponsors for a mobile e-health project are organizations looking to conduct studies on the use of wireless or mobile technology in healthcare delivery. These projects are typically funded through government research grants and the IT industry. In INET's case, an e-business acceleration project was used to deliver the mobile e-health project and was funded through the ICT-industry sector. However, in 2003, INET began to quickly understand that it would be difficult for the ICT industry to continue the sponsorship of research-originated projects without a much faster commercial payback period. INET clearly saw this as the need to rebalance funding from research to commercial sources. With this in mind, INET started working in collaboration with others on a strategy to incorporate the e-business acceleration project as an INET mobile e-health project and accelerate commercial successes. The INET mobile e-health project's objectives include the following:

1. Accelerate consensus building with an e-health solution that is focused on a disease state and driven by the medical model, with the primary objective to streamline communications and information exchange between patients and providers of community or home care, primary care, and acute care.

2. Acquire commercial funding early with a compelling business case. For instance, enhancing therapeutic compliance can improve patient quality of life with significant healthcare cost savings. It is well documented that regarding diabetes, this will have immediate and high-impact benefits for healthcare consumers, pharmaceutical firms, governments, insurers, and employers.

3. Avoid risk by reengineering large-scale healthcare-delivery processes in small, manageable pieces. Today, organizations can harness a rigorous method to incrementally enhance a process one step at a time, a way to achieve quick wins early and frequently.

4. Rapid development of simple-to-use, low-cost, private, and secure ICT solutions through wireless application service providers (ASPs). In addition to achieving rapid development, a wireless ASP can easily connect and bring together many independent healthcare information systems and technology projects.

To actualize the mobile e-health project, INET is looking to the mobile e-health project delivery model as a framework (Figure 1). For INET, this will support a mobile e-health project-management office (PMO) to manage the costs, quality, and ICT vendors; deliver many small projects; and replicate projects for local and international distribution. As a first case scenario for the model, INET is proposing an INET diabetes mobile e-health project with the leadership of a family physician. The INET PMO is provisioning a project manager to support this physician-led project to meet both research and commercial sponsors' interests and objectives in diabetes. A detailed description of the key attributes of the INET diabetes mobile e-health project follows.

Problem Statement

There are many communication and information exchange bottlenecks between patients and their family physicians that prevent the effective treatment of diabetes. A fundamental problem today is the inability to have a private and secure way to manage, search, and retrieve information at the point of care. For diabetes, physicians cannot quickly and easily respond to patients with high glucose levels. They need to wait for people to come to the office, respond to phone calls, reply using traditional mail delivery, or never receive the patient information.

Solution Mandate

The goal is to implement a diabetes monitoring program to enhance therapeutic compliance, such as releasing a program to enhance the usage of oral hypoglycemic agents (drugs) and/or the usage of blood-sugar monitoring devices.

Everyone wins when enhancing patients' ability to follow instructions in taking prescribed medication. The patient's health, safety, and quality of life improve with significant healthcare cost savings. However, it is well documented that many patients do not stay on treatments prescribed by physicians.[5] This is where wireless technology may have the greatest impact to enhance compliance.

One solution may be as simple as using a cell phone and installing a secure wireless application for patients to monitor glucose levels, and provisioning a physician to use a PDA (connected to a wireless network) to confidentially access, evaluate, and act on the patient's data

Business Case

In Ontario, the cost savings may represent almost $1 billion over 3 years. INET uses a simple calculation to determine the $1 billion savings. This can be found at http://www.inet-international.com (select the INET mobile e-health project section to review the calculations). The business case can be backed with additional data on how the cost of prevention (drugs) is far less than the cost savings associated

Figure 8. Impact form reducing AIC levels

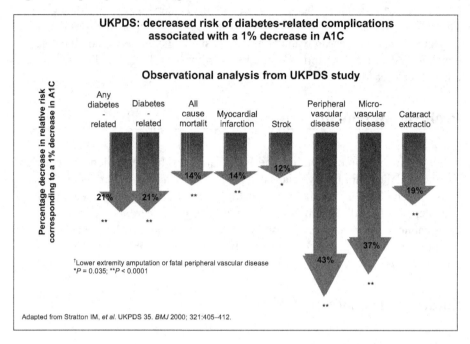

Figure 9. Economic burden

Economic Burden		
	Direct medical costs*	**Mortality costs***
Diabetes	$573	$455
Chronic Complications	921	619
- Neurologic disease	148	63
- **Peripheral vascular**	63	N/A
- Cardiovascular	637	545
-Renal	49	11
-Ophthamologic	6	N/A
-Chronic complications	17	N/A
General Medical	1,133	N/A
Total	2,627	1,074

* In millions

The economic cost of diabetes in Canada. 1998
University of B.C., Drs. Dawson, Gomes, Gerstein, Blanchard & Kahler

with reducing the risk of complications associated with diabetes. For instance, the impact of a 1% decrease in A1C is significant and evident in Figure 8.

The Canadian healthcare costs savings is presented in Figure 9, and when matched with the impact of reduced complications in Figure 8, the cost savings are significant.

More data are available to support the business case for the prevention of type 2 diabetes such as through lowering the incidence of end-stage renal disease (ESRD).

In summary, there are plenty of data today to quickly build consensus to fund and implement a national and international wireless diabetes program that enhances patients' quality of life with significant healthcare cost reductions.

Systems Development Life-Cycle Project Delivery

An INET diabetes mobile e-health project will be used to localize a wireless diabetes program led by a physician. Each project can easily and simply customize a program to quickly meet the unique needs of a rural and urban healthcare delivery setting taking into account age, ethnicity, income, language, and culture. These are small manageable projects and may cost $100,000 to $250,000 with a per-user operational cost of $1,200 to $1,500 per year. Each project collects data on patient and health-care-provider relationships, wireless medical informatics, therapeutic-compliance business cases, and ICT usability to accelerate the acceptance of a wireless diabetes program. The program may include a cellular network and application usage, support, PDAs for healthcare providers, and consulting fees for family physicians and other healthcare providers. However, it is expected that the costs may not include items such as consumer cell phones, medication, or blood-sugar monitoring devices and supplies. It is recommended that commercial and/or research sponsors pay for an INET project and help subsidize the user costs.

In summary, INET's research data indicate that using the mobile e-health project delivery model will increase ICT project success in healthcare. To realize and test this, INET plans to map the players from an INET diabetes mobile e-health project (use-case scenario) to the model. To show how this may work, please review the mapping exercise. The bold text represents a project player, and the text in brackets relates to the sections of the model presented in Figure 1.

- **Physician Mobile E-Health Project Lead** [Actualize Physician-Led Innovations in Disease Management]. Physicians provide the linkage to the medical model to enhance disease-management programs (wireless diabetes program) that provide better patient care and safety, improve research and education, increase healthcare quality, and reduce healthcare costs; these are the project deliverables [Healthcare Deliverables].

- **Commercial sponsor(s)** [Wireless & Mobility Commercialized Funding Needs/Requirements]. The project delivers information and communication solutions for the following:

  o Consumers wishing to improve their quality of life with an enhanced relationship with their healthcare providers, for example, family physicians.

  o Pharmaceutical firms looking to increase revenues with e-compliance programs.

  o Government or insurers investigating ways to significantly reduce administration and healthcare costs, and shorten healthcare delivery time cycles (wait times).

  o Employers wanting to increase productivity and avoid absenteeism with a healthier workforce.

- **Research sponsor(s)** [Wireless & Mobility Research Data Collection Needs/ Requirements]. The project develops intellectual property for researchers in the following fields:

  o Patient and healthcare provider relationships.

  o Wireless medical informatics.

  o Therapeutic-compliance business cases.

  o Wireless information-technology usability.

- The project also develops an INET mobile e-health project delivery team.

  o **Healthcare delivery team** [Field a Wireless Application]. For a wireless diabetes program, the players may include the following:

    • Healthcare consumer: People with diabetes.

    • Community care: Nurse specializing in diabetes.

    • Primary care: Family physician.

    • Acute care: Endocrinologist.

  o **Business-process analyst** [Validate a Business Case]

  o **Privacy and security consultant** [Confirm Mobile Trust Policies]

  o **Programmer using a wireless ASP** [Utilize a Wireless IT Infrastructure]

    • Wireless network and devices.

    • Device and application transcoding.

- Application service provider.
- Back-end connection.

In conclusion, INET is looking forward to further advancements in the mobile e-health project delivery model to meet the following goals:

- Achieve rapid advancements in healthcare delivery.
- Improve diabetes management.
- Enhance therapeutic compliance.
- Realize significant healthcare care cost savings[6].

INET is planning to continue its role as a source of use-case scenarios for the model with the delivery of mobile e-health projects.

Critical Success Factors

The preceding has served to outline all the critical aspects that must be considered when trying to actualize a mobile health initiative. This has been done by describing the major stages of a wireless implementation project by INET. What this case-study data highlight is that at one level, m-health initiatives must guard against the challenges and problems faced by most IT implementation projects, including failure of SDLC methodology, incompatibility with existing IT infrastructure, and a whole host of people issues such as resistance to the technology. However, there is also a unique set of complications that impact wireless initiatives in healthcare that also threaten the success of these initiatives. One of the leading challenges is the lack of acceptance of wireless technology as an appropriate facilitator of healthcare delivery. To mitigate and limit both sets of challenges (those shared by all IT implementation initiatives and those unique to m-health initiatives), we have developed the integrative framework presented in Figure 1 and utilize the methodology of adaptive mapping to realization (AMR), the mapping of case facts to the model, and subsequent and ongoing analysis of present-state outcomes to prescribe changes in order to ensure a superior future state. This is indeed a rigorous process and unique in regard to m-health implementations.

Clearly, mobile e-health or m-health projects are complex and require much planning and coordination within and between the webs of healthcare players. Success is never guaranteed in any large initiative; however, in order to realize the four major healthcare deliverables depicted in Figure 1 (enhance patient care and safety, improve research and education, increase healthcare quality, and reduce healthcare costs), it is vital that any m-health initiative focus on the key success factors of people, processes, and technology. Specifically, the technology must be correct and function as desired. Furthermore, it must integrate seamlessly with existing ICT infrastructure and enable the processes. The processes must be well defined and at all times ensure that they are of a high quality and error free. The Institute of Medicine in America (Committee on Quality of Healthcare in America, 2001) identified medical errors as the fourth leading cause of death. In trying to prevent such errors, it has identified six key quality aims, namely, healthcare should be (a) safe, avoiding injuries to patients from the care that is intended to help them, (b) effective, providing services based on scientific knowledge to all who could benefit and refraining from providing services to those who will not benefit (i.e., avoiding under use and overuse), (c) patient centered, providing care that is respectful of and responsive to individual patient preferences, needs, and values, and ensuring that patient values guide all clinical decisions, (d) timely, reducing wait times and sometimes-harmful delays for both those receiving care and those who give care, (e) efficient, avoiding waste, and (f) equitable, providing care that does not vary in quality based on personal characteristics. Finally, arguably the most critical key success factor is the web of healthcare players. Any m-health project must consider the impact and role of each of these players, and the interactions of such an initiative both within one group of the web of players as well as between groups of players. As discussed, and based on the findings from INET's longitudinal studies, it is critical to the ultimate success of these projects and the ability to realize the healthcare deliverables that they are in deed physician led.

Discussion and Conclusion

Healthcare in the United States and globally is at a crossroads. It is facing numerous challenges in terms of demographics, technology, and finance. The healthcare industry is responding by trying to address the key areas of access, quality, and value. M-health, or mobile e-health, provides a tremendous opportunity for healthcare to make the necessary evolutionary steps in order to realize its goals and truly achieve its value proposition. What is important is to ensure m-health excellence. By an in-depth analysis of the rich and longitudinal data of INET, we have developed an integrative macro-level model to facilitate the achievement of m-health excellence. To the best of our knowledge, it is the first such model, and while it is certainly not a panacea, it does help to set the stage and outline the key issues that must be addressed in a systematic fashion so that a successful m-health initiative might ensue. Healthcare globally must embrace the advances of wired and wireless ICT use to provide superior healthcare delivery. In such an environment, an imperative becomes the successful implementation of these wired and wireless initiatives. The wireless INET integrative model serves to fill this critical void. We close by strongly urging for more research in this area that will further test our framework.

References

Blair, J. (2004). Assessing the value of the Internet in health improvement. *Nursing Times, 100*, 28-30.

Boyatzis, R. (1998). Transforming qualitative information thematic analysis and code development. Thousand Oaks, CA: Sage Publications.

Committee on Quality of Healthcare in America, Institute of Medicine. (2001). *Crossing the quality chasm: A new health system for the 21st century.* Washington, DC: National Academy Press.

Dawson, Gomes, Gerstein, Blanchard, & Kahler. (1998). *The economic cost of diabetes in Canada.* Canada: University of British Columbia.

Eisenhardt, K. (1989). Building theories from case study research. *Academy of Management Review, 14*, 532-550.

European Institute of Medicine. (2003). *Health is wealth: Strategic vision for European healthcare at the beginning of the 21st century.* Salzburg, Austria: European Academy of Arts and Sciences.

Fadlalla, A., & Wickramasinghe, N. (2004). An integrative framework for HIPAA-compliant I*IQ healthcare information systems. *International Journal of Healthcare Quality Assurance, 17*(2), 65-74

Frost & Sullivan. (2004). *Country industry forecast: European Union healthcare industry.* Retrieved from http://www.news-medical.net/print_article.asp?id=1405

Goldberg, S. (2002a). *Building the evidence for a standardized mobile Internet (wireless) environment in Ontario, Canada: January 2002 update.* INET.

Goldberg, S. (2002b). *HTA presentational selection and aggregation component summary.* INET.

Goldberg, S. (2002c). *HTA presentation rendering component summary.* INET.

Goldberg, S. (2002d). *HTA quality assurance component summary.* INET.

Goldberg, S. (2002e). *Wireless POC device component summary.* INET.

Health Insurance Portability and Accountability Act (HIPAA) privacy compliance executive summary. (2001). Protegrity Inc.

HIPAA security requirement matrix. (2002). Retrieved from http://www.hipaa.org

INET. (2004, October). *Enhance therapeutic compliance using wireless technology.* Proceedings of the WNY Technology & Biomedical Informatics Forum, Niagara Falls, NY.

Kavale, S. (1996). *Interviews: An introduction to qualitative research interviewing.* Thousand Oaks, CA: Sage.

Kulkarni, R., & Nathanson, L. A. (2005). *Medical informatics in medicine: E-medicine.* Retrieved from http://www.emedicine.com/emerg/topic879.htm

Kyprianou, M. (2005). *Healthcare : Is Europe getting better?* The New European Healthcare Agenda: The European Voice Conference. Retrieved from http://www.noticias.info/asp/aspcommunicados.asp?nid=45584

Lacroix, A. (1999). International concerted action on collaboration in telemedicine: G8sub-project 4. *Sted. Health Technol. Inform., 64,* 12-19.

Lee, M. Y., Albright, S. A., Alkasab, T., Damassa, D. A., Wang, P. J., & Eaton, E. K. (2003). Tufts Health Sciences Database: Lessons, issues, and opportunities. *Acad. Med., 78,* 254-264.

Moore, T., & Wesson, R. (2002). Issues regarding wireless patient monitoring within and outside the hospital. *2nd Hospital of the Future Conference Proceedings.*

National Center for Health Statistics. (2002). *Health expenditures 2002.* Retrieved from http://www.cdc.gov/nchs/fastats/hexpense.htm

National Coalition on Healthcare . (2004). *Building a better health: Specifications for reform.* Washington, D.C.: Author.

Organisation for Economic Co-operation and Development (OECD). (2004). *OECD health data 2004* [CD-ROM]. Retrieved from http://www.oecd.org/health/healthdata

Pallarito, K. (1996). Virtual healthcare . *Modern Healthcare,* 42-44.

Plunkett's healthcare industry almanac (2005 ed.). (2004). Houston, TX: Plunkett Research, Ltd.

Russo, H. E. (2000). The Internet: Building knowledge & offering integrated solutions to healthcare. *Caring, 19,* 18-20, 22-24, 28-31.

Standish Group International Inc. (1994). *The CHAOS report.* Retrieved from http://www.standishgroup.com/sample_research/chaos_1994_1.php

Stratton, I. M., et al. (2000). UKPDS 35. *BMJ, 321,* 405-412.

Von Lubitz, D., & Wickramasinghe, N. (in press). Healthcare and technology: The doctrine of networkcentric healthcare. *Health Affairs.*

Wickramasinghe, N., & Goldberg, S. (2004). How M=EC2 in healthcare . *International Journal of Mobile Communications, 2*(2), 140-156.

Wickramasinghe, N., & Mills, G. (2001). MARS: The Electronic Medical Record System. The core of the Kaiser galaxy. *International Journal of Healthcare Technology Management, 3*(5/6), 406-423.

Wickramasinghe, N., & Silvers, J. B. (2003). IS/IT the prescription to enable medical group practices to manage managed care. *Healthcare Management Science, 6*, 75-86.

Wickramasinghe, N., et al. (in press). Assessing e-health. In T. Spil & R. Schuring (Eds.), *E-health systems diffusion and use: The innovation, the user and the user IT model.* Hershey, PA: Idea Group Publishing.

World Health Organization. (2000). *Health systems: Improving performance.* Geneva, Switzerland: Author.

World Health Organization. (2004). *Changing history.* Geneva, Switzerland: Author.

Endnotes

[1]   Harris Poll

[2]   *MercuryMD* (2003)

[3]   http://www.ctia.org. On December 18, 2004, at 9:48 a.m. eastern standard time, there were 172,984,768 current U.S. wireless subscribers.

[4]   This documents a consensus to begin the ICT industry's transition with a five-step wireless IT-infrastructure deployment road map. This document and other reference material are available at http://www.itacontario.com.

[5]   Fourteen to 21% of patients never fill their original prescription, and 30% to 50% of patients ignore or otherwise compromise their medication instructions (http://www.managedhealthcare executive.com/mhe/article/articleDetail. jsp?id=105388).

[6]   In Ontario, this may save $1 billion over 3 years (INET, 2004).

Chapter III

Healthcare Organizations and the Internet's Virtual Space:
Changes in Action

Stefano Baraldi, Catholic University, Italy

Massimo Memmola, Catholic University, Italy

Abstract

For some years now, the opportunity of innovating business models has basically been linked to continual progress in ICT. Healthcare is no exception; information and communication technologies are generally considered the most effective driver for changing organizations, improving quality, optimizing resources, and so forth, at least in theory. In practice, it is not clear which and how many of these opportunities are really exploited by organizations operating in healthcare. This chapter presents the results of a research project aimed at understanding to what degree and how Italian healthcare organizations make use of the virtual space made available to them by the Internet.

Introduction

"Who doesn't know the Internet?!"

This is certainly one of the most common answers when hospital patients are asked about their knowledge of the Internet phenomenon.

In fact, it is likely that the word Internet is one of the most common and widely used in Western societies. Whether walking in the streets of New York, Paris, London, or Rome, if passersby are asked the simple question, "What is the Internet?" the answer is always the same: The Internet is something that enables us to communicate, read, learn, play, purchase goods and services, make transactions, and more.

The answer does not vary. Often, in fact, people do not know the technology behind the Internet, but they know what it can be used for. In other words, people know how to use technology even if they do not know how it works (which can be said for many new technologies).

Certainly, new technological developments and their awareness can only increase further thanks to the spread in recent years (or rather in recent months) of mobile communication devices. Equally, it is easy to predict the same happening in developing countries: Because of its low cost, the Internet is often preferred to the telephone for communication with these countries.

The Web, therefore, has changed our lifestyle, our habits, and our way of working, interacting with our acquaintances, and, in few words, dealing with so many aspects of our daily life. We no longer have to physically go to the bank but can transfer money and check our account balances by logging on to our bank's Web site. In the same way, booking our holiday is easier when we can click on the site of a tour operator. The list goes on and on: We can shop online instead of at our local supermarket, or buy books and DVDs from a company abroad.

In the world of healthcare, this revolution has yet to be realized, at least fully. In recent years, literature has often indicated how the digitalization of clinical, organizational, and management processes of health structures brings undeniable benefits both for the efficiency and effectiveness of the company, as well as improving the quality of service offered to patients (Coile, 2002; Goldstein, 2000; Nicholson, 1999). More particularly, the Internet considerably affects the entire process of the creation of the value of a healthcare organization (HCO) if we consider the following:

- It gives the organization a global presence.
- It means it can offer more services that are more readily available to more users.
- It allows the collection and elaboration of a greater amount of information.

- It provides a new channel, the World Wide Web, through which it is possible to transfer information, provide services, make transactions, and create a privileged area of interaction between physicians and patients.

The possible applications in a healthcare organization, limited only by the state of the art of technological development, can be schematized according to their functional characteristics.

The Internet as a Tool for Reducing Geographic Distances and Aiding the Distribution of Services of a Healthcare Organization

The opportunity to provide services in inaccessible areas is one of the aspects of the Internet that has multiple applications for healthcare:

- **Home monitoring:** This service allows the patient to be cared for at home, or in the most convenient place, either through prearranged or automatic transmission of signals and vital parameters, or through the sending of alarms activated by predefined emergency situations. Healthcare personnel can also call back and give instructions to the patient or to semiautomatic equipment.
- **Telemedicine:** This is the provision of a treatment service or remote assistance in the form of tele-consultation, tele-presence, second opinions, Electronic Medical Record, (EMR), and so forth. Generally, these services are offered via the Web and may involve videoconference connection between the patient and the doctor or between medical personnel.

The Internet as a Tool for Sharing Information

The Internet is a special way to exchange information:

- The Internet can be used to share clinical information, both for applications developed to be shared totally or to be shared partially and securely. This can include information contained in clinical records (electronic health records, EHRs) to be shared with healthcare personnel belonging to different organizations who treat or have treated the same patient using secure electronic post services (encoded and authenticated).

- It can be used to research and develop new diagnostic-therapeutic processes or protocols aimed at providing welfare assistance through the use of applications similar to those already developed.
- It can also be used for epidemiologic observation through the development of an information flow for the rapid collection and distribution of epidemiologic information for healthcare companies and organizations.

The Internet as a Support Tool for Management and Training of Human Resources

The technologies of information and communication provide an effective tool the following:

- Running and managing the organization (business intelligence, data warehouse systems, etc.).
- Distance training of operators, providing considerable possible savings in costs and significant benefits for the levels of services offered.
- Managing the organization's information system and integrating, through the Internet, departments and organizational units.

The Internet as a Tool for Reducing Informative Asymmetry

Informative asymmetry is characteristic of any service-providing organization. The transmission of clinical information through the Web allows patients to be guided and informed about their health, and provided with expert advice about illnesses, prevention, and treatments available.

In this way, patients become more aware of the progress of their treatment and the traditional physician-patient relationship is transformed. However, there is still the problem of appointing an authority to guarantee the quality of the information provided.

The Internet as a Tool for Reducing the Cost of Transactions

The most popular service offered is that for making appointments online for diagnostic examinations or specialist treatments through an SBC (single booking centre) set up for a single healthcare organization or for a larger catchment area such as a city or even a region. Other possible applications include the possibility to access the results of diagnostic investigations online, to search for a medical specialist for the treatment of a pathology, to download required documentation for requests and reports, and so forth.

The Internet as a Multipurpose Tool with Virtually Infinite Capacity

The opportunity to provide services through the Internet to several users simultaneously without the physical presence of the doctor has meant that many other operations in the chain of value of a healthcare organization have been revised; this has been shown in many of the solutions described previously (telemedicine, home treatment, online booking, clinical information services, etc.).

In light of this, it is clear that the Internet offers the world of healthcare enormous potential, which can be exploited to improve both efficiency (helping reduce some of a country's highest public costs) and effectiveness (with a consequent improvement in the level of the quality of services offered). There are, however, factors that may limit the spread of Internet solutions or, at worst, lead to the failure of projects under way.

For patients, a highly negative factor is the still-limited access to the Web for many potential users (even in economically advanced countries) and what can be defined as electronic illiteracy, particularly for low-income families. The difference in terms of the needs of those who have access to the Web and those who do not, which is destined to increase, will make it even more difficult for healthcare organizations to decide what to offer. Furthermore, some patients are reluctant to use the Web in so far as they are unable to perceive the qualitative level of the information and services available or simply because they prefer to maintain a more personal relationship with their own physicians.

It is physicians themselves who can be the pioneers of the new technological standards of healthcare when they accept the Internet and ICT solutions in their work and consequently redefine their own competencies. It is not impossible to foresee that in the future the teaching of medicine will involve computers and Web tools for distance learning.

Finally, there are still healthcare organizations and managers who run them who are unable to understand how or why to use available technology, or, more importantly, how to evaluate what the returns for investments could be. In particular, whoever is in charge of an organization (production or healthcare) should be aware that introducing the technology is never an end in itself, but should be regarded as the result of a careful strategic plan through which a thorough evaluation of the real, physical, and virtual consequences of some activities of the chain of value is produced. The benefits, in cost reductions, that may come from the introduction of virtual processes in an organization might, in fact, be cancelled out by the greater costs that other units in the organization accrue.

Moreover, it is also relatively easy to find examples of failure, often caused by a lack of strategic vision or knowledge of how the Internet (and more generally ICT technologies) can change the traditional basic rules of treatment services (Given et al., 2001a; De Luca & Enmark, 2000; Fattah, 2000; Flory, 1999; Glaser, 2002; Minard, 2001, 2002).

On the other hand, several factors (including health demand, medical developments, reduced resources, and increasing competitive pressure) lead us to consider ICT as one of the fundamental means for solving the paradox of doing (ever) better with less. In healthcare, too, the key question seems to be not whether to deploy Internet technologies, but how to deploy them (Given et al., 2000; Porter, 2001).

Starting from this point of view, we carried out a research project (named Health. Net) in order to understand to what extent and how Italian HCOs actually exploit and use the virtual space offered by the Internet.

Objectives and Research Methodology

In our research, we have tried to focus on the real ability of HCOs to use the Internet to support their organizational, clinical, and management processes (Given et al., 2001b; Alemi, 2000; De Luca & Enmark, 2000; Solovy & Serb, 1999), and to create value (Glaser, 2000) not only for their own users, the patients, but also for all their stakeholders (physicians, staff, students, other HCOs, etc.).

In particular, we tried to shed some light about these issues:

- To what extent and how do Italian HCOs inhabit the Internet?
- What kind of strategy is guiding Italian HCOs in creating their own visibility on the Web?
- What are the results achieved? How much of the opportunities offered by the Internet have been actually exploited?
- What progress is being made by Italian HCOs in the exploitation of Internet potential?
- What are the latest trends and the scope of change?
- Is there any difference or delay in regard to other national healthcare systems?

The project has been carried out in two separate steps: the first in 2002 and the second in 2005. Both in 2002 and in 2005 we tried to accomplish the following:

1. Position the Italian healthcare system in an international context.
2. Analyze the public presence of Italian HCOs on the Internet.

In each year, the first stage of the project (aimed to position the Italian healthcare system in an international context) has been carried out by comparing Italian HCOs with the following:

*   A sample of 121 companies in six foreign countries (United States, Canada, United Kingdom, France, Germany, Spain); we paid attention to the biggest structures (i.e., those having over 200 beds) in the capital city or in a similar city in terms of healthcare facilities (Boston, Toronto, London, Paris, Frankfurt, and Milan). HCOs were identified thanks to data displayed on the Internet sites of national ministries.
*   Nine HCOs acknowledged as benchmarks in the international literature; benchmarks were identified according to the conclusions drawn in a study carried out in the United States (Goldsmith, 2000). The benchmark Web sites achieved the best performance in terms of contribution to business growth and patients' satisfaction.

During the second stage, we analyzed the presence on the Web of all the Italian HCOs (both public and private) that are officially registered in the database of the Italian Ministry of Health.

In order to give greater significance to the information gathered, it was decided, nevertheless, to consider only autonomous structures as much from a legal point of view as an economic one, in other words, only those organizations that have registered (or can register) a domain with their own name. Thus, for organizations dependent on a local health authority, we considered only the site of the latter; equally, the same approach was used for private structures that are subsidiaries to larger companies.

1,002 organizations were identified in 2002, and 990 in 2005. The reduction in the number of organizations usable for research depended on the following:

*   The mergence of some hospital structures by local health authorities decided by some Italian regions.
*   The setting up of a process of concentration in large national private healthcare organizations.

Each organization under study was classified as one of five main types: local health unit (LHU, 20%), independent hospital (IH, 10%), scientific institute for research and care (SIRC, 4%), university polyclinic (UP, 1%), and private structure (PS, 65%). The percentage of the total for each type did not change between 2002 and 2005.

For each of these organizations, a preliminary investigation accomplished the following:

a.   Identified the HCOs' Web domains by means of the most common (national and international) search engines; in case of broken (no) links, companies were contacted by phone.

b.   Took note of Web site status (active, not active, under construction, partially under construction).

At the end of this preliminary phase, 473 sites were identified (47% of the total) in 2002 and 574 sites (58% of the total) in 2005.

In the first year, successive elaborations of the research concerned all the organizations present on the Web. In 2005, we decided to take a sample of 136 sites of the 574 originally identified. The statistical technique used to describe the research sample is that of stratification (Frosini, Montinaro, & Nicolini, 1999).

Using this technique, it is possible to obtain an increase in the efficiency of a sampling plan as it reduces the sampling's order of greatness of error without increasing the number of samples.

The population was separated into strata internally as homogeneous and heterogeneous as possible. In other words, the population was divided into n underpopulations. From each stratum or underpopulation, a single random sample was taken; there were therefore as many samples as strata. The numerousness of each stratum was decided according to the variance that the stratum presented with reference to the data from the first year, for which a higher number of samples of strata with a greater internal variability were considered. The total number of the samples is obviously equal to the sum of the numbers of each single stratum.

To evaluate not only how many HCOs are on the Web but, above all, the features of their Net presence, we used the ICDT model (Information, Communication, Distribution, Transaction; Angehrn, 1997). Angehrn states that companies should look at the Internet not as a simple communication tool by which information can be exchanged (Nicholson, 1999), but rather as a space in itself, virtual of course, which they can "colonize" in different ways and to different levels, following a strategic vision or not. In this sense, the ICDT framework (Figure 1) effectively works to evaluate which kind of strategic approach is actually guiding HCOs in the conquest of their space on the Web, and, to sum up, how and to what extent, and how successfully, they exploit the opportunities offered by the Internet.

Figure 1. The ICDT model

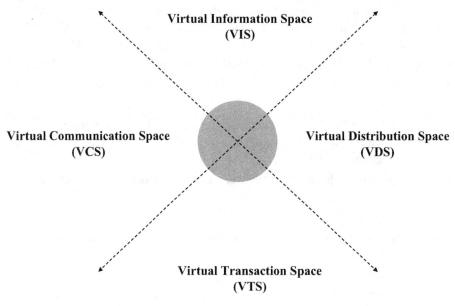

[Angehrn, 1997]

The model takes its name from the segmentation of Internet virtual space into four areas: a virtual information space (VIS), a virtual communication space (VCS), a virtual distribution space (VDS), and a virtual transaction space (VTS). This segmentation emphasizes that the "Internet has extended the traditional market space by providing new spaces in which economic agents can interact by exchanging information, communicating, distributing different types of products and services and initiating formal business transaction" (Angehrn, 1997, p. 362). Some particularities are as follows:

- The VIS consists of the new Internet-based channels through which economic agents can display information about themselves, and the products and service they offer. A hospital can provide patients and any other user with different kinds of information, such as its structure and history, information about how to reach the hospital, available services and clinical specialties, access modes, research projects, and so forth.

- The VTS consists of the new Internet-based channels through which agents can exchange formal business transactions such as orders, invoices, and payments. An HCO can refer a patient to remote transactions such as checkup or hospital-admission online booking, diagnostic-reports consultation, service payments, or medicine purchase.

- The VDS represents a new distribution channel suitable for a variety of products and services. It is a Web space by which companies can provide e-services or e-products. Using such space effectively definitely calls for modifications in the traditional physician-patient relationship as it hosts a wide range of services in telemedicine (tele-consultations, tele-diagnoses, remote access to health-improvement or health-monitoring programs), in prevention, in education, and in information on medicines.

- The VCS is the extension of traditional spaces in which economic agents meet to exchange ideas and experiences, influence opinions, negotiate potential collaboration, and so on. It is an opportunity to create a virtual community in which users can exchange ideas and opinions. Forums, chat lines, and news-groups can promote debate and discussion among patients and/or between physicians and patients about their own experience, pathologies, and related issues.

We mapped the HCOs' presence on the Web according to their approaches and modes to actually use these different spaces (Figure 2). In particular, we paid attention to the following variables:

Figure 2. Mapping the Internet presence

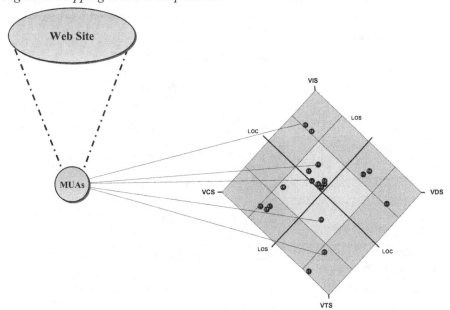

- **Complexity**, that is, how far HCOs turn to the Internet to support their institutional activity; an intensive and extensive use of such technology is likely to result in a rich and highly informative Web site that can provide users with corporate information (VIS), services (VDS), interaction and communication (VCS), and transactions (VTS).

- **Typology**, that is, the more or less balanced way in which HCOs exploit the Internet virtual space, achieving a presence in some rather than in all the different kind of spaces (VIS, VTS, VDS, VCS) of the IDCT model.

- **Quality**, measured by considering the level of sophistication (LOS) and the level of customization (LOC) of their Web sites. These two indicators were chosen in order to evaluate the concept of quality objectively and concretely rather than subjectively and vaguely. LOS refers to the technological level (indicating any multimedia elements), while LOC determines the degree of customization of the site content in accordance with the user's specific needs.

- **Effectiveness**, in terms of HCOs' Web site visibility (hits obtained using the most common search engines) and success (number of visitors).

Table 1. Methodology for analyzing the Internet presence

Step 1: Preliminary research	• identify Web-site domain by search engine or contacting the HCO by phone • verify Web-site status (active, not active, under construction, partially under construction)
Step 2: Internet presence complexity	• examine Web-site contents and classify them in minimum units of analysis (MUAs) • assess the existence of other structural features (foreign-language version, native-language search engine, Web-site map, extranet)
Step 3: Internet presence typology	• position retrieved MUAs in corresponding virtual space areas (VIS, VDS, VCS, VTS) • verify Web-site use of Internet potential
Step 4: Internet presence quality	• assign the LOS of each MUA • assign the LOC of each MUA
Step 5 : Internet presence effectiveness	• verify Web-site visibility by means of the most common search engines • measure average Web-site traffic

Thus, the presence of HCOs on the Internet has been evaluated by the study of the MUAs retrieved from their Web sites (see Table 1). An MUA is a Web site area featuring homogeneity in content and thus embodying a precise territory in the Internet virtual space. Each MUA may spread over more than a single Web page or share the same Web page as other MUAs. Each MUA has been positioned with regard to the following:

- Within its own virtual space of competence (VIS, VDS, VCS, VTS).
- The closer to the margins, the higher the LOS and LOC indexes, which appear on the axes in the graph.

This method has enabled us to evaluate to what degree and how the HCOs that are officially online are really enhancing their presence on the Web. Figure 3 gives as an example the maps relative to the Internet presence of HCOs. Each HCO exhibits one of the following characteristics:

Figure 3. The Internet strategy of an HCO

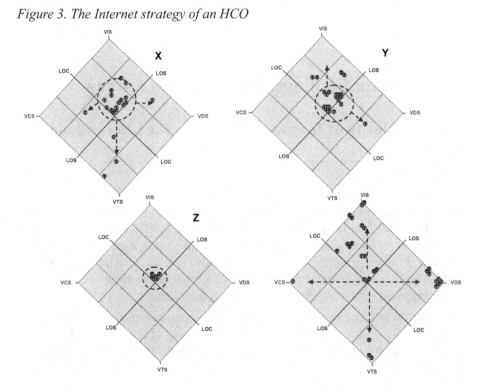

- Although starting with the use of the Internet as a simple, cheap means of communication, it is beginning to expand and significantly improve its presence on the Web, "attacking" other types of virtual space (e.g., X).

- It seems to be interested only in some of the opportunities offered by the Internet in various virtual spaces (e.g., Y).

- Although present on the Internet, it is unable to fully exploit the Web's potential (e.g., Z).

The different types of behavior shown by HCOs in occupying the Internet's virtual space have been grouped into three main clusters:

- The **pioneers**, whose results are better than the national average. These are the HCOs that are more convinced of the potential offered by the Internet and therefore exploit it to a greater extent.

- The **followers**, whose results are in line with the national average. These are the HCOs that, compared with the other members of the sample, are starting to expand and improve their presence on the Web to a greater extent.

- The **latecomers**, whose results are below the national average. These are the companies that, although present on the Web, only exploit its potential to a very limited extent.

Main Findings: International Survey

The Complexity of Internet Presence in the International Sample

There is a considerable distance between the benchmark and the other organizations that make up the international sample. Organizations in the United States and Canada perform best both for the quantity of contents offered on their sites and for the index of complexity. From 2002 to 2005 there was a substantial increase of values assumed from the variables investigated (particularly in the number of MUAs), with a reduction of the gap between European healthcare organizations when compared with those in the United States and Canada.

Through this analysis, we wanted to examine the consistency of the Internet presence of healthcare organizations, not simply according to the number of pages of content on the site, but rather for the value of the content offered in order to achieve the following:

- Give information in the VIS area
- Deliver services in the VDS area
- Provide a virtual communication space in the VCS area
- Accomplish transactions in the VTS area

Evidence shows that richer content in the different areas is limited to a higher Web site complexity as measured by two indicators: (a) the number of MUAs on the Web site and (b) the structural complexity index (SCI).

The number of retrieved Web site MUAs measures the richness of the content of HCOs' Web sites and ranks their complexity.

SCI is related to a series of structural features intended to improve Web site user friendliness. It is indeed likely that richer content requires greater skill of the user (i.e., patients, physicians, students, or any other stakeholder) in searching for desired information: In such cases, a native-language search engine and a Web site map are, no doubt, helpful. The SCI, then, results from the ratio between actual Web site structural features and research-relevant ones, including the following:

- Native-language search engine
- Web site map
- Foreign-language version of the Web site
- Extranet access

An extranet network can be described as a nonpublic presence on the Internet of an organization. Unlike an intranet, which presupposes a totally private exploitation of the network inside the organization and is reserved for the people who work there, an extranet enables information and content to be offered securely outside the organization. In fact, it can be used to share treatment protocols or, more generally, clinical information between hospitals or physicians, or to provide suppliers with information. Therefore, although not being a tool for improving the readability of a site's contents, an extranet does directly affect the degree of complexity of the Internet presence.

In general, what clearly emerges from examination of the complexity of Internet presence is the considerable distance that separates the group of the benchmark organizations from the others making up the international sample. Moreover, this gap, which is mostly consistent in other areas of research analysis, further increased in 2005.

Regarding the quantity of informative contents or services offered by the site, in the year, only the organizations in the United States and Canada performed in any way comparable with those of the benchmark organizations: The average number of MUAs identified was 16.79 for the United States and 17.35 for Canada compared with an average for the benchmark organizations of 23.56. The European average was considerably lower (12.71 MUAs) with the United Kingdom (12.96), Italy (15.47), and Spain (13.2) being more advanced than France (11.17) and Germany (10.75).

Analysis of data from 2005 shows that there was considerable growth in the quantity of content available on sites of healthcare organizations in both Europe and America, but not enough, however, to bridge the gap with respect to the benchmark organizations, which continued their steady growth. Once more, it was the U.S. and Canadian organizations, with an average number of MUAs of around 30 units that were nearer the benchmark 53.22. European organizations, nevertheless, markedly improved their position with an average number of MUAs of 26.32; in contrast to 2002, there was greater homogeneity with no major differences between European countries (Figure 4).

Analysis of the SCI, apart from the theme of the gap between the benchmark and the rest of the international sample (particularly as regards the European organizations), showed for 2002 a particularly high result for U.S. organizations (49%) compared to Europe (23%) and Canada (36%). In Europe, the highest values were in Italy (29%), Germany (33%), and the United Kingdom (24%). Scores for France and Spain were considerably lower at 13% and 17%, respectively.

Here, too, in 2005, there was greater homogeneity of values with an average European figure of 54%, perfectly in line with U.S. organizations at 50% and those in Canada at 54%; still, however, it is a long way from the benchmark of 75%.

Further interesting points emerge if we examine individual organizational features (foreign-language versions, site maps, internal search engines, access to an extranet).

We now find that the Web sites most likely to have an international slant are those in Italy, Germany, France, and, to a lesser degree, Spain: In these countries, in fact, a foreign-language version of the site is present in 35% of the sites analyzed in 2005. It is also the case that, because of the international importance of the English language (the official Internet language), organizations in the United Kingdom do not much concern themselves with presenting the content of their sites in another language. This factor is probably mitigated in the U.S. and Canadian organizations by the presence of a multiracial society: 29% of U.S. organizations and 15% of those in Canada offer foreign-language versions.

The site map and internal search engine are the two factors that most influence the user friendliness of a site, and nowadays are almost universally present on organizations' Web sites in the international sample; the high number of MUAs activated in sites inevitably creates difficulties in searching for information or services.

The extranet, finally, continues to have a low presence, even at an international level. It is present in little over half of the benchmark organizations (56%), in about

Figure 4. The complexity of Internet presence of the international sample

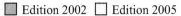

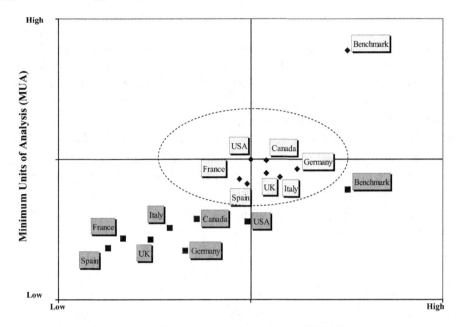

Structural Complexity Index (SCI)

a third of those in Canada (33%), a quarter of those in the United States (24%), and a lowly 13% of European ones.

To indicate the complexity of Internet presence, Figure 4 shows three distinct clusters:

- Pioneering organizations that constitute the benchmark and are clearly more advanced than those in the international sample.

- U.S. and Canadian organizations that, despite European organizations making progress, continue to lead, but are still some way behind the benchmark organizations.

- European healthcare organizations, the late arrivals, whose reference values were more homogenous in 2005 and are approaching U.S. and Canadian values.

The Typology of Internet Presence in the International Sample

Only the organizations that make up the benchmark have a widespread and homogenous presence on the Internet, fully exploiting its potential. Despite the progress made in 2005 compared with 2002, the organizations that make up the international sample only partially exploit the Internet, limiting themselves, in the case of European organizations, to a Web strategy with a low to medium profile.

Corporate Web presence enables companies to take advantage of virtual space, whose different areas (VIS, VDS, VCS, VTS) can be variously exploited.

The typology of Internet presence (Step 3) maps HCOs' strategies in colonizing such space in terms of the number of colonized virtual spaces and the actual exploitation of the different kinds of colonized virtual spaces.

The indicators employed in measuring such elements are, respectively, as follows:

- The colonization index, obtained by listing colonized areas (VIS, VCS, VDS, VTS).
- The coverage index, that is, Internet virtual space coverage, both globally and by single area.

The colonization index (how many virtual spaces are actually used by HCOs) is a rather simple, but very useful, indicator for measuring the evolutionary stage reached by an HCO's Internet strategy. However, as areas can be colonized even by a single MUA, this indicator does undoubtedly provide only an approximate value, which has to be completed with the analysis of the coverage index.

While the colonization index measures corporate colonization of single Internet virtual areas, the coverage index ascertains how far companies benefit from it. More specifically, the coverage index by single area is given by the ratio between (a) the number of MUAs in each area of the Web site and (b) the total number of research-retrieved MUAs for each area. The global coverage index is thus the arithmetic average value of all single areas; it measures the overall, though approximate, dimension of how far companies are taking advantage of the Internet's potential.

Regarding the colonization of different sectors of the Internet, it is clear that only the benchmark organizations have a widespread and homogenous presence, fully exploiting the many opportunities that virtual space provides (Figure 5).

The USA and Canada, as usual, are at a higher level than the European average, in both 2002 and 2005. Organizations in these countries, in fact, occupy three or more of the four areas (VIS, VDS, VCS, VTS) that the Web offers: The average numbers of areas colonized in 2005 were 3.67 and 3.00 respectively. European organizations, on the other hand, adopt a more traditional or low-profile strategy to exploit the Web; the colonization of Internet space remains on average limited to less than three areas (2.66).

Organizations in the United Kingdom (average colonized areas was 3) have a more pervasive Internet presence than the rest of Europe (about 2.5 areas).

Therefore, as often happens, numbers only are not enough to fully describe a phenomenon. Even if, in several cases, development of the presence occurs rather casually depending on contingent circumstances, the logical sequence of colonization of the different spaces, as happens in other sectors, is in this case too, tied to the need to resolve particularly serious organizational crises.

A typical path of development for Internet presence can therefore be outlined as follows:

- Entry onto the Internet is by means of the VIS area, which in strategic terms is that which is first attacked by organizations in the initial phase of their presence on the Internet, generally without clear rethinking of their marketing policy, and featuring a traditional approach as regards the media (the traditional "shop window" site); none of the international organizations studied were of this type.

- Later developments involve the VTS area, with the aim of facilitating interaction with patients (online booking of appointments, distance medical reports, complaints and suggestions), suppliers, staff, and so forth; 32% of HCOs have colonized the virtual space made available by the Internet in terms of VIS and VTS.

- Only later, and so far to a limited degree, do HCOs try to provide some of their services through the Internet (VDS area), especially LHUs; 44% of HCOs have managed to increase their presence on the Internet beyond VIS and VTS into a virtual distribution space.

- The VCS area is clearly that which is considered least important, partly because it does not correspond to any of the traditional organizational roles, nor does it provide immediate advantages in terms of optimization of existing organizational processes; 24% of HCOs are currently present in all four types of virtual space provided by the Internet.

The information that emerges from analysis of the coverage index average (Figure 5) shows that, with the exception of the benchmark organizations, the Internet still offers the potential for enormous developments. This is supported by the fact that even with the marked improvements seen in 2005 with respect to 2002 (with

Figure 5. The typology of Internet presence of international sample
■ Edition 2002   □ Edition 2005

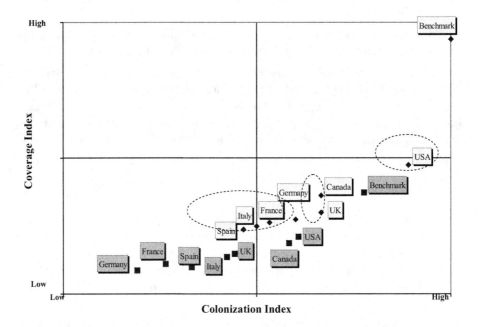

percentage variations equal to or over 200% for the United States and Canada, and Europe too), the coverage index remains overall rather low.

There is an abyss between the benchmark organizations and the rest of the world: The ratio of the exploitation of the potential of the Internet is 1:2 for organizations in the United States and Canada (the coverage index in 2005 was 38% and 29% with respect to 75% for the benchmark organizations) and a low 1:3.3 for European organizations (which had an average coverage index of 21%). In Europe, the figures for 2005 were again more homogenous than those for 2002.

An examination of the typology of the Internet presence for organizations in the international sample puts them into three groups (Figure 5):

- The pioneers, comprising the benchmark organizations.
- The chasers, representing organizations in the USA and Canada.
- The latecomers, the other organizations in Europe.

The Quality of Internet Presence in the International Sample

If the benchmark organizations are excluded, once more it is the U.S. and Canadian organizations that perform best for the quality of Internet presence. However, for the user friendliness of informative content and services available on the Web sites, the period of 2002 to 2005 showed a partial closing of the gap by European organizations.

In evaluating the qualitative dimension of Web site content (Step 4), we paid attention to both technical and user-related aspects, such as user-friendliness in information retrieval. Among the several options at hand, research ascertained the quality of Internet presence using the following criteria:

*   Level of sophistication.
*   Level of customization.

Both are calculated in average terms on Web site-identified MUAs.

The LOS varies according to the technology (in its widest meaning) employed on Web site content editing; its index ranges from 1 to 5 in order to account for intermediate situations between the following:

*   MUAs in static Web pages, edited without Internet-specific tools (using, for instance, MS Word, Powerpoint, and similar editors).
*   MUAs in highly interactive Web pages, edited using the most recent multimedia tools (such as Flash View animations, audio and video files, pictures, etc.).

The LOC calculates how far users judge Web site content to fit their needs. In this case, too, the index ranges from 1 to 5 in order to account for intermediate situations between MUAs, as follows:

*   MUAs in static Web pages that enable no customization of content and display no additional services.
*   MUAs in highly customizable Web pages offering user-profile-related additional services such as e-mail, forums, chat lines, newsgroups, and so forth.

Overall, from the point of view of the level of technological sophistication, there emerges a substantial difference between the organizations that constitute the bench-

Figure 6. The quality of Internet presence of international sample

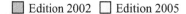

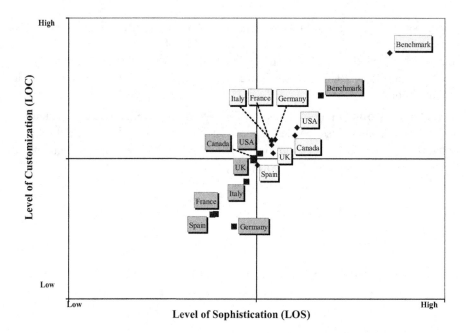

mark and the other healthcare organizations of the international sample, which, as has already been seen in other areas of research, grew further in the period from 2002 to 2005 (Figure 6).

In 2002, in fact, the LOS average for the benchmark organizations was 3.68 against an average value for the United States, Canada, and Europe of 3.04, 2.97, and 2.74 respectively. In Europe, sites of the healthcare organizations in the United Kingdom had values of 2.96, more in line with those of the United States and Canada than with those of the rest of Europe.

In 2005 there was, as usual, a general improvement in values, but it was more consistent for the benchmark organizations (average improvement of 20%) compared with the rest of the sample (average improvement of 15%). More specifically, in the second year of the project, the benchmark organizations had an LOS average of 4.41, which was considerably higher than the average values for the USA (3.43), Canada (3.41), and Europe (3.14). These values indicate that the distance between America and Europe remains largely unchanged. In Europe, the most consistent improvements were in organizations in France (from 2.57 to 3.16), Germany (from 2.76 to 3.20), and Spain (from 2.53 to 3.01).

The same can be said for user friendliness (LOC; Figure 6). Here, from 2002 to 2005 there was greater homogeneity in average values in the USA and Canada compared to Europe. In fact, the average improvement percentage of American and Canadian organizations was 11.5% compared with 30% for Europe. More particularly, the LOC for benchmark organizations rose from 3.90 in 2002 to 4.50 in 2005, for USA organizations from 3.08 to 3.44, for Canadian organizations from 3.01 to 3.33, and for those in Europe from 2.43 to 3.14. Analysis in 2005 did not indicate major differences between European organizations.

Neither were there major variations for the quality of Internet presence (Figure 6):

- Benchmark organizations had an established position.
- Organizations in the United States and Canada were classified as chasers.
- The remainder of organizations in the international sample, mostly European, lags considerably behind the first two groups.

The Effectiveness of Internet Presence in the International Sample

The visibility of the sites of the foreign sample by the main search engines is very high. Generally, the organizations with the best performances for traffic recorded are those that best exploit the potential the Internet offers and consequently their results are higher in all areas of analysis.

HCOs are likely to turn to the implementation of a Web site for basically two main different reasons:

- To maximize their visibility on the Web, that is, rendering themselves available to the highest number of potential users.
- To maximize Web site use by their stakeholders.

As a consequence, their Internet presence effectiveness (and success) depends on reaching such goals. Thus, we decided to measure HCOs Internet effectiveness (Step 5) by jointly considering the following:

- A Web site visibility index, obtained using the most common national and international search engines.

- A Web site traffic ranking as measured by Alexa.com, which offers an international list of sites ranked according to average Web site traffic.

The first indicator measures how simple it is to find Web sites using the most common search engines. All HCOs were searched by name in Italian and international search engines (Virgilio, Lycos, Excite, Iltrovatore, Google, Yahoo, Search.msn, Altavista). Indexes were the ratios between positive responses and search engines' total numbers.

Alexa.com is a site ranking Web sites all over the world according to their traffic. Lists are updated quarterly, measuring both Web site and Web-page access: The former quantifies Web site visitors per day (displaying its home page), and the latter is calculated considering the number of Web site URLs (uniform resource locators) accessed by the same user. The site hosting the highest Web site and Web-page traffic is ranked first.

It should be pointed out that Alexa.com can only provide an evaluation for the principal domain (e.g., http://www.domain.org or http://www.domain.com). For pages inside the domain (e.g., http://www.domain.org/page.html) or for sub-domains (e.g., http://www.hco.domain.edu), the value given is that of the principal domain. A zero value is given, however, in cases where traffic is so light compared to the average that it is not detectable (usually below position number 6,000,000 in the list).

Finally, Alexa.com's evaluation includes all sites on the Web. A process to classify Italian healthcare organizations appropriately has therefore been undertaken.

By evaluating the visibility of organizations in the international sample with the selected search engines, the factor that emerges most clearly is the high traceability both for benchmark organizations and the rest of the sample (Figure 7).

The organizations studied have an almost total visibility in that they are reachable using any of the search engines considered. This, in light of the results of 2005, is the case for the benchmark organizations, for those in the United States and Canada, and almost always for those in Europe. In the latter case, exceptions are Italy and France, although visibility is still high at 94% and 98% respectively.

As regards the ranking of site traffic, generally, those organizations that best interpret the role and potential of the Internet get greater success in return: Figures from Alexa.com for the benchmark organizations are, in fact, considerably higher than for the rest of the organizations in the international sample, both in 2002 and 2005 (Figure 7).

This general trend does not seem to be applicable in results recorded in 2002 for organizations in some countries, as, for example, Canada, Spain, and Italy. Ranking for site traffic, rather low for Canada and on the contrary high for Spain and Italy, are not in line with results obtained in other areas of analysis. In 2002, we attributed this to the following:

- The low population density for Canadian organizations.
- The fact that some sites of Spanish and Italian organizations were guests in the domain of a university or regional organization, and HCO site constituted only a section of the host site; in these cases, Alexa.com's evaluations were influenced by the volume of traffic recorded by the principal domain.

The 2005 research does seem to confirm these suppositions. Unlike in 2002, in fact, sites of organizations in Italy and Spain more often have their own domain that does not affect Alexa.com's evaluations. Recorded performances are therefore substantially in line with those obtained in other areas of analysis with the three types of HCOs:

Figure 7. The effectiveness of Internet presence of international sample

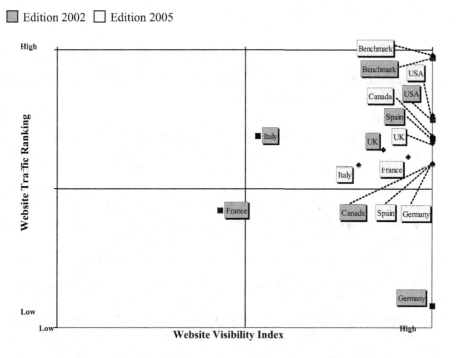

Figure 8. The ranking variation for Alexa.com vs. the Internet presence profile

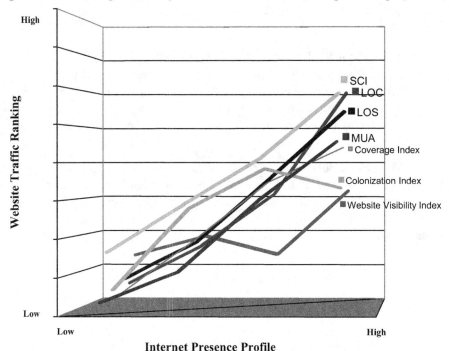

- The pioneers, or rather the group of benchmark organizations that obtain the best results both in terms of visibility with the search engines and ranking for traffic recorded for their site.

- The chasers, made up of the organizations in the USA and Canada, to which can be added, different from in other areas of research, the organizations of the United Kingdom, which have a markedly better performance for traffic recorded than the average of other European countries.

- A group of latecomers, basically the remainders of the European organizations (France, Germany, Italy, Spain) that are different from the best particularly for traffic recorded for their Web sites rather than for their visibility with search engines.

Finally, in light of major differences existing between the benchmark organizations and the rest of the foreign sample, it is useful to evaluate whether there is a relationship between the extent to which a site is used and the way in which the organization develops its Internet presence. In this respect, Figure 8 shows the ranking variations for Alexa.com vs. the indicators previously used to evaluate the Internet presence.

As a general, probably predictable, indication, it can be stated that the organizations that best make use of the potential of the Internet are rewarded. This is particularly evident as regards the number of MUAs, the SCI, and the possibility of personalizing the contents of the site.

Main Findings: National Survey

Italian Hcos and their Presence on the Web

The Internet is not yet a phenomenon that is present throughout Italian healthcare: Only 58% of all the organizations have a Web site.

The second part of the project presents a picture that is somehow contradictory. In fact, if the number of structures that are on the Web has not significantly increased (the rise over 3 years was only 11%, from 47% in the first year to 58% in the second), there is now a more homogeneous geographical distribution.

One of the striking facts about the first part of the study was that in the north, 63% of HCOs were on the Web, clearly distancing the centre (43%) and the south (33%). Today, there is still a difference, but the gap has narrowed: The north remains a benchmark with 68% of HCOs online, but the centre (53%) and the south (51%) are catching up.

This kind of heterogeneity between regions is not surprising and is confirmed by other studies concerning the spreading of a managerial culture in Italian HCOs (Anessi Pessina & Cantù, 2003; Baraldi & Monolo, 2004).

This general information has more relevance if analysis is undertaken by company type rather than by geographical area. For public organizations (local healthcare organizations and hospitals), SIRCs, and UPs, the full exploitation of the potential of the Internet is now a reality, regardless of geographical situation (almost everywhere penetration is around 90%); this is not the case for private organizations. These, therefore, seem to hold in much minor consideration the importance of a Web strategy as an essential component of their institutional strategy (the number of organizations online increased from 30% in the first year to 41% in the second).

Probably, public structures have a wider awareness of the potential offered by the Web than private ones because of (a) the recommendations provided by regional healthcare plans, (b) the need to provide information to the general public (e.g., emergency services, waiting lists for specific services), and (c) the attempt to simplify administrative procedures (e.g., booking appointments and giving medical exam results, such as blood tests, online). It should also be borne in mind that the private Italian HCOs are generally small (often with less than 300 beds) and have a limited budget for implementing a high-profile Web strategy.

The Complexity of Internet Presence

There is an increasing depth of presence on the Internet by Italian HCOs: The information available and services activated are increasing as are the tools that allow their content to be best exploited.

If we had to assess the tendency of the phenomenon under discussion based only on this aspect of analysis, we would have to affirm that the Internet is becoming increasingly important for Italian HCOs. The growth of the average amount of information or services offered through the Internet by Italian HCOs has gone from 10.22 to 17.80, which is a considerable rise of 74%.

As always, however, more information can be obtained by further analysis. Geographically, despite an overall improvement, there remains a substantial difference between organizations in the north and centre of the country (an average MUA value of 19.2 and 17.33) and those in the south (where the average MUA value is 13.32).

Analysis by type of organization is even more interesting as it makes it possible to outline, in some way, a taxonomy of approaches to the Web strategy according to the individual characteristics of each structure. It affirms that LHUs, IHs, and SIRCs have a presence on the Internet that can be described as established and definitely consistent (the average number of MUAs is respectively 21.76, 23.32, and 25.36, with variation with respect to 2002 of 95%, 72%, and 97%).

The situation is different for UPs and PSs. Regarding the former, of particular interest in the first year was the rather weak, improvised, or inadequate role played by these organizations in the overall picture. The MUA figure of 8.8 was closer to that of private organizations than any other type previously seen. In light of the results of the second year for this and the rest of the analyses, it is confirmed that the presence of UPs on the Net has been transformed: The average MUA value in this sector has increased by 148% reaching 22.00, much more in line with the other main organizations.

For PSs, the trend that emerged in the preliminary analyses already reported in previous pages is confirmed. In brief, considering the results obtained from other aspects of the analysis, private organizations are 3 years behind the other HCOs: The situation shown in the second year of the project (MUA value of 9.70 on average with an increase of 26%) is generally in line with the average figures for the first year. Similar observations to those expressed in earlier lines can be made here regarding the index of structural complexity. Here, too, the average increase is consistent (a

Figure 9. The complexity of Internet presence of Italian HCOs

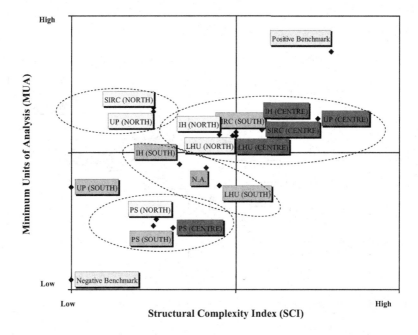

variation in percentage terms of 128%) by reason of a general improvement in those devices, in particular, site maps and search engines, which allow faster and easier access to the content available.

Considering the combination of the two variables (minimum unit of analysis and index of structural complexity), it is possible to position Italian healthcare organizations regarding the complexity of their presence on the Internet (Figure 9). Included in the HCOs are the following groups:

- A pioneer group of organizations (which includes the LHUs and IHs of the north and the centre, and the SIRCs and UPs of the centre) that invested early in the Internet and that hold that a Web strategy is an integral part of their business strategy and therefore make a substantial quantity of information available, with averages considerably higher than the national average.

- A chasing group of LHUs and IHs in the south that have average figures mainly in line with the national average.

Figure 10. The typology of Internet presence of Italian HCOs

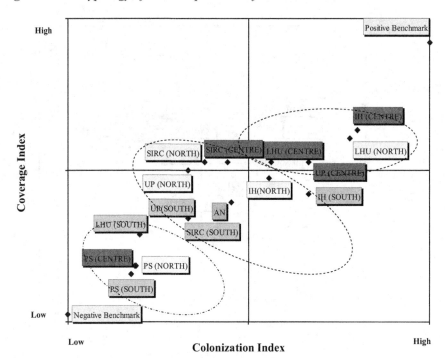

- An anomalous group of organizations comprising the SIRCs and UPs of the north, which analysis reveals has an unbalanced position: The contents of their sites are substantial (above the national average), but they are not backed up by an adequate index of structural complexity.

- A group of latecomers, mainly private HCOs, that have a rather light presence on the Internet, with sites offering basic and essential content, mainly of an informative nature, easily accessible even when lacking navigational aids (measured by the index of structural complexity).

- The negative and positive benchmarks representing the average of 5% of HCOs, presenting, respectively, the best and the worst scores.

The Typology of Internet Presence

The Web strategies of Italian HCOs are still mostly traditional, tending to provide information rather than services: Of the four areas in which the Internet's virtual space can be articulated, in over half of cases only two are used. So, inevitably, utilization of the Internet remains low. Despite steady improvements in recent years, Italian HCOs utilize less than a fifth of the potential that the web offers.

If we consider the results of the second year of the project for this sector, it emerges that Italian HCOs would appear to have some difficulties in getting over this hurdle. In other words, the go-to Internet has undoubtedly occurred and in some instances progress made since 2002 has been consistent (also because in several cases, the starting objectives were on a small scale). In most cases, nevertheless, a clear strategic design is still lacking that sees the Internet not only as a sophisticated instrument of communication, but also a lever through which to carry out company strategy, involving the re-planning of, if necessary, operational methods of healthcare services.

So, progress recorded in the level of colonization of the Internet's virtual space (with an increase of 23%) is undoubtedly a positive, but equally, it should be noted that around 60% of Italian organizations still have a traditional Internet presence based on the distribution or completion of information (VIS).

Figure 11. The quality of Internet presence of Italian HCOs

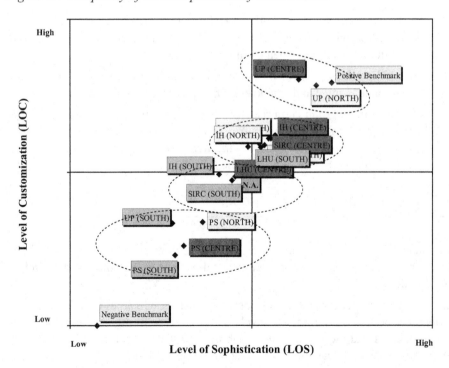

Investigating the analysis for each type of organization in more depth reveals that there has been consistent progress for LHUs, IHs, and UPs (growth percentages of 33%, 29%, and 43% respectively), rather less for the SIRCs (11%), and absolutely no progress for PSs.

The consistent increase of available content, even if this occurs as previously shown in a rather traditional way, has resulted in a clear improvement in the coverage index with an average increase of 88%.

The percentage variations are really surprising: 133% for LHUs, 90% for IHs, 111% for the SIRCs, a striking 183% for Ups, and, finally, a more modest 40% for PSs.

A glance at the absolute values, though, enables us to put the phenomenon into proportion. The starting values were in fact very low: a national coverage index of 8%, varying between 9% and 10% for LHUs, IHs, and the SIRCs; 6% for UPs; and 5% for PSs.

So, if following the progress made (the national coverage average is 15%), it is easy to see how much more improvement remains to be made.

The overall evaluation of Internet presence typology is thus reached by considering both the colonization index and the global coverage index. Figure 10 shows the following:

- The pioneers, in this case the IHs, LHUs, and the UPs of the centre to which can be added the LHUs in the north, exploit the Internet in a more homogeneous and broader way when compared with the national average.

- The SIRCs, IHs, and UPs in the north and the south perform mostly in line with the national average and can be classified together with the chasing group.

- The latecomers group comprises LHUs of the south and PSs, which have a less balanced presence (on average in fewer than two areas) and a much lower level of Internet use (particularly in the areas offering greater services such as the VDS, the VCS, the VTS) than the national average.

The Quality of Internet Presence

The quality of Internet presence of Italian HCOs is mainly good regarding both the degree of technological sophistication of the content available and the level of user satisfaction.

In 2002, we recorded for this dimension of analysis good values of reference with differences both at a geographical level and for the type of organization, basically due to budgets and the consequent possibility or not of utilizing specialized consultants. The second year shows the inevitable improvements following a general pattern common to the other dimensions of analysis.

The degree of technological sophistication (LOS) increased by 16% nationally (from 2.44 to 2.82), with the best performance coming from the UPs (a rise of 45%, from 2.15 to 3.12) and the worst from the PSs (no change).

The improvements for LOC are more consistent because the starting values were more modest. Progress on a national scale was 49%, but performances varied in a range from the highest of the UPs (85%, from 1.92 to 3.56) to the lowest of the SIRCs (23%, from 2.71 to 3.33). PSs did have an increase of 33% but recorded the worst overall performance with a LOC of 2.16.

Figure 11 positions Italian HCOs according to the quality of their Internet presence. Overall values do align with results from the other research areas, showing the following:

- The UPs of the north and the centre are very close to the positive benchmark.

Figure 12. The effectiveness of Internet presence of Italian HCOs

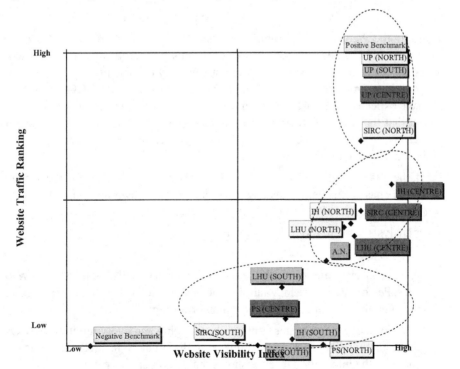

- The LHUs, SIRCs, and IHs of the north and the centre take the lead for national averages regarding quality of Internet presence.

- The chasers comprise the IHs and the SIRCs of the south, with a quality of Internet presence that, even if well below that of the leaders, is still in line with average national values.

- A final group of latecomers (mainly the UPs of the south and PSs), regarding this dimension of analysis, is in a generally weak position.

The Effectiveness of Internet Presence

Italian HCOs' Web sites have quite a good visibility on the Net: They can be reached by the most common search engines around 8 times out of 10. Their Internet effectiveness, however, is still limited in terms of overall Web site traffic.

Regarding the Web site visibility index, good results were already seen in 2002. In the national average, improvement was 27%, with a visibility index for search engines rising from 60% to 76%. Regarding the other dimensions of analysis, the best performance was from the UPs (with an increase of 72% and visibility with all the search engines used). The rise for the SIRCs was only 2%, although the visibility index was one of the highest (82%). PSs, despite a positive increase of 19%, are in line with the average values of the first year (visibility index of 63%).

Regarding the Web site traffic ranking, the second year of our research produced results that could be described as somewhat curious in that they are not in line with the results of previous analysis. There being generally consistent improvements in the complexity, typology, and quality of Internet presence of Italian HCOs, we would also have expected an improvement in the Alexa.com ranking. In other words, we thought that the real endeavors of the Italian HCOs to develop their presence on the Internet would have been translated into a greater success in terms of visitor numbers and traffic on their Web sites.

In fact, there is not a great difference between the results of the first and the second years. There has been, even if only slight, a reduction in the average national ranking (-3%) and increases only for the SIRCs and UPs (1% and 11% respectively).

Closer analysis leads to the following conclusions:

- Alexa.com evaluates traffic to the site, providing a global classification of Web sites. Performances are inevitably influenced by the rankings of other sites, be they healthcare related or not, or Italian or foreign.

- The world of the Internet is continually changing and evolving. Italian HCOs have undoubtedly progressed compared to the situation in 2002, but comparison with the international scene shows that the others have not been standing still and that Italian organizations, in several areas, are still behind the USA, Canada, and the United Kingdom (not to mention the abyss that distances Italy from international benchmarks).

Figure 12 positions Italian HCOs according to the effectiveness of their presence on the Internet. It shows the following:

- A first group of organizations, comprising UPs and the SIRCs of the north, in a leadership position with results in line with or very close to national benchmark references.
- A second group of companies, representing IHs and LHUs in the north and centre and the SIRCs of the centre in a chasing position, with results for effectiveness in line with national average values.
- A third group of organizations comprising the remaining Italian HCOs, which, at present, are in a relatively weak position as regards the effectiveness of their Internet presence.

Limitations and Further Research Areas

While aiming to provide a complete summary of the presence of Italian HCOs on the Web, our study nevertheless inevitably suffers from some structural limitations, which can be summarized as follows:

- The research project was carried out to evaluate, through external analysis, how and to what degree healthcare organizations currently exploit the potential of the Internet through their Web sites. The aim was to understand the strategy of colonization of Web space through analysis of the informative content or services offered. In this phase, we did not consider the internal aspects of the organization in the definition of the Web strategy (definition of the strategy, evaluation of costs, evaluation of performance), which does constitute an essential part of the research we are undertaking.
- The Internet is constantly evolving. Through our research, we attempted to take a picture of the state of the art of healthcare online, nationally and internationally, at two different times, first in 2002, then in 2005. This enabled us to identify the changes that took place over this period, but it was not our aim to define trends, find a way forward, or predict future developments.
- The aim of the research project is that of providing as objective an evaluation as possible of Internet presence. Although accurate scales have been elaborated

for the evaluation of LOS and LOC, there remain in this phase some inevitable and unavoidable sources of subjectivity.

- The evaluation of traffic recorded on the Web site was carried out using a site classifier Alexa.com. When sites are hosted by a main domain site (the sites of some LHUs on the regional site, or some PUs on the relevant university site), the figure refers to the main domain.

- An international comparison has been developed through the analysis of HCOs in the capital or similar city of each nation, similar in terms of the structure of the health service. As such, it inevitably provides only a general outline of the state of the art of online healthcare in each nation.

- The results obtained from the international sample, as for those from the survey regarding Italian structures, can be compared and confirm hypotheses only indirectly. This is because at the moment, and at least at a European level, we did not find similar studies about the actual level of exploitation of the Internet by healthcare providers. In this sense, it can be said that the lag shown by Italian structures compared with international benchmarks is basically in line with the limited attitude revealed by Italian companies to invest in technology (in 2003, only 2% of GDP [gross domestic product]) compared with the European average (3% of GDP) and the United States (3.8% of GDP).

Furthermore, the research activity carried out so far represents the first phase of a wider project that is planned for the carrying out of an in-depth analysis of a sample of organizations that will permit the analysis of the path followed in the definition of a Web strategy; its relative impact on organizational, clinical, and management processes; and the results in terms of cost reduction, quality improvement, customer satisfaction, and so forth.

Conclusion

"Who doesn't know the Internet?!"

The research undertaken leads to this conclusion: Nowadays, everyone, even in healthcare, knows the Internet; few, however, use it to its full potential.

In a few short years, the Internet has become one of the most widely used technologies in everyday life. In healthcare, however, despite the progress that has been made between the first and second years of our research project, it seems that the full potential of this technology is often not realized.

About the international survey, we can assert the following:

- There is a considerable distance between the benchmark and the other organizations that make up the international sample. HCOs in the United States and Canada perform best, but the European HCOs have reduced the gap in 2005.
- Only the organizations that make up the benchmark have a widespread and homogenous presence on the Internet, fully exploiting its potential. The HCOs that make up the international sample only partially exploit the potential of the Internet.
- If the benchmark organizations are excluded, once more it is the U.S. and Canadian organizations that perform best for the quality of Internet presence.
- Generally, the organizations with the best performances for traffic recorded are those that best exploit the potential the Internet offers and consequently their results are higher in all areas of analysis.

In Italy, many organizations are still unable to exploit this potential, not having activated a Web site. This is particularly true for PSs and those in the south. Public organizations, the SIRCs, and UPs have gradually become more aware of the Internet. The reasons for this change can be found in national health policies, in the increasing simplification of bureaucratic procedures and consequent reduction in costs, and, finally, in the growing awareness of the Internet's importance for the common good (improvement of relations with patients for healthcare organizations, and the facilitation of research and didactics for the SIRCs and UPs).

This has all led to a continual increase in the quantity of information and services offered by Web sites and greater focus on the tools (site maps, search engines, foreign-language versions) that facilitate access to their content. The gap in respect to international reference benchmarks and overseas organizations remains, however, substantially unchanged; this, moreover, is the same for all dimensions of analysis.

The approach to the Internet is rarely strategic, but often improvised, utilizing it simply as a means of communication. There rarely seems to be an awareness that investment in this technology should be in line with the institutional strategy of the organization (Minard, 2001), leading to a rethinking of the main ways of operating and assisting in healthcare.

This inevitably means that the possibilities that the Internet offers are not exploited as fully as they could be.

The quality of the contents available, in terms of technological sophistication and availability of access are, however, of a good level.

Like in the international survey, it is undeniably true that the best performance in terms of traffic recorded on the sites was achieved by the organizations that have invested more effort in this technology through the development of a high-profile Web strategy.

References

Alemi, F. (2000). Management matters: Technology succeeds when management innovates. *Frontiers of Health Service Management, 17*(1), 17-30.

Anessi Pessina, E., & Cantù, E. (2003). *L'aziendalizzazione della sanità in Italia: Rapporto Oasi 2003.* Milan: Egea.

Angehrn, A. (1997). Designing mature Internet business strategies: The ICDT model. *European Management Journal, 15*(4), 361-369.

Baraldi, S., & Monolo, G. (2004, July). *Performance measurement in Italian hospitals: The role of the balanced scorecard* [Working paper]. Proceedings of the 2004 Performance Measurement Association Conference, Edinburgh, Scotland.

Coile, R. C. (2002). *The paperless hospital: Healthcare in a digital age.* Chicago: Health Administration Press.

De Luca, J. M., & Enmark, R. (2000). E-health: The changing model of healthcare. *Frontiers of Health Service Management, 17*(1), 3-15.

Fattah, H. (2000). Failing health. *Media Week, 10*(17), 100-103.

Flory, J. (1999). Healthcare communications approaches for an online world. *Marketing Health Services* (pp. 25-30).

Frosini, B. V., Montinaro, M., & Nicolini, G. (1999). *Il campionamento da popolazioni finite.* Torino, Italy: UTET.

Given, R., et al. (2000). *Winning the loyalty of the e-health consumer.* Deloitte Research. Retrieved from www.dc.com/research

Given, R., et al. (2001b). *Understanding the NHS market for eHealth.* Deloitte Research. Retrieved from www.dc.com/research

Given, R., et al. (2001a). *Strategy and e-health.* Deloitte Research. Retrieved from www.dc.com/research

Glaser, J. P. (2000). Management response to the e-health revolution. *Frontiers of Health Service Management, 17*(1), 45-50.

Glaser, J. P. (2002). *The strategic application of information technology in healthcare organizations, (The Jossey Bass Health Series).* San Fancisco: John Wiley & Sons.

Goldsmith, J. (2000). How will the Internet change our health system? *Health Affairs, 19*(1), 148-157.

Goldstein, D. E. (2000). *E-healthcare: Harness the power of the Internet e-commerce & e-care.* MD: Aspen Publication.

Malcolm, C. (2001). Making a healthcare Web site a sound investment. *Healthcare Financial Management* (pp. 74-79).

Malec, B., & Friday, A. (2001). *The Internet and the physician productivity: UK & USA perspectives.* Working paper, EHMA Congress. Granada: EHMA Congress.

Minard, B. (2001). CIO longevity: IT project selection and initiation in the healthcare industry. *IT Healthcare Strategist, 3*(12), 1-12.

Minard, B. (2002). CIO longevity and IT project leadership. *IT Healthcare Strategist, 4*(1), 3-7.

Nicholson, N. (1999). *The Internet and healthcare.* Chicago: Health Administration Press.

Porter, M. E. (2001). Strategy and the Internet. *Harvard Business Review, 79*(3), 63-78.

Rodgers, J., et al. (2000). *HealthCast 2010: Smaller world, bigger expectations.* PricewaterhouseCoopers. Retrieved from http://www.pwcglobal.com/healthcare

Solovy, A., & Serb, C. (1999). Healthcare's 100 most wired. *Hospitals and Health Network, 73*(2), 43-51.

Section II

Challenges and Opportunities

Chapter IV

An e-Healthcare Mobile Application:
A Stakeholders' Analysis Experience of Reading

Niki Panteli, University of Bath, UK

Barbara Pitsillides, Nicosia, Cyprus

Andreas Pitsillides, University of Cyprus, Cyprus

George Samaras, University of Cyprus, Cyprus

Abstract

This chapter presents a longitudinal study on the implementation of an e-health mobile application, DITIS, which supports network collaboration for home healthcare. By adopting the stakeholders' analysis, the study explores the various groups that have directly or indirectly supported the system during its implementation. The system was originally developed with a view to address the difficulties of communication and continuity of care between the members of a home healthcare multidisciplinary team and between the team and oncologists often hundreds of kilometers away. DITIS evolved to be much more than that and even though it was introduced 5 years ago, it is considered a novel application. Despite this, its implementation has been slow, and several challenges, including the system's sustainability, have to be faced. This chapter aims to understand these challenges and the results of the study point to a diversity of interests and different degrees of support.

Introduction

Healthcare is an environment that has been experiencing dramatic progress in computing technology in order to process and distribute all relevant patient information electronically and overall to improve the quality of care. In particular, mobile e-health involves a spectrum of information and telecommunication technologies to provide healthcare services to patients who are at some distance from the provider and also to provide supporting tools for the mobile healthcare professional. The benefits of such mobile applications are numerous, with the main one being improvements in access to medical resources and care.

Recently, the healthcare and related sectors have been found to embrace mobile technology in e-healthcare applications. Though there have also been cases of mobile workstations being implemented at small medical units to facilitate easier access to specialist medical advice (e.g., Salmon, Brint, Marshall, & Bradley, 2000), most of the applications have been introduced to support patients at home. These could either be patient centered where patients and/or caretakers are given direct access to a mobile phone for communicating with the provider (e.g., nurse, doctor, counselor, etc.), or nurse centered where nurses who visit and care for patients at home have direct access to mobile applications for communicating with other medical staff.

It follows that the practice of e-health projects is often a collaborative activity requiring extensive and interactive communication within and between members of specialized occupational groups to coordinate patient care services. This becomes necessary when dealing with patients requiring a multidisciplinary team approach to their care, and who are treated outside the hospital environment. In such a case, the team is mostly geographically dispersed and rarely sees the patient together. This requires the creation of virtual multidisciplinary teams of care whose management and coordination can be supported by technology. In the study, we aim to explore the role of diverse stakeholders in an e-health application involving virtual multidisciplinary teams of care. Diverse stakeholders get involved at different stages of the project implementation and may experience different degrees of knowledge about the system itself, its significance, and its novelty. These along with their different backgrounds, interests, and expectations may contribute to different meanings and understanding about the system, its role, and its significance, which will ultimately affect system implementation.

Background

Stakeholders' Analysis

The role of stakeholders in IS implementation has long been recognized in the literature, though it has only been during the last few years that the identification of different stakeholders as well as the roles and interrelationships between them was found to be important for uncovering some of the complexity in system implementation (Pouloudi & Whitley, 1997).

Despite this, researchers have given different definitions to stakeholders. Sauer (1993), for example, makes reference to stakeholders as supporters, those who provide funding, information, and influence, whilst Beynon-Davies (1999) argues that there is a need to broaden this definition. As he puts it, "…not all groups with an interest in the development of an information system necessarily support that development. Some stakeholder groups may have a definite negative interest in the success of a given project" (p. 710). Following from these, in this chapter, in an attempt to keep a broad definition, stakeholders are defined as those with a direct or an indirect interest in a project.

According to Mitchell, Agle, and Wood (1997), stakeholders can be distinguished in terms of three relationship attributes: power, legitimacy, and urgency. Power is the ability to impose influence on the relationship; legitimacy is "a generalized perception or assumption that the actions of an entity are desirable, proper, or appropriate within some socially constructed system of norms, values, beliefs, and definitions" (Suchman as cited in Mitchell et al., p. 574). Finally, "[u]rgency is based on time sensitivity and criticality" (Mitchell et al. as cited in Howard, Vidgen, & Powell, 2003, p. 31). A combination of these three attributes contributes to different types of stakeholders who have different roles and expectations.

Empirical Study

In this section, we present the case of an e-health mobile application and adopt a stakeholders' analysis to understand its implementation process and the challenges faced.

Mobile applications are an increasingly important technology for improving the quality of health services, especially at the point of care. They enable the formation of virtual teams of care, and timely, effective, and quality patient management are the expected outcomes. The role of stakeholders in supporting such innovative applications is vital.

Figure 1. DITIS system architecture

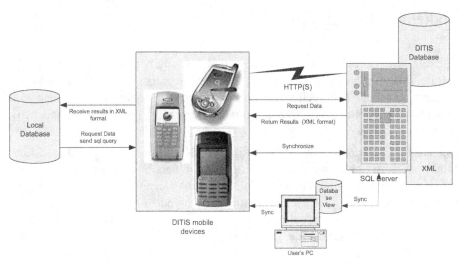

The System

DITIS: Virtual Collaborative Teams for Home Healthcare (http://www.ditis.ucy.ac.cy) is an Internet- (Web) based group collaboration system with secure fixed and mobile (GPRS [general packet radio service], GSM [Global System for Mobile Communication], WAP [wireless application protocol]) connectivity (see Figure 1).

It enables the effective management and coordination of virtual collaborative healthcare teams. It provides a secure access to e-records from anyplace and anytime via desktop computers (at work) or a variety of mobile devices (when on the go). It includes a set of tools for effective scheduling and coordination of team members, with features including automatic notification and alerting. It makes use of supportive tools relevant to home care that improve efficiency and minimize errors. The collaboration platform is based on identified roles and scenarios of collaboration, analyzed using the unified modeling language (UML).

The DITIS project was initiated in 1999 to support the activities of the home healthcare service of the Cyprus Association of Cancer Patients and Friends (Pasykaf). The goal of DITIS is to deliver a product that can improve the quality of the citizen's life. Contrary to today's health processing structure, which is, in all practical terms, facility-based care, this project aims to shift the focus onto home-based care, where everything is moving around the patient, supported by a team of multidisciplinary healthcare professionals. Given that the team cannot be by the side of the patient at all times, DITIS developed a collaborative software system to support dynamic virtual healthcare teams, customized for the differing needs of each patient at different times. The virtual healthcare team is supported in its provision of dedicated, personalized, and private service to the home-residing patient on a need based and timely fashion, under the direction of the treating specialist. Thus, it is expected that chronic and severe patients, such as cancer patients, can enjoy optimum health service in the comfort of their home (i.e., a focus on wellness), feeling safe and secure that in case of a change in their condition, the healthcare team will be (virtually) present to support them. The present users of the system include the healthcare professionals treating cancer patients (home-care nurse, oncologist, treating doctor, psychologist, physiotherapist, social worker, etc.) and the Pasykaf administration. It is expected that the system will be extended to other paramedical professionals, as, for example, the Pharmacist and the Cancer Registry, currently located at the Ministry of Health. Furthermore, the system can be adapted to cater to other home healthcare needs, as, for example, cardiac, renal, or diabetic patients.

DITIS deploys a novel networked system for tele-collaboration in the area of patient care at the home by a virtual team of medical and paramedical professionals, implemented using existing networking and computing components (the novelty of the system and competing approaches are briefly discussed in A. Pitsillides et al., 2005; B. Pitsillides et al., 2004). The system was originally developed with a view to address the difficulties of communication and continuity of care between the home healthcare multidisciplinary team (Pasykaf) and between the team and the oncologist often many miles apart. DITIS has through its database and possibility of access via mobile or wire-line computers offered much more than improved communication. Its flexibility of communication and access to the patient's history and daily record at all times, anywhere (in the case of home patients, outpatients, or an emergency hospital admission), has offered the team an overall assessment and history of each symptom. DITIS thus has the potential to improve the quality of life of the patient, for example, by offering the nurse the possibility of immediate authorization to change prescriptions via mobile devices, and the oncologist the possibility of assessment and symptom control without having to see the patient. It also offers the home-care service provider the opportunity to plan future services and lobby for funding by offering audit, statistics, and performance evaluation, and with these in place, the possibility for research.

The User Organization

Pasykaf, the user organization, was founded in 1986 to provide support to cancer patients and their families during their period of rehabilitation and is manned with highly qualified medical, paramedical, and nursing staff. In 1992, it started a home-care service for cancer patients. Specially trained palliative-care nurses in close cooperation with doctors (general practitioners and oncologists), physiotherapists, and psychologists attend and care for patients at home, focusing on maintaining the best possible quality of life, including medical care and psychological support.

In the context of home care, home-care professionals visit patients at home. Traditionally, the team of professionals was (loosely) coordinated by weekly meetings, or in case of some urgent event, information was exchanged by telephone calls or face-to-face meetings. Often, the same information is requested from the patient so each professional can build his or her own medical and psychosocial history and treatment notes (handwritten). Traditionally, these handwritten notes were filed at the Pasykaf district offices once the healthcare professional returned to the office. On a scheduled visit, the file had to be removed from the office and taken with the healthcare professional to the patient's house. This was inflexible and restrictive as there was no possibility of access by another healthcare professional at the same time. Furthermore, after hours, on-call professionals had to make a special visit to the office to collect the patient file (even if there was no other business with the office). For a patient visit to the hospital, especially in an emergency, there was no possibility of immediate access to the patient file from the attending home-care nurse. Therefore, there was limited possibility for continuity of care.

As with every manual system, there was limited possibility for audits and statistics, research was difficult, evidence-based medicine was not supported, dynamic coordination of the team was almost impossible, and communication overheads were very high and extremely costly in human and monetary terms. DITIS aims to address these problems in the provision of home-care services by a team of professionals.

Generally, given the limitations of the existing home-care delivery models, the need for improved ICT-supported practices emerged. Even though the context of health reform may vary across countries, major objectives are similar and include the following:

- A move toward people-centered services.
- A commitment to healthy public policy and a desire to improve the health status and quality of life of individuals and communities.
- Increased emphasis on knowledge- and evidence-based decision making, and efficiency and effectiveness in service delivery.
- A shift from facility-based health services and a focus on illness to community-based health services and a focus on wellness.
- The integration of agencies, programs and services to achieve a seamless continuum of health and health-related services.
- Greater community involvement in priority setting and decision making.

DITIS aims to support the above healthcare reform objectives. We focus our analysis on home healthcare of cancer patients, but expect our results to be applicable to home healthcare in general as well as cross-cultural and cross-border interoperability. Thus, through DITIS, we expect to assist in the delivery of better home care by offering the healthcare team services that are aimed at achieving a seamless continuum of health and health-related services despite the structural problems of home care as compared to facility-based care.

The system was initially deployed in one district, District A, and was gradually implemented in another three districts. The study has adopted the longitudinal approach in data collection for the first 5 years of the system from 1999 to 2004.

Methods and Data Collection

This research is interpretive as our aim is to capture stakeholders' interpretations of the system itself and their use of the system. To this end, our research method is qualitative in nature, examining "humans within their social settings" (Orlikowski & Baroudi, 1991, p. 14).

The fieldwork has taken place in various district sites of Pasykaf. Each site is served by a number of palliative-care nurses who visit patients regularly in their house to offer support. Data on DITIS were collected on different stages of the implementation process.

Phase 1: The preliminary part of the research has studied the use of mobile telephones by a group of palliative-care nurses during the period of August to September 2000. Interviews with three nurses and one doctor in District A have enabled the study of nurse-to-nurse interactions and nurse-doctor interactions via the use of mobile telephones, whilst also contributed to gathering information on their level of awareness about DITIS and its potential use in palliative care.

Phase 2: This part of the study took place in May 2001 and involved the use of a structured questionnaire that was sent to DITIS developers and potential users. It aimed to explore stakeholders' expectations regarding DITIS. A copy of this questionnaire appears in the appendix.

Phase 3: The third phase of data collection took place in April 2003. By this time, DITIS was implemented in four district sites. During this phase, current users of the system in three district Pasykaf offices were interviewed: one psychologist and three nurses. The main issues explored during interviews included the participants' actual use of DITIS, their own explanation of why they use DITIS the way they do, and their understanding of what users' and others stakeholders' role should be for achieving effective DITIS use.

Phase 4: This final part of the study took place in April 2004. Interviews included users (nurses and psychologist) as well as members of other stakeholder teams. During this phase, it was found that even though DITIS has been implemented

Table 1. Summary of the adopted data-collection approach

Phase	Time Period	No. of Interviews	Purpose
Phase 1	August-September 2000	4	Understand context of work, users' awareness about DITIS
Phase 2	May 2001	7	Users' expectations
Phase 3	April 2003	4	Level of usage & explanations given
Phase 4	April 2004	5	Stakeholders' own evaluation of DITIS

effectively with the right support from the project team, anxiety was identified at different levels with regard to its future.

Table 1 summarizes the data-collection approach adopted.

Results

Overall, the data reveal that DITIS offers innumerable opportunities for palliative-care nurses and other cancer-care practitioners. DITIS is currently widely accepted as an invaluable tool in palliative care. Nurses, psychologists, and doctors acknowledge that DITIS has numerous advantages and that they are willing to incorporate it in their work activities. DITIS can improve communication, coordination, and collaboration among members. Due to the huge amount of data regarding new and old patient records that need to be handled on a daily basis, DITIS enables users to access data quickly either from their office or remotely. Furthermore, it can be used as a statistical tool for producing internal reports for the district offices and the head office as well as external reports required by the Ministry of Health and other government departments.

"Pasykaf will be able to extract more information and statistics about cancer symptoms. Information about cancers and their occurrence by region will help to detect possible reasons that may be responsible about cancer (e.g. factories in the areas, etc.)" (Developer, Phase 2).

"Life will be so much easier with DITIS to fill in the gaps from unknown to known" (Nurse, District B).

Interestingly, even though technophobia was identified in Phase 1 of this study as a possible negative factor in the effective implementation of the system, it was later expressed by nurses across different district sites that participated in Phase 3 that users are generally willing to adapt the system in their day-to-day work because they expect that their tasks will be executed faster and easier, saving time and effort.

It was agreed among all participants in the study that the project was a novel one, that it was a radical departure from what existed previously, and that it was designed to be at the core of Pasykaf's activities (i.e., the provision of healthcare to home-based cancer patients).

Implementation Problems

During the first three phases of the study, there was a general feeling that DITIS had not yet been sufficiently incorporated in the daily work activities of the healthcare workers and that this would be a slow process. The main problems identified were with regard to the implementation process. A nurse in District B clearly said that this process "has been slow from all points of view" (Phase 3).

It has been widely recognized that the effectiveness of the system implementation was jeopardized due to financial resources being constrained or at times becoming unavailable. The limited budget that the project development team had to work with mainly had two major implications. First, only a small number of mobile handsets could be acquired. As a result, only nurses in one district office (District A) were given a mobile device with DITIS application, whilst other district offices have to rely only on PC- (personal computer) based applications of DITIS. The latter, however, restricts the use and the potentials of DITIS, which has been developed as a wireless application to promote virtual collaboration in cancer care. Second, limited financial resources have influenced staff availability on the project implementation team. The project has been experiencing staff discontinuities since the early stages of DITIS development. Mainly graduate students have been used for this project under the guidance of two computer-science academics, both members of the DITIS project team. Even though the latter remained the main drivers of the project, DITIS was only one of several projects that they were involved in and therefore could not give their full attention as required by the criticality of the project nature. To quote a nurse, "The system was done on borrowed time" (District A, Phase 3), indicating that for most of the implementation period, there were no full-time project members; rather, even though the system had gained the enthusiasm of several people who committed themselves to the system, none of them could make a full-time commitment. Instead, there were several temporary project members throughout the duration of the implementation process leading to staff turnover. Therefore, the high staff turnover and the lack of full-time staff brought inconsistencies and delays in the project development.

As the psychologist who participated in the study (Phase 3) put it, "There is no person in charge and this leads to communication problems." The same person suggested that there was a need for frequent meetings to keep users informed about the state of the implementation process, alleviate doubts, and improve coordination among cross-groups (e.g., nurses, doctors, psychologists).

Based on the results of the longitudinal assessment, corrective measures were taken, including the creation of a more stable team due to the commitment of all relevant actors and availability of funding. These corrective actions were acknowledged by all the users interviewed in Phase 4 of the study. During this phase, there was a general feeling of satisfaction about the use of DITIS in the day-to-day work practices as users have by now begun seeing the benefits of the system.

"The system is more reliable."
"It is 100% better."

Despite these positive results, there appears to be some anxiety among users about the sustainability of the system. DITIS has been successful in attaining the initial goals set. However, several difficulties, some non-technical in nature, were encountered during the development and deployment of DITIS in Phase I. These are outlined below:

- Underestimation of the workload involved in order to populate the DITIS database, and misjudgment of effort for encouraging users who are used to a paper-based system to switch to a new system, which was in essence still incomplete and under construction

- Network technology limitations (e.g., WAP over GSM). The migration to new technologies (GPRS/UMTS [Universal Mobile Telecommunications System] and ADSL) is resolving many of the original technical problems: Service is always available and bandwidth is much higher.

- Deployment with mobile devices has been limited mainly due to the cost related with having each member of the home-care team have his or her own mobile device, and the pace of new device launches with enhanced functionality. Recently, 30 mobile devices were acquired and are being deployed. This will allow a number of multidisciplinary teams to operate.

- Sustainability of DITIS due to uncertain financial support. Potential funding may derive from the government through its spin-off company initiative. Another one is the Ministry of Health for its planned community-care service.

Stakeholders: Relationships and Roles

The main stakeholders of DITIS derive from inside as well as outside the user organization. A key stakeholder group consists of the nurses and other members of the medical team in the district offices, for example, psychologists who directly use the system. Gradually, the nursing team has begun to embrace DITIS and overcome the initial resistance that was mainly caused due to the delays in implementing the system.

A second group involves the university computer scientists who designed the original system and currently manage the project development and implementation. They have the intellectual property of the system and are the ones who actively promote the system to national and international organizations in order to attract funding that would enable them to continue developing and improving the system. Their interest is in "the expansion of the collaborative system for usage in other fields (e.g., cardiac home care, insurance sales, etc.) and its eventual commercialization" (interview with project leader). The cost of setting up such an infrastructure and supporting it vs. the benefits, such as quality of life and time saved, are difficult to justify in monetary terms; also, there is the potential benefit of using GPRS and ADSL (Asymmetric Digital Subscriber Line) (always connected, higher speeds) vs. the earlier GSM/WAP mobile telephone device and ISDN (integrated services digital network) for the fixed computer lines (dial-up, low bandwidth), and the costs of maintaining such a telecom infrastructure to consider. Another barrier is the high cost of handheld devices and rapid change of technology, which have hindered projects development Cyprus-wide.

An independent commercial software organization that supported the initial idea and design of DITIS has also been a key stakeholder. This organization has recently withdrawn from the project after the first stage of its implementation as they could not see the financial viability and profitability of the project and were no longer interested in investing in mobile-based applications. Another commercial software organization with a focus on mobile e-services has recently joined the project.

Figure 2. Stakeholder typology (Mitchell et al., 1997, p. 874)

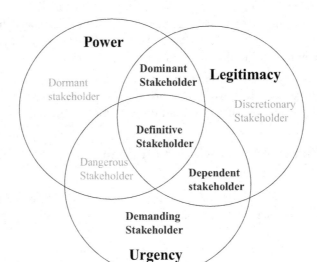

Finally, an important stakeholder is Pasykaf as the user organization that hosts the system in its district offices as well as in the headquarters. Pasykaf, a charity organization, is interested in the project as it can see that the system can predominantly enable it to produce national and regional statistical information on cancer and cancer patients, which are required by the government. Our study, however, has shown that the involvement of Pasykaf management in this project has remained limited. Several users recognized that Pasykaf in particular should undertake a more active role in the implementation of the project by investing more time and administrative support.

Accordingly, the stakeholders derive from diverse sectors, and even though they all want DITIS to succeed, their expectations are different. From our data, it was found that all stakeholders agree on the important role that the system could have in cancer support.

In what follows, we describe these different types of stakeholders and identify their relevance to the DITIS project. Based on Mitchell et al.'s model (1997), Figure 2 depicts the stakeholders' relationship attributes and the different types of stakeholders.

Definitive stakeholders possess power, legitimacy, and urgency. The project initiators at the university fall in this group. Due to their active involvement in the design and promotion of the system, they possess both power and legitimacy whilst simultaneously are aware of the risks involved if no funding is secured. For this, they have a clear and immediate mandate to give priority to the project and have considered several options, one of which is commercialization.

Demanding stakeholders exist where there is urgency but no other relationship attribute. Within this category, we place the commercial organization that became involved in the initial design of the system. It did so for commercial interests. Due to the financial difficulties that the project faced, the company decided to withdraw their demands for commercialization and thus profitability has not been met.

Dormant *stakeholders* **are the** *stakeholders* who possess power to impose their will, but by not having a legitimate relationship or an urgent claim, their power remains unused…Dormant *stakeholders* have little or no interaction with the firm. However, because of their potential to acquire a second attribute, management should remain cognizant of such *stakeholders*, for the dynamic nature of the stakeholder-manager relationship suggests that dormant *stakeholders* will become more salient to managers if they acquire either urgency or legitimacy. (Mitchell et al., 1997, pp. 874-875)

In the case of DITIS, the dormant stakeholder is the government and other funding bodies that have shown an interest in the system. Because of their potential role in the future of the system, the project management team (university) should remain cognizant of such stakeholders. It is expected that the government's power will be exercised when there is a better recognition of the potentials of the system not only in cancer care but in the other health areas, too.

Dependent *stakeholders* **are those** "who lack power but who have urgent legitimate claims as 'dependent,' because these *stakeholders* depend upon others [e.g., other *stakeholders*] for the power necessary to carry out their will" (Mitchell et al., 1997, p. 877). Nurses represent this type of stakeholders. Their voice has mainly been represented through Pasykaf (management) itself, and, as it was found in the study, their interest in the system is predominantly for efficiency and accuracy in statistical analysis.

Dominant stakeholders are both powerful and legitimate. Their influence in the relationship is assured since by possessing power and legitimacy they form the dominant coalition. In our case, we find that Pasykaf itself belongs to this group of stakeholders as it represents the only host of the system. However, as a charity organization, it has been unable to fund the project itself and therefore its position has remained weak and the management team needs to look elsewhere for financial support. It is expected that when other funders are found, for example, the government, or with the commercialization of the system, the power of Pasykaf will be reduced as other host organizations will emerge, such as those for cardio-care support.

Other stakeholder groups that were identified in Mitchell et al.'s (1997) framework and shown in Figure 2 but not found in our own study are as follows:

- Discretionary stakeholders possess legitimacy, but have no power for influencing the firm and no urgent claims.

- Dangerous stakeholders possess urgency and power but no legitimacy; this may result in the use of coercive power.

- Non-stakeholders possess none of the attributes and thus do not have any type of relationship with the rest of the group.

Accordingly, it is found that in the case of DITIS there were five different distinct types of stakeholders. The diversity in their views, which is expressed in the reasons they give to the legitimacy of the system, as well as the perceived urgency and their ability to influence the system have all contributed to different degrees of support. In this study, apart from the commercial organization, all the other stakeholders have been positively supportive of the implementation process. They have been doing so differently, however, with some, such as the university team, taking a more active role in securing funding and hosting the project team whilst others, such as Pasykaf, remaining more passive. This passiveness is not, however, due to a lack of interest but rather due to other priorities.

Future Trends

The potentials of mobile applications in e-health are tremendous. They could be used for supporting health professionals in offering care through improved e-tools (e.g., for improved access to patient records by all health professionals, improved collaboration, and streamlined workflow), which are especially useful for the community-care environment. However, even though mobile technology is a key factor for enabling the formation of geographically proximate medical teams at the point of care, their effective implementation depend not only on the level of support provided to users, but also by the extent to which there is a shared understanding and support among diverse groups of stakeholders.

An important contribution of this study is an examination of the role of diverse stakeholders on the implementation of such a novel e-health initiative. With a growing recognition that e-health can make an impact on the provision of healthcare as well as that e-health applications are becoming even more global, investigating the diversity that may exist among the various stakeholders is becoming vital for their success.

Conclusion

This chapter presents a longitudinal study on the implementation of an e-health mobile application, DITIS, which supports network collaboration for home health-care. Our study has found that users' support has gradually improved over the last years as they have been increasingly exposed to the system capabilities and have recognized the advantages of the system in their day-to-day work for both administrative and consultation purposes. Another reason for this is that the nurses have gained participation in the project team with periodical meetings with the project manager and developers. Yet, the future of the system is uncertain as future funding to gain sustainability may not be available. Such a complex and novel system has not gained shared support by all parties concerned, with one company dropping out (while another one joined) and others not taking an active role. The long-term solution is commercialization, which is currently pursued, but as with any new ideas and products, there is considerable risk involved. The study has adopted the stakeholders' analysis (Mitchell et al., 1997) and found that there are different relationship characteristics among the key stakeholders who show diversity in interests, expectations, and levels of involvement in the system implementation.

DITIS has appeared to act as a useful fuel for improving patient records and promoting an integrated approach that has a direct impact on the quality of treatment and healthcare support to home-based cancer patients. However, even though this is a novel application and despite the fact that it was introduced 5 years ago, the implementation has remained slow and the system has not yet been able to secure its place and its future in the health sector; rather, it has gradually been making an impact on the healthcare support provided by some nurses and medical staff. Lack of organization ownership makes the future of the system uncertain. What has enabled it to survive was the enthusiasm of some key individuals, mainly the university team and those users who could see the direct benefits of the system on the quality of cancer care. Speedy commercialization of the system seems to be the solution for its long-term survivability, and this is the current focus. In the meantime, DITIS is at present being deployed for its healthcare collaboration and patient-management aspects in the context of two EU- (European Union) funded e-TEN (deploying Trans-European e-services for all) market validation projects (HealthService24 and LinkCare) involving trials for cardiac-patient monitoring.

References

Beynon-Davies, P. (1999). Human error and information systems failure: The case of the London ambulance service computer-aided dispatch system project. *Interacting with Computers, 11*, 699-720.

DITIS: Virtual Collaborative Teams for Home Healthcare. (n.d.). Retrieved July 2005 from http://www.ditis.ucy.ac.cy

Howard, M., Vidgen, R., & Powell, P. (2003). Overcoming stakeholder barriers in the automotive industry: Building to order with extra-organizational systems. *Journal of Information Technology, 18*, 27-43.

Mitchell, R., Agle, B., & Wood, D. (1997). Towards a theory of stakeholder identification and salience. *Academy of Management Review, 22*(4), 853-887.

Orlikowski, W. J., & Baroudi, J. J. (1991). Studying information technology in organizations: Research approaches and assumptions. *Information Systems Research, 2*(1), 1-28.

Pitsillides, A., Pitsillides, B., Samaras, G., Dikaiakos, M., Christodoulou, E., Andreou, P., et al. (2005). DITIS: A collaborative virtual medical team for the home healthcare of cancer patients. In *M-health: Emerging mobile health* (pp. 247-266). Kluwer Academic/Plenum Publishers. Springer: New York.

Pitsillides, B., Pitsillides, A., Samaras, G., Christodoulou, E., Panteli, N., Andreou, P., et al. (2004). *User perspective of DITIS: Virtual collaborative teams for home-healthcare: Vol. 100. Current situation and examples of implemented and beneficial e-health applications* (pp. 205-216). IOS Press: The Netherlands.

Pouloudi, A., & Whitley, E. A. (1997). Stakeholders identification in inter-organizational systems: Gaining insights for drug use management systems. *European Journal of Information Systems, 6*, 1-14.

Salmon, S., Brnt, G., Marshall, D., & Bradley, A. (2000). Telemedicine use in two-nurse-led injuries units. *Journal of Telemedicine and Telecare, 6*(Suppl. 1), 43-45.

Sauer, C. (1993). *Why information systems fail: A case study approach.* Henley on Thames, UK: Alfred Waller.

Suchman, M. C. (1995). Managing legitimacy: Strategic and institutional approaches. *Academy of Management Review, 20*, 571-610.

Appendix

DITIS Implementation Project, May 2001

Please take a few minutes to answer the following questions. I would appreciate if you return the completed questionnaire to me by e-mail.

Thank you very much for your contribution.

Please tick that which most identifies your involvement in the DITIS project.

Designer/Developer: □
User (Nurse): □
User (Doctor): □
Other (please specify): □

What do you think the main benefits would be for implementing DITIS within Pasykaf?

After implementation, how do you expect DITIS to be used on a day-day to basis?

What would you expect the results of the pilot study to be? Please identify in your answer those factors that you think might enable or constrain the success of this pilot.

Chapter V

Behavior Change through ICT Use:
Experiences from Relatively Healthy Populations

Marieke W. Verheijden, TNO Quality of Life, The Netherlands

Abstract

New communication technologies have made an impact on several areas of our everyday life, including the areas of health and health promotion. The Internet provides opportunities for personalized interactive health communication at a much larger scale than is possible in face-to-face communication. It has been suggested that only interactive health-behavior-change Web sites that advise, assess, assist, provide anticipatory guidance, and arrange follow-up have the potential to lead to successful behavior change. Additional factors that may affect the success rate of behavior-change programs are the reach of and the exposure to such programs. This chapter elaborates on all of these factors.

Introduction

Rapid developments in information and communication technologies have made a plethora of computer-based applications affecting our everyday life available. Not surprisingly, these applications have also been taken up in health-related areas. Electronic medical records, computerized reminders for preventive services, and computer-aided diagnosis of tumors are examples of this. Examples can also be found in the area of health-behavior change. New communication technologies are increasingly used as a replacement for and supplement to traditional health education and health-behavior-change programs. In addition to searching for health information, large groups of Internet users have received information from tailored health-promotion programs through interactive health communication and/or have participated in online surveys (Sciamanna, Lewis, Tat, Napolitano, Fotheringham, & Marcus, 2002).

Evidently, the ultimate goal of Web-based health-promotion programs is to make an impact on public health. The extent to which such programs have an impact on public health is determined by two components: the reach and the efficacy (Abrams, Emmons, & Linnan, 1997). This chapter will discuss the following components in more detail: the Internet as a communication channel, potentials and minimum requirements of Web-based behavior-change programs, the delivery and reach of and exposure to Web-based behavior-change programs, and the feasibility and effectiveness of Web-based behavior-change programs. This chapter focuses specifically on relatively healthy populations as opposed to patient populations. Koelen and Van den Ban (2004) discussed that conventional mass media generally focus on new discoveries about diseases and their treatment. Much less attention is focused on disease prevention, health behavior, or early detection. This is in sharp contrast with the general idea that an ounce of prevention is worth a pound of cure, and stresses the need for a focus on applications for relatively healthy populations. This is particularly challenging because (otherwise healthy) at elevated risk for cardiovascular disease stated that they were more interested in using Web-based health-promoting programs for information when confronted with a direct medical condition than for prevention purposes (Verheijden, 2004).

Commonly Used Behavior-Change Theories

Changing health behavior is challenging, and it is widely assumed that to be successful one needs to understand the determinants of behavior and behavior change. To help increase this understanding, several models have been developed. It is beyond the scope of this chapter to provide a full overview of health-behavior (-change) models.

Therefore, this paragraph only briefly touches upon some of the most frequently used models: the theory of planned behavior and theory of reasoned action, the health belief model, the social cognitive theory, and the stages-of-change model. According to the theory of planned behavior, intention is the basis for any behavior change. Intention in turn is a function of attitude toward the behavior (change), the perception of the social pressure from other people (subjective norm), and perceived behavior control (self-efficacy). The health belief model postulates that health behavior is determined by the belief in a personal health threat combined with the belief in the effectiveness of a healthy behavior, provided that there is an internal or external cue to action, such as an article in a newspaper or a close friend having a heart attack. The social cognitive theory proposes that interrelated personal, environmental, and behavioral factors determine behavior. The stages-of-change model is one of the very few models specifically developed to address behavior change (as opposed to addressing merely the behavior itself). The stages-of-change model postulates that individuals can be classified in one of five stages of change based on their current behavior and on their intentions to change in the (near) future. Progressing through the stages of change thus reflects behavior change. The model also explicitly states that behavior change is not a watershed event. Instead, it is a spiraling progress and regression to earlier stages of change is possible (Ajzen, 1991; Bandura, 1986; Koelen & Van den Ban, 2004; Mullen, Hersey, & Iverson, 1987; Prochaska & DiClemente, 1983; Prochaska & Velicer, 1997). Especially in recent years, the models are increasingly used as a theoretical basis for all sorts of health-promotion practices, including Internet-based programs.

The Internet as a Communication Channel

Communication is essential to inform people about health and to keep health issues on personal and public health agendas. New technologies provide the means for interactive health communication. This was defined by the Science Panel on Interactive Communication and Health (SciPICH) as "the interaction of an individual—consumer, patient, caregiver, or professional—with or through an electronic device or communication technology to access or transmit health information or to receive guidance and support on a health-related issue" (Robinson, Patrick, Eng, & Gustafson, 1998). Electronic applications that focus exclusively on administrative, financial, or clinical data were excluded from this definition. The SciPICH identified six specific functions of interactive health communication: to relay information, enable informed decision making, promote self-care, manage demand for health services, promote peer information exchange and emotional support, and promote healthful behaviors. All aspects can be valuable in health-behavior-change programs.

Table 1. Characteristics of hybrid media for health-promotion programs

Reach	Potentially large
Push and pull	Pulled delivery of information, potentially large information overload. Limited insight in the actual reach of the intended target audience
Tailoring and personalization	Possible, but only using preprogrammed structure. Less direct than interpersonal communication
Social support	Particularly promising for stigmatizing topics when people prefer to be anonymous. Possibly difficult to establish relationships
Space and time dependence	At convenience of users, no geographical boundaries. Equipment is still largely tied to locations.
Speed	Content can easily be updated in the event of new developments.
Costs	Development costs (particularly of advanced applications) can be high. The cost per person may end up being relatively low because of potentially large reach.

The outcome of health communication is determined by characteristics of the source, message, channel, and receiver. The word channel refers to the way the message is brought across from the source to the receiver. Evidently, health-promotion messages can be transferred using a number of channels. These channels may vary in the extent to which the receivers are at liberty to interpret the message to their own views, in the extent to which feedback is possible, and in the extent to which group interaction is possible. Channels can also be placed on a continuum ranging from one-way flow of information (e.g., mass media) to active two-way exchange (e.g., an in-person conversation). Channels that typically possess characteristics from both sides of the continuum, such as interactive computer technologies, are usually referred to as hybrid media (Koelen & Van den Ban, 2004). These hybrid media potentially reach large numbers of people and support a level of interactivity that is impossible with conventional mass media. Table 1 (see also Leeuwis & Van den Ban, 2004, for a more detailed discussion) presents characteristics of hybrid media in applications for health-promotion programs.

Potentials of Behavior-Change Programs through the Internet

The increasing research interest in behavior-change programs through the Internet is due in part to the increasing numbers of people at risk for non-communicable lifestyle-related chronic conditions. While individual approaches have shown to be effective for many different health behaviors in many different target groups, they are very expensive, thus limiting their applicability for reaching large groups of people. When targeting large groups of people, most approaches use the same generic information for all members of the target group. This information is oftentimes action oriented and may thus fail to be of interest and relevance to the majority of the target group. As discussed previously, hybrid media may provide a solution to this.

There are also other potentials of new communication technology that may be used for health-promotion programs. In addition to the characteristics discussed in Table 1, using new communication technologies for health-behavior change also has several other advantages related to interactivity, appeal, and engagement. Tailoring and instantaneous feedback based on the responses of the participants, for example, may increase the interactivity and improve the personal relevance of the content. Wantland, Portillo, Holzemer, Slaughter, and McGhee (2004) also suggested that "interactivity may help reduce attrition and provide benefits in producing positive behavioral change." The appeal of health-behavior-change programs can be increased by giving convenient access to the program and by allowing participants to choose what material they receive, when, and how often. Some target groups may also prefer computer-delivered programs over other delivery methods. Finally, computer-based health-behavior-change programs may have advantages over traditional programs in terms of engagement, for example, by facilitating open communication, particularly on sensitive issues. Furthermore, multimedia interfaces may reduce the level of literacy required to work with the materials, and they allow for role-playing in a virtual environment (Owen, Fotheringham, & Marcus, 2002).

De Nooijer, Oenema, Kloek, Brug, De Vries, & De Vries (2005) conducted a survey among 15 experts involved in the development and evaluation of health-promotion interventions on the Internet. The experts unanimously stated that a solid theoretical base is essential for the success rate, however, the extent to which the large numbers of available Web-based health-promotion interventions are sufficiently theory based may be limited at the present time (Evers, Prochaska, Prochaska, Driskell, Cummins, & Velicer, 2003). Notably, most experts were involved in interventions based on frequently used theories in health-behavior change, such as the social cognitive theory and the stage of change model. Apart from the benefits for the users, the experts also benefits for designers and researchers. These benefits include easy access to information, the reach of potentially large groups of users, low costs for dissemination of information, the possibility to keep the information up to date with relative ease, the possibility for follow-up, integrated questionnaires and feedback, tailored questionnaires, and central data management.

The experts also mentioned several disadvantages of Web-based health-promotion programs. One of the major concerns is related to the actual reach of the target audience. As of yet, it is unclear how experts can assure that the intended target audience knows about the availability of the site, that they will actually access the site, and that they will stay on the site long enough to receive the message. Furthermore, the Internet provides health workers with much less influence on (or control over) what people do with the information that is presented to them. Another concern that was mentioned is the lack of quality control on Web-based information and the large numbers of sites containing false and possibly dangerous information. Limited personal communication and the possibility that health-related information may easily be overlooked or disregarded in view of the large numbers of e-mails and Web sites people are exposed to were also mentioned.

Minimum Requirements of Web-Based Behavior-Change Programs

Several instruments to rate the quality of Web sites are available (Commission of the European Communities, 2002; Gagliardi & Jadad, 2002; Gattoni & Sicola, 2005; Griffiths & Christensen, 2005; Risk & Deznowagis, 2001). However, very few of these instruments are applicable to the interactive Web sites that are used for health-behavior-change programs (Evers et al., 2003). It was therefore attempted to extrapolate clinical-practice guidelines for face-to-face counseling to guidelines for Web sites that provide interactive health information. These guidelines for Web sites can be summarized using six *A*s, which provide a minimum set of criteria for a Web-based program to have the potential to change behavior (Cummins et al., 2003; Evers, Cummins, Prochaska, & Prochaska, 2005). Notably, meeting all criteria does not guarantee successful behavior change.

The initial criterion in face-to-face counseling is to *ask*. In this step, the behavioral concerns of the patient are assessed. Because it was assumed that people only visit sites on topics they are interested in or concerned about (e.g., people visiting a site on smoking cessation want help to quit smoking for themselves or for someone close to them), this criterion was not included in the Health Behavior Change on the Internet (HBC-I) screener. The five criteria that were included are to advise, assess, assist, provide anticipatory guidance, and arrange follow-up (Table 2). These strategies will be discussed in more detail below.

Evers et al. (2005) reviewed almost 300 Web sites targeting alcohol use, diet, exercise, smoking, and three disease-management programs using the five *A*s. Less than 10% of the Web sites met all five criteria. The criterion that the fewest Web sites met was anticipatory guidance. The assess criterion was met most frequently (slightly over half of the sites). Sites on smoking cessation, diet, and exercise relatively frequently met four to five of the criteria.

Table 2. The five As for Web sites providing interactive health information

Advise	Create sense of urgency about the need to change
Assess	Assess (determinants of) behavior
Assist	Provide (tailored) feedback on assessments, suggest strategies for change, support behavior-change efforts
Anticipatory guidance	Manage tempting situations, provide relapse prevention, maintain motivation
Arrange follow-up	Specify time frame, encourage continuous participation

Advise

The first criterion includes advising individuals about behavioral risks and about the need to change this behavior. This may be done in part by using (self-) assessment approaches that will be discussed under "Assess." It is also recommended that the purpose of the Web-based program (i.e., to initiate and sustain behavior change) is made clear to the users.

Assess

People who are unaware of the problem behavior and the need to change will most likely not be interested in any information on behavior change. This makes (self-) assessment of the problem behavior to create a sense of urgency an obvious criterion. Assessment is also important because it provides the basis for tailoring feedback messages. Unfortunately, Anhøj and Jensen (2004) showed that patients were more likely to trust their own (oftentimes incorrect) perception of dietary and physical-activity habits than the outcomes presented to them by a computer program. Similar findings were found in other studies (Anhøj & Nielsen, 2004; Brug, 1999). This complicates the feedback that can be given on self-assessment tools. Yet, as Verheijden, Jans, and Hildebrandt showed that people's belief that their lifestyles are healthy already is one of their major arguments for limited interest in Web-based behavior-change programs, misconceptions about people's actual behavior need to be taken away as much as possible.

A variety of outcomes may be assessed. One may think of weight, BMI (body mass index), exercise levels, fitness levels, nicotine dependence, and so forth. Patients' willingness to change is also frequently assessed (Evers et al., 2003). Several theoretical frameworks to do so are available (e.g., the stages of change model, the precaution adoption process model, the theory of planned behavior). Bensley et al. (2004) presented three behavior-change projects based on the e-health behavior-management model. This model integrates several commonly used behavior-change concepts to assess and affect people's readiness to change behavior. The model subsequently directs the user to stage-matched information that is available on the Internet (as opposed to most tools designed for research purposes that refer people to information specifically designed for the study). Bensley et al. argued that using preexisting information sources reduces costs compared to developing new computerized expert systems. However, all existing sources people are directed to do need to be reviewed and approved for credibility and appropriateness. The ever-changing nature of the content of Web sites may therefore impose serious constraints on the applicability of this approach. The HBC-I screener also suggests to assess other psychosocial variables that can affect behavior change, such as attitude, self-efficacy, and subjective norms. Collecting data on these variables may be interesting from a research perspective, but their value in behavior-change practice may be limited, as research on dietary behavior thus far has shown no added value of tailoring to psychosocial variables in addition to feedback on food-consumption levels (Brug, Steenhuis, van Assema, Glanz, & De Vries, 1999).

Assist

The assist criterion includes providing support, reinforcement, and understanding; providing people with intervention options and negotiating an intervention approach; and providing assistance during the change process. One may think of increasing knowledge, offering nicotine-replacement therapy, setting realistic goals, developing skills, getting social support, and so forth. Interestingly, while Web-based behavior-change programs allow for interaction with health professionals as well as with peers, the evidence suggests that when given access to both, subjects tend to rely more on communication with health professionals than on peer support (Nguyen, Carrieri-Kohlman, Rankin, Slaughter, & Stulbarg, 2004).

Tailoring may play an important role in the assist criterion. In tailoring, the information that is presented to people is a selection of all available information based on their socio-demographic, behavioral, motivational, psychosocial, and physical characteristics. In this respect, communication through new interactive-technology applications mimics the level of tailoring that can take place when an individual is given a face-to-face consultation with a health worker. In addition, interactive health communication may also be personalized. The underlying assumption for tailoring the behavior-change program to individuals' characteristics is that this will increase the personal relevance of the message, thus increasing people's attention and the subsequent motivational and behavioral impact (De Vries & Brug, 1999; Dijkstra & De Vries, 1999; Kreuter, Farrell, Olevitch, & Brennan, 2000; McGuire, 1985; Petty, Barden, & Wheeler, 2002). Furthermore, the tailoring process assures that

Textbox 1. Computerized tailoring

The first-generation computer tailoring programs produced print communications. This print was in many ways similar to the traditional brochures that were used in non-tailored print approaches, but we presented people with a selection of the information as opposed to including all possibly relevant information. Later generations were available online and provided instant feedback. Interactive Web sites and the delivery of programs through personal digital assistants (PDAs) are examples of this approach (Owen et al., 2002; Redding et al., 1999).

Computer tailoring of health-promotion programs tends to take place in two steps. In the initial step, people complete an assessment instrument to assess (determinants of) the behavior of interest. In the second step, a computer program (sometimes also referred to as an expert system; see, for example, Redding et al., 1999) selects relevant message segments that are combined into tailored feedback on the outcomes of the assessment phase. The feedback often includes personal, normative, and ipsative feedback. This provides the respondent with information relevant to their current behavior (in itself and in relation to their prior behavior), and relevant to their behavior in relation to a reference group of peers (Owen et al., 2002; Redding et al.). Expert systems thus combine the individual matching that can be obtained in an individual clinic-based intervention with the relatively low costs associated with a public health approach. They also have the following additional benefits: ease of documentation, ease of transfer to multiple sites, increased consistency in decision making, increased potential for replicable results, permanence, and low costs (Waterman, 1986). A combination of personal contact and a computerized expert system is also possible, for example, when the health-promoting actions of a health professional are guided by information provided by an expert system (Velicer et al., 1993).

---

no redundant information is presented to the participant. This saves people time on deciding what they should and what they should not read (see Textbox 1 for a brief description of computerized tailoring).

Anticipatory Guidance

It is generally accepted that behavior change, once successfully initiated, may be difficult to sustain, especially when facing difficult situations. The HBC-I screener therefore includes anticipatory guidance: counseling and support for anticipated barriers and for relapse prevention. This may be done by specifically addressing frequently occurring barriers for behavior-change maintenance. Other possibilities include providing guidance on managing tempting situations or relapse prevention. Evidently, including anticipatory guidance in Web-based behavior-change programs does not guarantee successful maintenance of the desired behavior. However, the frequent relapses that occur in behavior change more than justify inclusion of this strategy in the HBC-I screener. Unfortunately, as was shown in the study by Evers et al. (2005), anticipatory guidance is infrequently included in health-behavior-change programs.

Arrange Follow-Up

Most researchers and practitioners agree that behavior change is a process that occurs over time rather than a watershed event. To maximize the likelihood of successfully affecting health behavior, repeated efforts are expected to be necessary. These continuous efforts may build on people's motivation levels, or they may help prevent relapse. As a result, the provision of a follow-up, either in the program itself or by referring people to other sources of follow-up interventions, is also part of the HBC-I screener (Evers et al., 2005). In practice, this may take place using daily or weekly e-mail reminders to keep users in touch with the program. Follow-up in person, for example, through telephone calls is also possible. Repeating the assessment tends to be a good trigger for people to participate in follow-up programs.

In addition to the *A*s, health-behavior-change Web sites should be designed to assure sufficient security of potentially sensitive health information by preventing unauthorized access. Password registration and members-only access are examples of this. Evers et al. (2003) found that many behavior-change programs had a privacy-policy statement and provided information on how the data collected from users was used.

Delivery of Web-Based Programs to Promote a Healthy Lifestyle

As Abidi and Goh (2000) suggested, "healthcare information should be personalized according to each individual's healthcare needs and it should be pro-actively delivered, i.e., pushed towards the individual." The need for pushed (as opposed to pulled) delivery may be particularly crucial for lifestyle programs. Yet, as was also discussed by Van Woerkum (1997), the main consequence of using new media for communication is that the role of senders and receivers in the process changes. Traditionally, health professionals and/or knowledge institutions sent messages they deemed useful to the intended users. The initiative for knowledge transfer was thus on the site of these parties. New communication technologies have gradually moved the initiative to the users, who are looking for information on demand to answer any question they may have. Many people have incorrect perceptions of their lifestyle; they underestimate their intake of energy and fat, and they overestimate their physical activity levels. As a result, people may mistakably perceive themselves to have a healthy lifestyle and may therefore not be interested in programs telling them how to improve their behavior. Pushed delivery of programs with fast and easy self-assessment tools may help to achieve the self-awareness that is crucial to initiate health-behavior change. Evidently, the challenge is to develop a strategy for pushed delivery of such self-assessment and counseling tools. Edutainment (education entertainment) may be a useful approach in this respect. Edutainment uses an entertainment approach (games, quizzes, TV shows, etc.) to make something of an otherwise possibly dull learning experience. *Sesame Street* is one of the clear examples of this strategy. Positive experiences with edutainment in health, for example, in diabetic children have been shown (Aoki et al., 2004). Fewer generally known examples exist for adults. Edutainment, however, has been successfully applied in several health-related areas such as drug abuse, drunk driving, HIV/AIDS, family planning, and nutrition. One of the examples that are currently available for adults in many countries throughout the world is the RealAge test (http://www.realage. com, RealAge, Inc.). RealAge provides personalized health information and health-management tools. Their flagship product, the RealAge test, compares biological vs. calendar age based upon people's answers to several questions on topics that affect the rate of aging. An interesting feature of the RealAge test is that it uses tactics based on the net-present value theory (see Allegrante & Roizen, 1998) to express expected health benefits later in life (a relatively abstract long-term benefit) in life years (a more concrete benefit that makes sense to people in the present). The RealAge test is available though the Internet and is supported by mass media activities such as TV shows.

Reaching the Target Group

One of the important challenges in health promotion is to reach the intended target audience. It has been postulated that the reach and impact of health-promotion programs are likely to be much larger when the target groups' desires are taken into consideration early in the development and planning process (e.g., in the precede-proceed model by Green & Kreuter, 1999). Indeed, as with any other health-promotion activity, the desires and needs of the target audience should be taken into consideration when designing, testing, and implementing Web-based health-promotion programs. At present, the content, style, and graphic design of the health-promotion programs are often based on the message health workers want to convey and on the personal preferences and skills of the Web designers. Yet in the end, the users decide what they like or dislike and can leave a site full of relevant information with one simple mouse click. When users visit a site, they decide in 5 to 10 seconds whether or not they will stay. This decision process tends to be based on the home page and possibly one additional page. This decision is personal and largely emotional, based on several satisfiers and dissatisfiers (ZBC Consultants B.V., 2004). Unfortunately, most programs tend to be designed top-down, which reduces the odds of programs meeting these satisfiers and dissatisfiers.

An alarming notion to add to this is the tendency that "to those who have (information), to those will be given" (Koelen & Van den Ban, 2004). Indeed, Verheijden (2004) conducted a non-response survey and showed that participants in a Web-based tailored nutrition counseling program were a relatively well-educated and healthy sub-sample of the target audience. It was also suggested by Brug, Oenema, and Campbell (2003) that interactive health communication may be particularly appealing to highly educated, motivated women. However, there is also evidence that interactive health communication can be as effective among people with lower education as among higher educated people. The appreciation and exposure may even be more positive among the lower educated people. This is in line with the idea that the ability to adapt the pace of learning to one's personal interests and skills (which characterizes interactive health communication) may be particularly beneficial to lower educated people. Evidently, a crucial factor related to access that needs to be addressed is the possibility that people do not use the Internet because they do not want to use it or because they see no need to use it (Bush, Bowen, Wooldridge, Ludwig, Meischke, & Robbins, 2004). Efforts to counteract this phenomenon need to be made simultaneously with current efforts to reduce the digital divide (see below), which are largely focusing on providing people with access to computers and the Internet and to training in the use of hardware and software.

One of the complicating factors related to the reach of the target group is the digital divide between those who do and those who do not have access to information technology. As discussed by Bush et al. (2004), the term Internet access is used in a variety of meanings in different contexts (see also Leeuwis & Van den Ban, 2004, for a discussion on factors determining the availability of information technology). Based on an extensive search of overlapping information sources, they suggested that issues of Internet and Web access can be either connectivity issues (for example, availability, capability) or human interface issues (for example, literacy; language; education; race, ethnicity, and culture; income; disability and age; experience and familiarity; skill and training). Although the socio-demographic characteristics of the advantaged and the disadvantaged groups may vary to some extent, younger, well-educated, employed men generally seem to be on the right side of the digital divide. Evidently, the digital divide also limits the rise of Web-based applications in less developed countries. Four successive phases in access to information technology have been suggested: (a) motivation, (b) possession of or access to hardware and software, (c) skills, and (d) actual use (Van Dijk, 2005). The digital divide is believed to be narrowing, but evidently access remains a barrier for some groups, such as the elderly, illiterate, and physically or mentally handicapped people. As a result of this, the use of certain media in health-promotion interventions can have political implications because they have benefit for some but are unavailable to others. Despite the advantages of interactive Web-based behavior-change programs, it has been suggested that communicative intervention programs should use a combination of different media, either simultaneously or in succession. Combining multiple media approaches may increase the likelihood of a program being effective because different members of the target audience may have preferences for and/or access to different media. Furthermore, different media have different qualities that may be combined to form an optimal intervention mix (Leeuwis & Van den Ban).

Data on the reach of Internet-based health information are not readily available. Over the past few years, almost 75% of the Dutch Internet users used the Internet to obtain health-related information. Estimates based on 2002 state that 73 to 110 million American adults used the Internet to look for health information for themselves or for others (Nguyen et al., 2004). U.S. data on use of the Internet for health information by the chronically ill showed that 46% reported using the Internet to seek for health information or advice in the past year (Wagner, Baker, Bundorf, & Singer, 2004). Over 10% reported at least monthly use. The likelihood of using the Internet for health information was higher among people with multiple chronic conditions than among people with hypertension, cancer, or heart problems only.

A recent study among members of a Dutch Internet panel showed that 28% of the respondents had participated in Web-based tailored lifestyle programs; 57% expressed an interest in such programs, and 15% expressed no interest (Verheijden, Jans, Hildebrandt, 2006). People who were interested in or had participated in such programs were younger and more frequently female than people who expressed no interest. People most frequently reported a general interest in their own lifestyle and in online tests as reasons to participate. Much less frequently, people were interested in possible improvements in their lifestyle. The main reasons for people not to be interested in Web-based tailored lifestyle programs was their preference to discuss lifestyle and health issues with (primary care) physicians or other health professionals. They also did not want to worry about their lifestyles, or believed there was no need for them to participate in Web-based tailored lifestyle programs because their lifestyles were already healthy.

Evers et al. (2004) and Evers et al. (2005) collected similar data in a longitudinal study among a sample of American Internet users on the use of Web-based health-behavior-change programs. At baseline, approximately 25% of the respondents used the Internet for health-behavior-change or disease-management programs. The vast majority of the remaining respondents had no intention to start using such programs. At follow-up 1 year later, most people who were initially using health-behavior-change programs were no longer doing so. This may be explained by the consistently high level of cons across all levels of motivation to use the Internet for health-behavior change that were found. Other explanations may be that some programs were no longer available through the Internet, or that the programs had a fixed duration of less than 1 year. Whatever the exact reasons, it is safe to conclude that even with continued use, the disadvantages of using programs on the Internet are prevalent, and that users are at risk for discontinuing use (Evers et al., 2004).

Exposure to Web-Based
Health-Promotion Programs

The central route of the elaboration likelihood model (Petty & Cacioppo, 1986) suggests that the extent to which people are persuaded to change their behavior depends on the time they spend elaborating on the topic. In Web-based health promotion, one might thus hypothesize that increasing the amount of time individuals spend reading or thinking about content on a Web site increases the likelihood of behavior change (Sciamanna et al., 2002). Future research and practice would therefore benefit from more data on exposure to (specific parts of) interventions. Sciamanna et al. and Leslie, Marshall, Owen, and Bauman (2005) suggested that people with low levels of motivation to increase their physical activity levels use other parts of intervention Web sites and have other preferences than people with high levels of motivation. Unfortunately, data on exposure to intervention programs are hardly ever presented despite the fact that the use of a Web site, visits to specific pages on a site, the use of links, and so forth can all be monitored in contemporary Web-based health-promotion programs. Only 5 of the 22 studies in the review by Wantland et al. (2004) reported measures of intervention exposure. Oenema, Brug, and Lechner (2001) increased the likelihood of high exposure to the intervention program by installing it on a local hard disk, thus preventing people from browsing the World Wide Web. Marshall, Leslie, Bauman, Marcus, and Owen (2003) conducted an 8-week intervention with repeated reminders to use the site. They found that less than half of the participants used the site. Only 26% of the participants used the site more than once. An 8-month intervention program on nutrition counseling (Verheijden, 2004) was used by only 33% of the study participants (median site-use frequency among users was one time). Participants from different study populations were asked to report reasons for their relatively infrequent site use. Barriers that were reported included lack of time, lack of interest in the topic, limited access to computers and/or the Internet, and low expected benefits (Sciamanna et al., 2002; Verheijden, 2004).

Current Web-Based Behavior-Change Programs

Large numbers of Web sites currently offer people the possibility to participate in behavior-change programs. These sites vary largely in content and quality (Evers et al., 2003). They also vary in the number of health behaviors that are addressed and in the freedom of choice participants have in terms of the topics they would like to work on. Verheijden et al. (2006) showed that possible users of health-behavior-change programs differ in their views about what lifestyle topics should be addressed. Roughly half of the participants said they would like to decide for themselves which behaviors to focus on. The other half stated that Web-based behavior-change programs should address all relevant lifestyle issues. Very little evidence exists about addressing multiple lifestyle issues simultaneously. Evers et al. (2003) found that the vast majority of Web sites cover more than one behavior. Some of these combined programs integrate the five As for all behaviors involved. Others offered a modular approach in which the five As were kept separate. Research by Vandelanotte, De Bourdeaudhuij, Sallis, Spittaels, and Brug (2005) showed that the simultaneous approach may be more effective in certain groups of people than a sequential approach.

The Feasibility and Effectiveness of Web-Based Health-Promotion Programs

As for any other health-promotion tool, a major prerequisite for large-scale implementation of Web-based health-promotion programs is a solid evidence base for their (cost) effectiveness. Conducting feasibility and pilot studies is an important first step in the evaluation of Web-based health-promotion programs (De Nooijer et al., 2005). Unfortunately, there is very little literature available on this topic. This is likely the result of journal editors being more interested in randomized controlled trail on the effectiveness of the final programs. As a result, publishing pilot studies is relatively low on researchers' priority lists. However, to prevent the wheel being reinvented over and over again, the field would benefit from the publication of the successes and failures in pilot studies, such as was done by Anhøj and Jensen (2004). They conducted a feasibility study on a Danish Internet application for lifestyle changes in diet and physical activity among five general practitioners (GPs) and 25 patients. Data were collected rather randomly from different sources and participants were likely a convenience sample. Nevertheless, the results from this study provide some valuable insights for the design of future behavior-change programs. One of the important outcomes of this study is that patients seemed to give a high value to personal relations with a health-care professional. All patients agreed that "the use of the program did not provide the necessary support in their struggle towards a healthy lifestyle." They said that the program itself was merely a tool and that it could never replace support from a health-care professional. It was also reported by both patients and GPs that the program needs to allow for frequent interaction, especially in the beginning, to prevent patients' loss of interest. Anhøj and Jensen (2006) suggested breaking up advice messages into small fragments delivered on a daily basis. They also suggested allowing patients to complete dietary assessment instruments on a regular (possibly even daily) basis, despite the fact that these instruments may have been developed to assess dietary intake patterns over the past month. Evidently, the latter limits the reliability of questionnaire results and the content of advice, which is one of the other key issues that needs to be addressed for similar programs to be successful in bringing about patients' lifestyle changes.

However, aside from a number of position papers indicating the potential of these programs and a few studies with varying external validity, very little evidence is available. Wantland et al. (2004) reviewed behavioral-change outcomes in Web-based vs. non-Web-based interventions for (amongst others) weight control, weight-loss maintenance, nutrition, physical activity, and the secondary prevention of heart disease. They concluded that there was substantial evidence that use of Web-based interventions improves behavioral-change outcomes. De Nooijer et al. (2005) conducted a systematic review on the effectiveness of Web-based health-promotion programs in the areas of smoking, physical activity, nutrition, weight loss (maintenance), safe sex, alcohol, and drugs. They found favorable effects (part of which were significant) on knowledge, awareness of actual consumption levels, self-efficacy, and intention to change behavior. The evidence on behavioral and biomedical outcomes was conflicting, and effect sizes were predominantly insignificant. Werkman, Kroeze, & Brug (2005) conducted a systematic review of the effectiveness of computer-tailored education on physical activity and dietary behaviors. They found most consistent evidence for tailored interventions on fat reduction. There was very limited evidence to support computer-based tailored programs for physical activity. Textbox 2 briefly presents examples of computerized health-behavior-change programs in the area of physical activity (Marshall et al., 2003), smoking (Velicer et al., 1993), and diet (Block, Block, Wakimoto, & Block, 2004). These examples were chosen because they represent different options in computerizing: Web sites, e-mails, and computerized expert systems. Several methodological drawbacks characterize some of the studies included in the reviews described above. For example, one needs to be cautious when interpreting studies in which a Web-based program (i.e., increasing access to information) is compared to a no-treatment control. The difference in exposure to intervention may be a co-intervention to such an extent that the effects of the use of new media cannot be assessed properly (Nguyen et al., 2004). A nice series of studies in which this was not an issue was conducted by Harvey-Berino, Pintauro, Buzzell, et al. (2002), Harvey-Berino, Pintauro, Buzzell, and Gold (2004), and Harvey-Berino, Pintauro, and Gold (2002). They compared

Textbox 2. Some examples of computerized health-behavior-change programs

---

Expert System for Smoking Cessation

Velicer et al. (1993) tested a computerized expert system for smoking cessation in 870 smoking volunteers. People were randomized into one of four groups: (a) standard treatment using self-help manuals and booklets, (b) individualized treatment using manuals based on all constructs of the trans-theoretical model, (c) interactive individualized treatment using manuals based on all constructs of the trans-theoretical model and three printouts of computer-generated reports on people's responses to mailed questionnaires, and (d) extensive interactive, individualized treatment using manuals based on all constructs of the trans-theoretical model, three printouts of computer-generated reports on people's responses to mailed questionnaires, and a series of short calls from counselors to provide personalized feedback. Self-reported smoking abstinence was measured in a pretest and after 6, 12, and 18 months. Individualized treatment and standard treatment led to similar abstinence rates through 12 months. At 18 months, individualized treatment outperformed the standard treatment. Interactive, individualized treatment outperformed standard and individualized treatment at all time points; quit rates were roughly twice as high. The extensive interactive, individualized treatment was slightly less effective than the regular interactive, individualized treatment.

E-Mailed Nutrition Intervention Program

Block et al. (2004) conducted an experiment in which employees at a corporate work site were offered the opportunity to work with the e-mailed Worksite Internet Nutrition (WIN) program. Some of the principles in the WIN program were relevance to the employee, tailoring, flexibility and individual choice, skill facilitation, commitment and goal setting, reminders and reinforcement, and multiple strategies and channels. Eighty-four of the 230 employees participated in the program. No control group was included in the study. The participants increased the frequency of fruit and vegetable consumption by 0.73 times per day and decreased the frequency of fat consumption by 0.39 times per day. Participants found the program helpful (93%) and all of the participants who admitted not to be health conscious at baseline would recommend the program to others.

Web-Site Physical-Activity Program

Marshall et al. (2003) conducted a study among 655 participants who were recruited from an Australian regional university. Participants were randomized to receive either a print-based or a Web-based active-living program. Most people preferred the Web-based program. However, 36% of the participants in the print-based group had kept the intervention materials for future reference, while only 12% of the participants in the Web-based group reported to have bookmarked the Web site. There was a trend for participants in both groups to participate in more physical activity after the intervention, but there were no statistically significant differences within or between study groups.

---

weight-loss maintenance programs, one of which was delivered via the Internet. The studies provided conflicting evidence on the effectiveness of Internet interventions for weight-loss maintenance; the Internet intervention was at best slightly (yet insignificantly) better than other treatment modalities. Generally, the attendance of meetings was lower in the Internet intervention and patients tended to prefer the in-person intervention.

A study by Consoli, Ben Said, Jean, Menard, Plouin, and Chatellier (1995) showed that knowledge scores among hypertensive patients increased significantly more when standard education by physicians, nurses, dieticians, and pamphlets were supplemented with one computer session using an expert system. Some positive experiences were also reported for other conditions, such as heart failure and asthma (Krishna, Francisco, Balas, Konig, Graff, & Madsen, 2003; Stromberg, Ahlen, Fridlund, & Dahlstrom, 2002). It has been suggested, however, that these findings cannot simply be extrapolated to the effectiveness of computerized behavior-change programs in general because the sense of urgency to participate in such programs is much larger for people who have been diagnosed with chronic conditions than for people who need to make behavioral changes in the prevention of chronic conditions. This may be partly the result of the fact that people can expect benefits from participating in such programs in the long term, while the effort needs to be put in at present (Allegrante & Roizen, 1998).

Potential for Harm

Initially, many interactive health communication systems were available under supervised conditions in academic research settings only. More recently, however, many systems have become available to the general public. There is even less evidence available on these publicly available products than on the academically available ones, and monitoring occurs rarely. As SciPICH (Robinson et al., 1998) already stated, this should raise legitimate questions about their quality, cost, and potential to cause harm. The risks of direct and indirect damage to people's health as a result of health-promotion programs through interactive health communication may be considerable. This may be the result of inappropriate treatment or delays in seeking necessary medical care. The expert panel in the study by De Nooijer et al. (2005) that was discussed previously also expressed concerns about places people can turn to when they need more information after they have read misleading or terrifying information on the Web. Other examples include damaged trust in health-care clinicians and misleading claims for health products. SciPICH strongly equal for the active involvement of health-care professionals in assessing and assuring the quality of interactive health-promotion tools and in contributing to the development and implementation of such tools. It advocated that this approach would lead to much better outcomes for both health-care providers and health-care consumers than an approach in which professionals ignore or disparage such tools.

Conclusion

New communication technologies provide many opportunities for health-promotion programs, but they also have some considerable limitations. The ultimate proof of the pudding, of course, is in the eating. As Evers et al. (2005) already stated:

Until the field solves the problem of helping significant percentages of populations toward effective action and maintain such action, Internet programs will not be able to realize their potential to be the lowest cost modality for delivering tailored communications that can have the highest impact on health promotion, disease prevention, and disease management.

It is a challenge for all those involved to make the best possible use of the new technologies for programs that were originally designed for face-to-face or print strategies, as well as to develop new programs specifically designed in a new media format. However, to make full use of the possibilities of face-to-face and computer-based interaction in health communication, the World Wide Web should never fully replace consultations and clinical examinations by health professionals. Depending on factors such as the available resources (time, space, staff, etc.) and the personal preferences of all individuals involved, an ideal mix of intervention approaches may be composed.

References

Abidi, S. S., & Goh, A. (2000). A personalised healthcare information delivery system: Pushing customised healthcare information over the WWW. *Studies in Health Technology and Informatics, 77*, 663-667.

Abrams, D. B., Emmons, K. M., & Linnan, L. A. (1997). Health behavior and health education: The past, present, and future. In K. Glanz, F. M. Lewis, & B. K. Rimer (Eds.), *Health behavior and health education: Theory, research, and practice* (pp. 453-478). San Francisco: Jossey-Bass.

Ajzen, I. (1991). The theory of planned behavior. *Organizational Behavior and Human Decision Processes, 50*, 179-211.

Allegrante, J. P., & Roizen, M. F. (1998). Can net-present value economic theory be used to explain and change health-related behaviors? *Health Education Research, 13*, i-iv.

Anhøj, J., & Jensen, A. H. (2004). Using the Internet for lifestyle changes in diet and physical activity: A feasibility study. *Journal of Medical Internet Research, 6*, e28.

Anhøj, J., & Nielsen, L. (2004). Quantitative and qualitative usage data of an Internet-based asthma monitoring tool: From the user's point of view. *Journal of Medical Internet Research, 6*, e23.

Aoki, N., Ohta, S., Masuda, H., Naito, T., Sawai, T., Nishida, K., et al. (2004). Edutainment tools for initial education of type-1 diabetes mellitus: Initial diabetes education with fun. *Medinfo, 11*, 855-859.

Bandura, A. (1986). *Social foundations of thought and action.* Englewood Cliffs, NJ: Prentice-Hall.

Bensley, R. J., Mercer, N., Brusk, J. J., Underhile, R., Rivas, J., Anderson, J., et al. (2004). The eHealth behavior change management model: A stage-based approach to behavior change and management. *Preventing Chronic Disease.* Retrieved December 5, 2006, from http://www.cdc.gov/pcd/issues/2004/oct/04_0070.htm

Block, G., Block, T., Wakimoto, P., & Block, C. H. (2004). Demonstration of an emailed worksite nutrition intervention program. *Preventing Chronic Disease.* Retrieved December 5, 2006, from http://www.cdc.gov/pcd/issues/2004/oct/04_0034.htm

Brug, J. (1999). Dutch research into the development and impact of computer-tailored nutrition education. *European Journal of Clinical Nutrition, 53*(Suppl. 2), S78-S82.

Brug, J., Oenema, A., & Campbell, M. (2003). Past, present, and future of computer-tailored nutrition education. *American Journal of Clinical Nutrition, 77*, 1028S-1034S.

Brug, J., Steenhuis, I., van Assema, P., Glanz, K., & De Vries, H. (1999). Computer-tailored nutrition education: Differences between two interventions. *Health Education Research, 14*, 249-256. Bush, N. E., Bowen, D. J., Wooldridge, J., Ludwig, A., Meischke, H., & Robbins, R. (2004). What do we mean by Internet access? A framework for health researchers. *Preventing Chronic Disease.* Retrieved from http://www.cdc.gov/pcd/issues/2004/oct/04_0019.htm

Commission of the European Communities. (2002). eEurope 2002: Quality criteria for health related Websites: Position Paper. *Journal of Medical Internet Research, 4*(3), e15.

Consoli, S. M., Ben Said, M., Jean, J., Menard, J., Plouin, P. F., & Chatellier, G. (1995). Benefits of a computer-assisted education program for hypertensive patients compared with standard education tools. *Patient Education and Counselling, 26*, 343-347.

Cummins, C. O., Prochaska, J. O., Driskell, M. M., Evers, K. E., Wright, J. A., Prochaska, J. M., et al. (2003). Development of review criteria to evaluate health behavior change Websites. *Journal of Health Psychology, 8*, 55-62.

De Nooijer, J., Oenema, A., Kloek, G., Brug, H., De Vries, H., & De Vries, N. (2005). *Bevordering van gezond gedrag via Internet: Nu en in de toekomst.*

De Vries, H., & Brug, J. (1999). Computer-tailored interventions motivating people to adopt health promotion behaviors: Introduction to a new approach. *Patient Education and Counselling, 36*, 99-105. Optima Grafische Communicatie in Rotterdam: The Netherlands.

Dijkstra, A., & De Vries, H. (1999). The development of computer-generated tailored interventions. *Patient Education and Counselling, 36*, 193-203.

Evers, K. E., Cummins, C. O., Johnson, J. L., Paiva, A., Prochaska, J. O., Padula. J., et al. (2004). *Towards maximizing Internet impacts on health promotion and disease management.* West Kingston, RI: Pro-Change Behavior Systems, Inc.

Evers, K. E., Cummins, C. O., Prochaska, J. O., & Prochaska, J. M. (2005). Online health behavior and disease management programs: Are we ready for them? Are they ready for us? *Journal of Medical Internet Research, 7*, e27.

Evers, K. E., Prochaska, J. M., Prochaska, J. O., Driskell, M. M., Cummins, C. O., & Velicer, W. F. (2003). Strengths and weaknesses of health behavior change programs on the Internet. *Journal of Health Psychology, 8*, 63-70.

Gagliardi, A., & Jadad, A. R. (2002). Examination of instruments used to rate quality of health information on the Internet: Chronicle of a voyage with an unclear destination. *British Medical Journal, 324*(7337), 569-573.

Gattoni, F., & Sicola, C. (2005). How to evaluate the quality of health related Websites. *Radiologia Medica* (Torino)*, 109*(3), 280-287.

Green, L. W., & Kreuter, M. W. (1999). *Health promotion planning: An educational and environmental approach* (2nd ed.). Mountain View, CA: Mayfield.

Griffiths, K. M., & Christensen, H. (2005). Website quality indicators for consumers. *Journal of Medical Internet Research, 7*(5), e55.

Harvey-Berino, J., Pintauro, S., Buzzell, P., DiGiulio, M., Casey Gold, B., Moldovan, C., et al. (2002). Does using the Internet facilitate the maintenance of weight loss? *International Journal of Obesity and Related Metabolic Disorders, 26*, 1254-1260.

Harvey-Berino, J., Pintauro, S., Buzzell, P., & Gold, E. C. (2004). Effect of Internet support on the long-term maintenance of weight loss. *Obesity Research, 12*, 320-329.

Harvey-Berino, J., Pintauro, S. J., & Gold, E. C. (2002). The feasibility of using Internet support for the maintenance of weight loss. *Behavior Modification, 26*, 103-116.

Koelen, M. A., & Van den Ban, A. (2004). *Health education and health promotion.* Wageningen: Wageningen Academic Publishers, The Netherlands.

Kreuter, M., Farrell, D., Olevitch, L., & Brennan, L. (2000). *Tailoring health messages: Customizing communication with computer technology.* Mahwah, NJ: Lawrence Erlbaum Associates.

Krishna, S., Francisco, B. D., Balas, E. A., Konig, P., Graff, G. R., & Madsen, R. W. (2003). Internet-enabled interactive multimedia asthma education program: A randomized trial. *Pediatrics, 111*, 503-510.

Leeuwis, C., & Van den Ban, A. (2004). *Communication for rural innovation: Rethinking agricultural extension.* Blackwell Science.

Leslie, E., Marshall, A. L., Owen, N., & Bauman, A. (2005). Engagement and retention of participants in a physical activity Website. *Preventive Medicine, 40*, 54-59.

Marshall, A. L., Leslie, E. R., Bauman, A. E., Marcus, B. H., & Owen, N. (2003). Print versus Website physical activity programs: A randomized controlled trial. *American Journal of Preventive Medicine, 25*, 88-94.

McGuire, W. J. (1985). Attitudes and attitude change. In G. Lindzey & E. Aronson (Eds.), *The handbook of social psychology* (Vol. 83, pp. 854-864). Mahwah: NJ: Lawrence Erlbaum Assoc.

Mullen, P. D., Hersey, J. C., & Iverson, D. C. (1987). Health behavior models compared. *Social Science and Medicine, 24*, 973-981.

Nguyen, H. Q., Carrieri-Kohlman, V., Rankin, S. H., Slaughter, R., & Stulbarg, M. S. (2004). Internet-based patient education and support interventions: A review of evaluation studies and directions for future research. *Computers in Biology and Medicine, 34*, 95-112.

Oenema, A., Brug, J., & Lechner, L. (2001). Web-based tailored nutrition education: Results or a randomized controlled trial. *Health Education Research, 16*, 647-660.

Owen, N., Fotheringham, M. J., & Marcus, B. H. (2002). Communication technology and health behavior change. In K. Glanz, B. K. Rimer, & F. Marcus Lewis (Eds.), *Health behavior and health education: Theory, research, and practice* (pp. 510-529). John Wiley & Sons.

Petty, R. E., Barden, J., & Wheeler, S. C. (2002). The elaboration likelihood model of persuasion: Health promotions that yield sustained behavioral change. In R. J. DiClemente, R. A. Crosby, & M. C. Kegler (Eds.), *Emerging theories in health promotion practice and research: Strategies for improving public health* (pp. 71-99). San Francisco: Jossey-Bass.

Petty, R. E., & Cacioppo, J. T. (1986). *The elaboration likelihood model of persuasion.* New York: Academic Press.

Prochaska, J. O., & DiClemente, C. C. (1983). Stages and processes of self-change of smoking: Toward an integrative model of change. *Journal of Consulting and Clinical Psychology, 51*, 390-395.

Prochaska, J. O., & Velicer, W. F. (1997). The transtheoretical model of health behavior change. *American Journal of Health Promotion, 12*, 38-48.

Redding, C. A., Prochaska, J. O., Pallonene, U. E., Rossi, J. S., Velicer, W. F., Rossi, S. R., et al. (1999). Transtheoretical individualized multimedia expert systems targeting adolescents' health behaviors. *Cognitive and Behavioral Practice, 6*, 144-153.

Risk, A., & Deznowagis, J. (2001). Review of Internet health information quality initiatives. *Journal of Medical Internet Research, 3*(4), e28.

Robinson, T. N., Patrick, K., Eng, T. R., & Gustafson, D. (1998). An evidence-based approach to interactive health communication: A challenge to medicine in the information age. *Journal of the American Medical Association, 280*, 1264-1269.

Sciamanna, C. N., Lewis, B., Tat, D., Napolitano, M. A., Fotheringham, M., & Marcus, B. H. (2002). Use attitudes toward a physical activity promotion Website. *Preventive Medicine, 35*, 612-615.

Stromberg, A., Ahlen, H., Fridlund, B., & Dahlstrom, U. (2002). Interactive education on CD-ROM: A new tool in the education of heart failure patients. *Patient Education and Counselling, 46*, 75-81.

Vandelanotte, C., De Bourdeaudhuij, I., Sallis, J. F., Spittaels, H., & Brug, J. (2005). Efficacy of sequential or simultaneous interactive computer-tailored interventions for increasing physical activity and decreasing fat intake. *Annals of Behavioral Medicine, 29*, 138-146.

Van Dijk, J. A. G. M. (2005). *The deepening divide: Inequality in the information society.* Sage Publications.

Van Woerkum, C. M. J. (1997). Media choice in nutrition education of general practitioners. *American Journal of Clinical Nutrition, 65*(Suppl.6), 2013S-2015S.

Velicer, W. F., Prochaska, J. O., Bellis, J. M., DiClemente, C. C., Rossi, J. S., Fava, J. L., et al. (1993). An expert intervention for smoking cessation. *Addictive Behaviors, 18*, 269-290.

Verheijden, M. W. (2004). *Nutrition counselling in general practice: The stages of change model.* Unpublished doctoral thesis. Wageningen University: The Netherlands.

Verheijden, M. W., Jans, M. P., & Hildebrandt, V. H. (in review). *Web-based tailored lifestyle programs: Exploration of the target group's interests and implications for practice.* Manuscript submitted for publication.

Wagner, T. H., Baker, L. C., Bundorf, M. K., & Singer, S. (2004). Use of the Internet for health information by the chronically ill. *Preventing Chronic Disease.* Retrieved from http://www.cdc.gov/pcd/issues/2004/oct/04_0004.htm

Wantland, D. J., Portillo, C. J., Holzemer, W. L., Slaughter, R., & McGhee, E. M. (2004). The effectiveness of Web-based vs non-Web-based interventions: A meta-analysis of behavioral change outcomes. *Journal of Medical Internet Research, 6*(4), e40.

Waterman, D. A. (1986). *A guide to expert systems.* Redding, MA: Addison-Wesley.

Werkman, A., Kroeze, W., & Brug, J. (2005, June). *A systematic review of the effectiveness of computer-tailored education on physical activity and dietary behaviors* (Abstract 1-50396). Proceedings of the Fourth Annual Conference of the International Society of Behavioral Nutrition and Physical Activity, Amsterdam, The Netherlands.

ZBC Consultants B.V. (2004). *E-commerce met de e van emotie.* Retrieved December 5, 2006, from http://www.zbc.nu/main.asp/ChapterID=2555

Chapter VI

Challenges, Opportunities and Solutions for Ubiquitous Eldercare

Paolo Bellavista, University of Bologna, Italy

Dario Bottazzi, University of Bologna, Italy

Antonio Corradi, University of Bologna, Italy

Rebecca Montanari, University of Bologna, Italy

Abstract

A non-negligible number of elder citizens, who represent a growing fraction of the population in developed countries, have to face a number of daily-life problems stemming from their partial and progressive loss of motor, sensorial, and cognitive skills. That often makes it difficult or impossible to live autonomously and, in today's small families, often forces elder hospitalization. Device miniaturization and ubiquitous connectivity can provide the technological support for valid alternatives to hospitalization, capable of reducing welfare costs, elder sense of loneliness, and elder exclusion from social relationships. On the one hand, wired and wireless sensors and actuators can improve elder life independence, for example, by transforming homes in smart eldercare environments with remote health-status monitoring, remote diagnostics, and facilitated house activities. On the other hand, pervasive wireless computing enables novel opportunities for caregivers, elders, and their family members, friends, and neighbors to collaborate and coordinate in an impromptu way to provide eldercare and social support anytime and anywhere. The chapter overviews the state-of-art of solutions for elder assistance, typically at home, and for coordinated-care networking by pointing out the need for advanced context-aware frameworks to properly establish ubiquitous and spontaneous communities of helpers when needed.

Pervasive Computing to Reshape Eldercare

The world population is aging rapidly and at a growing rate (United Nations Population Division, 2006). As the elderly people percentage is mounting, health and social costs are increasing, too: Medicare costs for elder individuals hosted in nursing homes are demonstrating to be significantly higher than for people continuing to live in their homes (Helal et al., 2003). However, the transformation of traditional families into small or single-family units makes it difficult to guarantee sufficient healthcare for elders at their homes, thus often forcing aging people's hospitalization, with both a notable emotional and economic impact.

Technological advances and cost reduction in computing devices and network solutions play an important role in enhancing elderly people's independence in day-to-day life by limiting their need for hospitalization, increasing the quality of medical care and technical assistance, and reducing elder health and social costs. Unquestionably, the Internet and the World Wide Web have been the first crucial driving forces in several changes that occurred in elder lifestyles: In the last years, several elder-related Web sites have emerged to link professionals in aging with their peers to provide aging adults and their families with direct links to suitable caregivers, and to allow them to input and transmit daily health records of elderly patients; moreover, several care agencies have started to use Web-based multimedia streaming to broadcast health programs (*Seniors-Site.Com*, 2006; Elder Services Network Links, 2006).

In addition to the above advantages basically stemming from Internet connectivity, recent developments in wireless technologies, sensors, and actuators are enabling new classes of eldercare applications available anywhere and at anytime, that is, *ubiquitous* eldercare services. The common guideline behind ubiquitous eldercare is the complete shift of the locus of health control from hospitals to pervasive systems deployed close to where elderly users live and move, with the main goals of increased independence, safety, and quality of life on the one hand, and of care cost savings on the other hand. For instance, a wearable device that appropriately alerts an elder with motor impairments of potential risks while he or she is walking around the home, possibly by coordinating with other wireless devices knowing the obstacles in that environment, could help the elder avoid bad falls and save consequent hospitalization costs.

We claim that recent and ongoing research activities about ubiquitous eldercare can be classified into two main categories: ubiquitous assistance and ubiquitous care networking support.

Ubiquitous assistance services have the primary goal to enhance living environments, primarily elder homes, with ubiquitous and possibly unobtrusive networked devices in order to support and facilitate daily-life activities of elderly people. For instance, wearable embedded diagnostic systems can collect and analyze health conditions, with the purpose of detecting diseases before the more severe symptoms become readily apparent. Sensors can help elders with mental and physical impairments to carry out their daily activities independently, for instance, by reminding them about routine activities and guiding them through their home environments (Wilson et al., 2003). Display devices, in the form of digital picture frames, can provide visualizations of elder daily-life activities to geographically distant family members by balancing the desire of family members to keep their elders safe and the willingness of seniors to age in place (Mynatt, et al., 2001). Several prototypes of smart homes with embedded sensors and networked devices are starting to emerge as examples of integrated ubiquitous assistance environments that tend to bridge the gap between aging people's homes and professional healthcare systems available at hospitals, hence postponing elder transfer into hospitals and nursing homes (Helal, et al., 2005).

Another primary goal of ubiquitous assistance services is to provide not only home-based but also anytime, anywhere assistance and support, for example, to guarantee safety requirements and prompt assistance to senior people when they are moving outside their usual living environments. Anytime and anywhere outdoor assistance is a challenging technological and organizational task and is probably the hottest topic in the ubiquitous assistance research: It requires several support mechanisms and tools, from wireless-enabled monitoring that enables healthcare professionals to control health indicators independently of an elder's physical position, to location systems that permit one to identify where elders are when in need of help.

The design of eldercare assistance services usually tends to assume an exclusively medical definition of care and to address only elders' physical decline in isolation from other dimensions of the aging experience. However, focusing only on medical-related issues may limit the positive impact of pervasive technologies in ubiquitous eldercare. The risk is of delivering advanced eldercare services that enhance life independence but increase the sense of loneliness of elders and promote their physical isolation into their homes, with the potential danger of senior estrangement from the physical spaces where social interactions typically occur. We claim that designers of ubiquitous eldercare solutions should carefully explore all the possibilities of pervasive technologies to increase elder people's opportunities for social engagement and intellectual stimulus. The underlying observation is that life quality and safety depend on a rich set of social relationships among a variety of individuals. Not only medical personnel contribute to care giving for elders, but also people passing by, friends and neighbors can play a significant role in eldercare and in enhancing elders' feeling of connection with the rest of the world (Hirsh, et al., 2000).

Some relevant research efforts are starting to address that crucial issue of reducing elder social barriers by supporting and promoting interactions between members of so-called eldercare networks. An eldercare network consists of all the people providing assistance to an elder: family members, friends, neighbors, and paid help, such as professional caregivers, doctors, pharmacists, and house cleaners. eldercare networks should be focused on the elder and created ad hoc, largely based on opportunity, with the possibility to rapidly and effectively accommodate changes in members, elder's needs, priorities, and so forth. As a consequence, it is impossible to assume that members statically know each other, together with their responsibilities and skills at any time. Pervasive computing offers new opportunities to enable effective coordination between the different members of an eldercare network. Wireless ubiquitous technologies have the potential to encourage novel forms of anytime and anywhere care assistance based on impromptu interactions and collaborations. For instance, individuals in the vicinity of an elder person with a heart emergency could be alerted and collaborate together to provide immediate help by exploiting Bluetooth-enabled portable devices used to monitor the heart situation of the elder and to broadcast the alert message in a peer-to-peer way, for example, by favoring the discovery of neighbors with medical skills. Novel ubiquitous care networking services are required to extend assistance applications, such as health monitoring, data collection, and analysis, with new functions that can properly handle both the formation of ubiquitous ad hoc first-response groups whenever an emergency event occurs and timely group interactions to deal with unexpected situations, possibly requiring helpers to make rapid decisions with only a partial perspective of the circumstances.

A relevant common guideline that is emerging in the design and development of ubiquitous eldercare services, for both individual assistance and social networking support, is context awareness. In the following, we use the definition of context as the collection of any information useful to characterize the run-time situation of one user involved in an eldercare service session, for example, the elder's or helper's location, the elder's or helper's profile (including physical and cognitive characteristics of the elder to be assisted and skills of the co-located potential helpers), and any additional data describing the current eldercare situation, for example, the elder's pulse, the state of water-flow valves, and medicine consumption (Dey, 2001).

Novel ubiquitous eldercare solutions should exploit the full awareness of context information to shape accordingly elder assistance, the formation of ad hoc helping communities, and the management of caregiver interactions.

The rest of the chapter is organized as follows. First it introduces the primary requirements of ubiquitous support to the aging population, structured according to a largely accepted classification framework. Next it is devoted to the overview of the most relevant recent activities about pervasive eldercare by distinguishing between ubiquitous assistance services and ubiquitous care networking ones. Among novel, promising, and challenging care networking solutions, the section will focus on the main design choices of AGAPE, our middleware for context-aware caregiver coordination in highly dynamic environments, with the main aim of exemplifying the main concerns and solution guidelines of the field. Primary open issues and expected directions of evolution of the ubiquitous eldercare area, together with conclusive remarks, end the chapter.

The Aging Population and Its Needs

To better position the past and ongoing research activities about ubiquitous eldercare, let us first introduce a solution classification widely accepted in the medical area. The U.S. National Center on Medical Rehabilitation Research (NCMRR) has proposed a model for categorizing the research and development efforts in technology for aging in well conditions (Mann, 2004). The NCMRR model identifies five main working directions at different levels, each focusing on a specific problem: cellular, organ, action, task-role, and social limitations. The cellular level relates to aberrations in normal physiological processes and in the cellular structure, with the consequent diseases and/or genetic abnormalities. The organ level focuses on solutions to impairments of organs, for example, the heart, or of whole organ systems, for example, the cardiovascular system. The action level refers to a person's inability to properly perform some actions due to functional limitations of the responsible organs. The research at the task-role level is addressed to people's disabilities in properly performing tasks and activities in specific physical and social contexts. The highest research area in the NCMRR model is about social barriers that are obstacles to people's engagement in a rich socio-emotional environment.

Each NCMRR level involves specific technological requirements and calls for proper solutions to enable elder autonomy in daily living, with less dependencies on family, friends, and caregivers. Let us note that only in the cellular area, the research focus is more on the curing and recovery of aging people rather than on technological solutions to improve quality of life. In more detail, research efforts in the organ and action levels seek to design and develop devices to compensate for limitations in common physical functionality, such as movement or hearing. For example, advanced signal-processing technologies have permitted us to significantly improved the sound quality provided via digital hearing aids. Technologies at the NCMRR task-role level aim at reducing the impact of disabilities on aging people's lives. For instance, hearing aids should also allow aging people with hearing disabilities to participate still actively in usual conversations. Research efforts in the social-limitation level, instead, analyze the social attitudes of aging people to derive the guidelines for the development of technological solutions that can improve the social interactions of elders with the external world.

Pervasive computing offers relevant and challenging opportunities to design and implement ubiquitous eldercare solutions, especially at the task-role and social-limitation levels. In fact, on the one hand, pervasive technologies enable the realization of wearable devices with ubiquitous connectivity that can assist elders in their activities and address their disabilities anywhere and anytime. On the other hand, pervasive solutions can also permit us to overcome social limitations by providing elders with a plethora of communication artifacts and services specifically tailored to set them within the context of a rich social and emotional framework and to reduce their sense of loneliness.

Technology Requirements and Impact on Elder Life

Identifying the guidelines to follow in order to evaluate the suitability of an eldercare technology is an extremely challenging task. Elder needs typically vary from person to person and evolve during time as the elder's physical and cognitive conditions decline.

The design of technological solutions for ubiquitous assistance requires carefully taking into account the specific type of organic, psychosocial, and environmental factors that affect elder daily functions. Eldercare-related technologies must be unobtrusive, easy to use, and appropriately tailored to the possible types of impairments and social limitations affecting their users. The main impairment categories that eldercare technology research is addressing are movement, vision, hearing, and cognition. Movement impairment can derive from injuries or diseases and can affect different capabilities, for example, walking, using hands, and moving the trunk and the neck. A non-negligible fraction of elderly people have vision impairments, from poor vision to total blindness, that limit their free movements in both indoor and outdoor environments. Similarly, the partial or total loss of hearing is usual, thus negatively affecting elder communication, with a consequent sense of isolation and depression. Hearing impairments can also relevantly reduce safety: For instance, failing to hear a fire alarm or not being able to clearly understand a pharmacist's direction about taking medications could lead to serious consequences. Cognitive impairments, often due to Alzheimer's disease, affect elders' ability to complete everyday tasks independently. Several elders experience a progressive decline in cognitive performance, with a sense of confusion, disorientation, limited attention, and memory impairment. That determines the decline in elders' ability to meet safety, self-care, household, leisure, social-interaction, and vocational needs, eventually resulting in them losing even the ability to perform basic activities, such as eating, dressing, and grooming.

Effective technologies for ubiquitous eldercare also require considering networking support aspects, necessary to favor the needed complex interactions among the elder's social, psychological, and environmental dimensions. Socio-psychological aspects deeply impact the elder's will to accept assistance technologies and the role elders can play (or attribute to themselves) within the community where they live. Pervasive technologies should promote elder engagement, that is, the elder's degree of connectivity with the world, together with the ability to communicate and share experiences and friendship. The balance between independence and engagement primarily determines the elder's quality of life; pervasive technologies should provide an appropriate trade-off between the two. Technologies promoting excessive elder independence at the expense of engagement may make elders feel isolated and alone; assistance technologies excessively relying on external people can make elders feel an important focus of attention but may make them feel infantilized.

The design of eldercare solutions is further complicated by the fact that technologies should not only overcome physical limitations and address social requirements, but also appropriately match elders' tastes and requirements. Devices for elder assistance can increase functional capabilities but can also be perceived as items that emphasize and stigmatize limitations. Therefore, eldercare technologies also have to enhance elders' perceptions of self-worth by carefully avoiding elders feeling ashamed, powerless and depression (Pirkl & Pulos, 1997). Older people tend to resist interaction outside their homes or limited environments because of embarrassment of their reliance on assistance supports; devices that positively contribute to user self-image are more readily adopted. Aesthetic considerations such as look and feel, size, and materials, as well as product function and underlying technologies, are essential components of eldercare solutions, thus often deciding the success or failure in their adoption.

Finally, designers of eldercare solutions at the role-task and social limitation levels should carefully consider also the elder's perception of his or her own abilities, just as important as elders' actual abilities. On the one hand, elders' functional abilities may decrease in such a slow and incremental manner that it is hardly noticeable in everyday life: That can result in the elder's perceived ability remaining at a much higher level than actual skills. Overconfidence may cause an elder to refuse the adoption of helpful devices. Senior people may attempt movements and actions that were once well within the range of ability, not realizing that they now threaten their health and safety. On the other hand, when the decline of functional ability is due to a traumatic event, such as a fall, the elder-perceived ability typically decreases to a much lower level than the real one.

Ubiquitous Eldercare Solutions for Aging in Place

The impact and use of technology for aging in place has been the subject of extensive research in both academia and industry over the last years. Various technologies, such as robots, smart phones, and wearable devices, are starting to be widely employed in elderly living environments to provide the kind of support required to effectively assist elders in routine life activities. The solution proposals in the literature tend to address two main needs. A primary requirement is helping elders to maintain their independence in basic activities (eating, bathing, dressing, toileting, etc.) while aging in place and also providing support for the constant monitoring of health conditions and prompt alerting in case of emergency. The other main requirement is permitting elders to be set within a rich social and emotional context by promoting their interactions with their family and friends and to support activity coordination between the different actors involved in ubiquitous eldercare. This section overviews state-of-the-art ubiquitous eldercare solutions by examining their evolution along two research mainstreams: ubiquitous assistance and ubiquitous care networking support.

Ubiquitous Assistance

Supporting elder life independence is a challenging task and requires addressing several technological and social issues. Motor, sensory, and cognitive decline of elders relevantly complicate the design of ubiquitous assistance services: There is the strong need of highly usable solutions, possibly tailored to the type of impairment of the final user; in addition, assistance services should appropriately and easily fit into everyday activities and devices by limiting the perception of intrusiveness that elders and family members could associate with assistance technologies (also taking into account privacy motivations). To further complicate the scenario, designers of assistance services should carefully consider the social effects of computer-supported eldercare by adopting the suitable trade-off between the need for assistance and the need for autonomy. For instance, over reliance on sophisticated technology can negatively affect elder self-confidence in autonomous living and might even lead to a decline in capability (Hirsh et al., 2000). With cognitive disabilities, it is critical that ubiquitous technologies do not prematurely replace the human's own capacity to act.

By considering the experiences and the literature about ubiquitous assistance services, it is possible to recognize some common requirements emerging: context awareness; context monitoring; management, filtering, and aggregation of heterogeneous context data; context distribution; and elder-caregiver communication.

A primary requirement is context awareness, that is, the need of full visibility of the elder's context (elder's physical gestures and health status; physical environment conditions, such as temperature; spatial layout of the objects in the elder's living environment; etc.) to dynamically determine and perform eldercare support activities accordingly. For instance, the visibility of context is fundamental to design effective solutions for people affected by Alzheimer's disease: An assistance service should adapt its functions depending on whether the target to be supported is the elder (when still in relatively lucid conditions) or the caregiver (when the elder has lost physical and/or verbal skills).

Context-information gathering is another key element in the development and deployment of ubiquitous assistance. There is the need to interact with the elder's living environment to collect heterogeneous monitoring data relevant for context determination, for example, the current location of elders and surrounding devices, the state of resources, and the elder's health conditions. In most recent ubiquitous assistance services, sensors such as RFIDs (Radio-Frequency Identifiers), accelerometers, pressure sensors, and video cameras are pervasively deployed in the environment where the elder lives and/or embedded in the devices she or he currently uses. Several technological challenges have to be faced notwithstanding the widespread availability of low-cost sensors of different types: Sensors currently exhibit a high heterogeneity in nature, accuracy, and performance; they should be unobtrusive and must not alter the existing living environment; and their monitored data should be as interoperable as possible so that different assistance tasks can exploit the context information derived from the same sensors (Mihailidis, Fernie, & Barbenel, 2001; Morris & Lundell, 2003). In addition, sensor technologies for eldercare should not place any additional load on final users by requiring explicit input or effort from older adults (Mihailidis & Fernie, 2002). Let us note that even remembering to wear a device or sensor can represent a cognitive load that a fraction of the elder population cannot tolerate, for example, elderly people affected by dementia.

Assistance services also need mechanisms and tools to turn raw monitoring data from sensors into context information at a higher level of abstraction, ready to be used by the application level. In particular, to improve software reusability and to accelerate assistance-service prototyping, it is necessary to decouple the design and implementation of context-aggregation solutions from the development and deployment of assistance services. This permits the designer to hide assistance services from the technological details and difficulties of sensing and monitoring. For example, an emergency response service generally requires knowing when an elder is calling for prompt help, for example, when he or she has fallen, rather than the sampled value from a position sensor. In addition, to recognize that a complex context situation has occurred, it is often necessary to compare sensed values from multiple sensors. Aggregation is the support functionality that allows one to interpret and reason about complex context situations by putting together and processing the raw monitoring data coming from sensors, generally increasing the abstraction level of context data. With the term intelligent inference, instead, the literature identifies the support function that exploits gathered context data (possibly after aggregation) to infer the current situation of the elder. Different technologies have been explored for context inference: Hidden Markov chains or Bayesian networks can model the elder's state, and planning technologies can help in promptly determining appropriate strategies to react to elder situations given that ubiquitous-assistance response times are often a critical factor.

Many-to-many context-data distribution is another important requirement to take into account in the design of ubiquitous assistance services. On the one hand, assistance applications need to exploit context that may come from different, and often heterogeneous, sources. On the other hand, the same context information should be distributed to many distributed service components. For instance, an integrated positioning system may consist of indoor (wireless fidelity or Wi-Fi based) and outdoor (Global Positioning System or GPS based) location detectors, possibly producing positioning data in different formats; the result of their work should be distributed to both an emergency service and an elder-tracking service for family members.

Finally, support for effective communication and presentation of service results to elders and caregivers is crucial. According to the gathered, aggregated, and inferred context, ubiquitous assistance services should provide consistent feedback to the elder by performing actions that can be classified in three categories. They should try to persuade elders to perform a specified task; for example, the assistance service should remind an elder to take his or her medicines or to wash his or her hands. That may require assisting the elder in the different steps necessary to complete the task, possibly requiring multimodal interfaces that exploit different communication channels (displays, speech outputs, etc.). A further line of action is performing the needed tasks on behalf of the elder; for instance, an assistance service may control gas sources, house heating, windows, and doors. Finally, in the case of emergency situations, the assistance service should alert elders' relatives, friends, and neighbors, together with emergency services, such as hospitals and fire brigades, by exploiting different communication mechanisms and solutions, for example, voice-based phone calls, text-based Short-Message Service (SMS), e-mails, and so forth.

An Overview of State-of-the-Art Ubiquitous Assistance Solutions

To the best of our knowledge, probably due to the number of complex technological and social challenges related to the design of ubiquitous assistance services, no solution in the literature already addresses all the design requirements briefly sketched in the previous section. The different industrial and academic proposals of ubiquitous assistance services tend to focus on a few specific eldercare aspects, primarily the treatment of specific disabilities or assistance with one specific task, for example, cooking or toileting. The consequence is that currently available elder-care solutions provide only a subset of the support functions needed for eldercare assistance, for example, alerting, monitoring, and context aggregation. Also, by following a historical line of evolution, the section surveys the most relevant assistance system experiences by pointing out how and to what extent they answer the identified design requirements for ubiquitous eldercare.

Personal Emergency Response Systems (PERSs) represent the first partial step in the idea of ubiquitous assistance. PERS are generally telephone-based services consisting of a manually activated help button, such as a necklace or wristband, and of a home communicator connected to a residential phone line (Canada Mortgage and Housing Corporation, 1988). PERSs have rapidly evolved by including simple remote health-monitoring devices that measure and track physiological parameters, such as pulse, skin temperature, and blood pressure (Miskelly, 2001). These solutions usually require elders to continuously wear monitoring devices and/or to manually enter data measurements, which are automatically transmitted to an evaluation or emergency station (Asada, Shaltis, Reisner, Rhee, & Hutchinson, 2003; Doughty, Lewis, & McIntosh, 2000). Most PERSs, however, have experienced limited applicability because they typically require a non-negligible utilization effort from elders, sometimes with long training times.

In addition to PERSs, the first assistance services have focused on the provisioning of monitoring and control functions to improve elder life independence in home environments. The Bath Institute of Medical Engineering, within the framework of the ENABLE Project, has developed several assistance devices tailored to people affected by dementia, primarily to compensate for their loss of memory (Bath Institute of Medical Engineering, 2006; ENABLE Project, 2006). For instance, a tap monitor is proposed to prevent bath or basin flooding: The system includes two modified taps, a level sensor, flow valves, and a central controller that can monitor the use of valves and vary water flow by commanding valve opening. The same project has also prototyped a cooker-monitoring system that can detect potentially dangerous situations on a gas hob and intervene by turning off the cooker knobs; if that automated intervention is insufficient, the monitoring system isolates the hob from gas supply and triggers an alert via SMS messages. The cooker-monitoring solution exploits infrared, smoke, and gas sensors to detect potential problems; standard cooker knobs are replaced with automatic knobs that use spring power, thus leaving the cooker immediately reusable after intervention. Another similar proposal focused on locating items that elders lose, such as home keys or glasses. The proposed solution consists of a central panel with a radio transmitter and a display showing a list of items, with each item associated to a button; a small fob is attached to the items commonly mislaid. When an elder looks for a lost item, she or he can press the associated button in the central panel, and the radio transmitter sends a coded message to the relevant fob, which replies by beeping.

More recent research activities provide monitoring-based assistance support to enable autonomous activities of daily living. For example, in McKenna, Marquis-Faulkes, Gregor, and Newell (2003), the authors demonstrate the benefits of motion tracking in eldercare, with an interesting extension of their proposal being to support people with severe vision deficiencies. In particular, the proposed system provides users with a notification of changes in the layout of familiar environments to avoid accidents during navigation. The system is based on the exploitation of computer vision to detect changes in the position of obstacles in a room between two consecutive visits of the same person. It can generate alert messages describing the occurred changes every time a person enters a room; alerts are notified via a speech output interface.

Other research exploits the location and tracking of elders to infer the current activity they are involved in and to determine the set of assistance tasks to perform accordingly. For instance, the Agent-Based Intelligent Reactive Environment Group at MIT (Massachusetts Institute of Technology) proposes a system that monitors the location of a person in a room and determines which activities are being completed and with which objects (Koile, Tollmar, Demirdjian, Shrobe, & Darrell, 2003). The space where the user lives is partitioned into zones by observing human activities. In particular, authors distinguish between physical and activity zones: Physical zones model locations and could be inferred from the observation of static furniture configuration; activity zones represent location and motion contexts and need to be identified through statistics of human behavior and activities. A multi-camera stereo-based tracking system is used to provide a history of 3-D information for every person in the observed space. A few research proposals track elder actions and behavior with the goal of obtaining relevant information about his or her status of wellness. The availability of data about elders' activities permits caregivers to evaluate the possible physical and cognitive decline of an elder and to tailor care activities accordingly. The Elite Care's Oatfield Estates Cluster in Milwaukie has developed several sensing devices to embed into home environments; for instance, wearable elder-locator tags based on periodic infrared pulses can continuously track an elder's location, while weight sensors can monitor sleeping disorders (Stanford, 2002). Narrator is another example of an assistance system that permits one to continuously track elders' actions and summarize daily movements (Wilson & Atkeson, 2003). The collected information includes data from motion detectors, pressure mats, drawer and door switches, and RFID tags. Monitoring data are summarized to represent eldercare-relevant daily events in a compact human-readable format.

Some recent work has focused on the remote health monitoring of elderly individuals. Wrist-worn ElectroMyography (EMG) sensors and portable ElectroCardiogram (ECG) systems are examples of commonly available devices that permit one to continuously monitor bio-signals. These health-monitoring data can trigger alerts to help professionals manage chronic diseases and to promptly respond to emergencies. For instance, Liszka, et al. (2004) present an ECG-monitoring solution that permits one to detect hazardous situations, such as heart arrhythmias or strokes, by integrating portable ECG sensors with a PDA (Personal Digital Assistant). The PDA processes acquired bio-signals by exploiting the R-peak detection algorithm; in the case of the detection of an emergency situation, the PDA triggers an alarm, communicates the corresponding ECG to health professionals, and alerts the elder's family members. Although most proposals focus on the monitoring of one kind of bio-signal, for example, ECG, the emerging trend is toward data fusion of bio-signals from different and heterogeneous sources. For example, some emergency situations can be easily detected by considering together monitoring data such as ECG, galvanic skin response, and accelerometer-based data detecting impacts such as from falls. In addition, a further recent research direction is devoted toward the engineering of innovative monitoring techniques to simplify data acquisition and make it as unobtrusive as possible; for instance, e-textile technologies are investigated to embed sensors directly into elders' clothes.

All the above prototypes and research proposals put in evidence the extreme heterogeneity of the monitoring indicators to be controlled and the need for higher level mechanisms to handle, filter, aggregate, and extract assistance-relevant information from raw monitoring data. That is the primary reason why significant research work has been recently directed to the design, implementation, and deployment of supports for acquiring, aggregating, and distributing context information in pervasive computing environments. In Dey, Abowd, and Salber (1999), the authors propose to apply a context-management infrastructure tailored to eldercare environments. Heterogeneous sensing sources acquire context information from the environment where elders live; collected context information is processed by context interpreters, responsible for monitoring data aggregation and interpretation; and special-purpose assistance services run in conjunction with the context-management infrastructure by benefiting from the visibility of aggregated context data. The main advantage of the approach results in the clear decoupling of assistance service implementation from all technical and deployment detail of sensing components.

Another interesting hot topic in eldercare research consists of exploiting context, for example, elder gesture monitoring, to infer the psychophysical status of elders by tailoring assistance services accordingly. For example, the smart environment prototype of Cognitive Orthosis for Assisting activities in the Home (COACH) can suggest to a person with severe dementia the tasks to perform to complete a daily-living activity, such as hand washing, with the aim of reducing elder reliance on caregivers. Switches and motion sensors track user actions during hand washing and cameras monitor whether taps were on, when soap was being used, and when hands were inside the sink basin (Mihailidis, Fernie, & Cleghorn, 2000). Artificial-intelligence learning and planning techniques are exploited to plan the proper sequence of steps to perform to complete the hand-washing task and to provide elders with appropriate help in the form of visual or verbal prompts when necessary. In addition, the Nursebot Project is investigating novel ways for caregiver robots to assist in eldercare activities (Pollack et al., 2003). Mobile robots can observe elder behavior via onboard sensors and can exploit the Autominder planning system to compute a suitable plan to complete a typical daily-living activity (taking medicine correctly, performing routine hygiene, keeping medical appointments, etc.). Autominder models elders' daily plans, tracks their execution by reasoning about elder observable behavior, and makes decisions about whether and when reminders have to be issued.

Most recent and articulated research efforts in the ubiquitous assistance field focus on the design and development of integrated homes instrumented with a variety of sensors and actuators to help elders in aging in place. For example, the Gator Tech Smart House is a pervasive environment tailored to fit elder needs (Helal et al., 2005). It uses heterogeneous sensing technologies, abstracts and manages context information, and integrates several assistance applications. In particular, a smart mailbox senses mail arrival and notifies the occupant. A smart front door permits keyless entry by residents and authorized personnel. Smart blinds can be preset or adjusted via a remote device to control ambient light and provide privacy. A smart bed can monitor elder sleep patterns and keep track of sleepless nights, while a smart mirror displays important messages or reminders to house occupants when needed. The smart bathroom includes a toilet-paper sensor, a flush detector, a shower that regulates water temperature and prevents scalding, and a soap dispenser that monitors occupant cleanliness and notifies a service center when a refill is required. In addition, smart-phone support integrates traditional telephone functions with remote control of all appliances and media players in the living room, with the possibility to convey reminders to home owners also while they are away. Moreover, the main peculiarity of the Gator Tech Smart House is to permit and facilitate the deployment and the discovery of new assistance services and their integration within the environment. The Smart House system core consists of an infrastructure layer that maintains leases of activated services. Basic services represent the physical world through sensor platforms, which store service bundle definitions for any sensor or actuator. Assistance application developers can create composite services by using a service-discovery protocol to browse existing applications and existing bundle services to compose new bundles. The support infrastructure also includes (a) a knowledge layer to maintain ontologies for available appliances, devices, and offered services, (b) a context-management layer to allow application developers to create and register contexts of interest, (c) an application layer to activate and deactivate offered assistance applications, and (d) a graphical-based integrated development environment to set and configure the smart environment.

A final relevant example of an integrated smart environment is the Honeywell Independent LifeStyle Assistant (ILSA; Haigh, Phelps, & Geib, 2002). It supports continuous data monitoring via home-installed sensors; these data are analyzed, processed, aggregated, and then possibly delivered to ILSA clients and their caregivers. The main peculiarity of ILSA is the adoption of common off-the-shelf hardware (home automation and control products from Honeywell), along with the adoption of a *de facto* standard for the support, development, and deployment of assistance services. In particular, assistance applications are implemented in terms of JADE (Java Agent Development Framework) agents that interact via standard communication protocols, such as FIPA (Footnotation for Intelligent Physical Agents).

Ubiquitous Care Networking Support

Ubiquitous care networks require appropriate technologies and solutions to promote the creation of ad hoc care communities and to support the various types of interactions needed to suitably coordinate care-related activities. The peculiar characteristics of eldercare networks, however, significantly complicate the development and deployment of effective support solutions for elder-caregiver communication and collaboration. Eldercare networks typically have an informal structure and are generally created on demand, dynamically, and largely on an opportunistic basis. Network members are generally loosely coupled, and roles and relationships between them often change, for example, depending on member presence, availability, skills, and elder conditions. In addition, it is impractical to assume that all network members statically know each other, that any member will take on any amount of responsibility, that members have homogeneous skills and share the same motivations to care for the elder, or that all members have the same access to the elder healthcare concerns. The case of heart emergency briefly sketched in the introduction exemplifies these characteristics.

Recent research work in the field has identified grouping, communication, and coordination as primary issues to address in the design of support mechanisms and tools for eldercare networks, and has consequently concentrated its efforts on the achievement of these three functions.

It starts to be recognized that it is suitable to base the design of eldercare networking services on the metaphor of group, team, or community. In fact, the interactions among care network members are facilitated by providing each one of them with a kind of group feeling that seems to increase the motivations to actively join a care network. In addition, from a technical perspective, restricting the scope of elder-related information dissemination and sharing to the only members of a group, with common goals and interests, simplifies the establishment of inter-member interactions and the process of care-related decision making. Various types of eldercare communities can be formed spontaneously to provide impromptu assistance without any prior planning. For instance, teams consisting of co-located users (people with no medical expertise or passing-by physicians) can be created on the fly to assist an elder on the street in the case of faintness; neighborhood communities, which group together people living in physical proximity, can also play a crucial role in both promoting social integration and supporting elder neighbors aging in place.

Ubiquitous care networking calls for the design and development of group-management supports that permit an elder, family member, friend, or caregiver to easily identify available groups sharing the same care goal and to dynamically create and join an eldercare group of interest by providing network members with the full visibility of all needed care-related information. Traditional group membership mechanisms provide the required support for the formation and the management of groups in distributed, open, and dynamic systems (Chockler, Keidar, & Vintenberg, 2001). However, the characteristics of eldercare networks undermine several assumptions of wired distributed systems: In traditional distributed computing, it is often possible to determine *a priori* the set of available groups and to statically identify group members; group membership does not vary frequently. Collaborating entities tend to operate from fixed devices with stable network connections and common computing capabilities in terms of available memory, computing power, battery, and so forth. Device failure and network partitioning rarely occur and it is feasible to organize solutions for deterministic recovery from them. On the contrary, in eldercare networks, group membership management cannot rely on preconceived knowledge about the set of available group members and their characteristics and device properties, but it is rather necessary to provide users with the possibility to compose, dissolve, join, and leave groups on demand from highly heterogeneous terminals with possibly different forms of wireless connectivity. In addition, members of an eldercare network may require belonging to several groups at a time and dynamically changing their played role within all or part of their groups.

Context awareness is emerging as a crucial property in the design of group-management solutions in the eldercare field. Novel group supports should exploit the full visibility of context information, such as elder conditions; the location of users, terminals, and resources; and user profiles of preferences and skills to promote the collaboration among members of eldercare communities anywhere and anytime. Among the various and heterogeneous data belonging to context, location plays a primary role. For instance, group-management supports could exploit the visibility of an elder's location to create an eldercare network with only the entities in the elder's vicinity to provide prompt help, especially in the case of emergency situations. The visibility of other context information may guide the formation of suitable ad hoc groups. For example, medical skills or the agenda of caregivers may be helpful to give priority to the choice of people with appropriate experience and time to perform the needed care-related activities.

Eldercare networks also need effective communication and coordination support to favor elder health-condition reporting and information dissemination among care network members. Traditional message-oriented and event-based models, or more advanced ones, such as blackboard-based or tuple-based ones, can be exploited for group-oriented communication. However, it is difficult to envision one single solution that fits all communication and coordination requirements of ubiquitous eldercare networking. An open challenge is to achieve rapid and prompt information dissemination to support medical-critical care services while minimizing, at the same time, issues of reporting reliability. For instance, prompt assistance may be the difference between death and life in the case of heart attacks: Circulation and ventilation should be initiated as soon as possible to avoid irreversible cerebral damage; however, high levels of stress and lack of time, typically characterizing emergency situations, could create untrained and unrehearsed reactions that cause critical delays in intervention and contribute to inconsistencies and omissions in the critical data to share. Novel communication and coordination solutions should proactively propose and stimulate collaboration among potential helpers co-located with the elder in need of help. They should clearly notify helper community members about the occurring situation, inform helpers about the activities to perform by supporting them in the collaborative definition and assignment of assistance tasks, and coordinate their activities via synchronization and data sharing, dynamically tailored to the group of participants and to the degree of criticality of the elder situation.

An Overview of State-of-the-Art Ubiquitous Care Networking Solutions

The research and development efforts in ubiquitous care networking are still quite immature, with only a few solution proposals. No current state-of-the-art solution exists that provides comprehensive support to all the requirements discussed in the previous section, while the few proposals in the literature focus on specific aspects of care networking, either grouping, communication, or coordination. This section surveys and discusses ubiquitous eldercare solutions at the NCMRR social-limitation level, moving from simple early attempts in care networking support to more recent advanced research.

Preliminary work in the field proposed systems with the simple goal of reinforcing emotional connections between elders and family members who live apart. Gust of Presence provides an easy-to-use, point-to-point communication support for enabling affective interactions between distant family members (Keller, Van der Hoog, & Stappers, 2004). The Living Memory Box (Stevens, Abowd, Truong, & Vollmer, 2003) and Digital Storytelling (Balabanovic, Chu, & Wolff, 2000) assist family members in preserving memories of their elders in a variety of media forms, such as photos, video, and audio, and support the sharing of that multimedia information.

A few recent research activities have already started to address advanced collaboration support for ubiquitous eldercare networking by handling grouping, communication, and coordination management issues (Bardram, 2005; Bottazzi, Corradi, & Montanari, 2005; Consolvo, et al., 2004; Munoz, Rodriguez, Favela, Martinez-Garcia, & Gonzales, 2003; Mynatt et al., 2001; Santana, Rodríguez, González, Castro, & Andrade, 2005). All this research bases networking on the abstraction of group (or team or community) by differing in the kind of group-management support provided. In particular, it is possible to distinguish three main types of group-management support. First of all, some systems exploit the grouping abstraction to restrict the scope of elder-related information to eldercare network members and, thus, to shape information distribution. The set of group members is a priori determined and only registered members can access the eldercare networking support (Consolvo, et al., 2006; Mynatt, et al., 2001; Santana, et al., 2005). For example, in Santana et al., users have to register themselves on a dynamic Web page in order to insert new content and access the already posted information.

A second set of more advanced solutions provide presence-aware group management (Bardram, 2005; Munoz et al., 2003). As in the first category of networking supports, these systems statically determine group composition, but with the additional ability to keep track of the group members who are simultaneously connected. For instance, in Munoz et al., (2003) an instant messenger has been extended to provide each group member with the full visibility of all potential collaborators available. Similar considerations apply to Bardram (2005), that rules group member visibility according not only to group belonging and online connectivity, but also to the current activities the group members are involved in.

A third research direction proposes to dynamically compose on demand ad hoc eldercare groups on the basis of more complex context conditions. In these systems, not only people's online availability, but also their current location is taken into account to rule the visibility of collaborating partners. For example, the AGAPE (Allocation and Group-Aasdre Perudsive Enviornment) system allows one to create, join, and leave ad hoc groups dynamically by exploiting the visibility of entity location, interest, and activities to define groups of collaborating partners (Bottazzi et al., 2005). In particular, user location determines the scope of group-member visibility, whereas user attributes govern the discovery and joining of groups of interest. As a key feature, AGAPE provides group-management support for both indoor and outdoor environments with a high degree of dynamicity; this permits the building of outdoor collaborative care applications, particularly useful in emergency scenarios. Whenever AGAPE detects, via its monitoring service, that an emergency situation is occurring, it promptly promotes and manages the formation of ad hoc first-response groups. Individuals in physical proximity with aging people (passing-by users or neighbors) are alerted and requested to join the ad hoc helping team and to collaborate together to provide elders with immediate help when alerted of an emergency.

Similar considerations apply when considering communication support. Current care networking supports tend to provide differentiated group communication functionality. Santana et al. (2005) support asynchronous communication to permit group members to not have to be online at the same time to communicate. In particular, users exchange messages or documents with collaborating partners by posting them on a centralized server; message delivery occurs when group members connect to the server. For example, in Santana et al., (2005) a blog-like Web page is the medium enabling collaboration activities. In systems like those in Munoz et al. (2003), Bardram (2005), and Bottazzi et al. (2005), the authors recognize the need also for synchronous forms of collaboration and provide context-aware mechanisms to identify appropriate partners for message delivery. In particular, these solutions recognize the possibility to identify collaborating partners on the basis of their played role in collaboration (Bottazzi et al., 2005; Munoz, et al., 2003), on their currently performed activities (Bardram, 2005; Bottazzi, et al., 2005), or simply on the basis of their attributes (Bottazzi, et al., 2005). AGAPE further pushes the exploitation of context awareness in communication by proposing to rule message scheduling and presentation format based on context information: User preferences and current activities influence message scheduling, whereas user-device characteristics and elder capabilities determine how to adapt message formats.

Coordination is the less investigated aspect in eldercare networking supports. Some basic solutions support only professional caregiver coordination by exploiting statically defined protocols (Bardram, 2005; Munoz et al., 2003). In particular, in Bardram (2005), only users involved in the same care activity can coordinate their tasks, while Munoz et al. (2003) permits the coordination of users via their access to shared resources. On the contrary, systems such as that in Mynatt et al. (2001) support differentiated forms of coordination for different user categories, and the visibility of care activities to perform is propagated to all members. For example, an agenda reminds elders' relatives and friends about relevant care tasks, for example, taking the elder to the doctor, and permits one to distribute caring duties. AGAPE extends the idea of coordination through a shared agenda to allow not only caregivers or family members, but also passing-by users to coordinate care tasks, even in outdoor emergency scenarios. Passing-by users with their wireless portable devices are provided with a to-do list describing the specific operations to do and a how-to list of basic instructions about the most appropriate ways to perform care tasks.

Conclusion and Open Issues

The growing percentage of aging people puts the welfare systems of our developed-world societies under relevant stress. There is a strong and increasing need to provide novel forms of eldercare services capable of controlling costs that raises relevant and challenging social and technological problems. Internet technologies at first and pervasive computing more recently, are playing a major role in reshaping the eldercare assistance model. In particular, pervasive technologies are opening up novel possibilities for anytime and anywhere assistance to elderly people. On the one hand, ubiquitous connectivity, miniaturization, and the mobility of devices have fostered the development of novel assistance solutions that can constantly monitor and compensate for motor, sensory, and cognitive difficulties of elders while moving both in indoor and outdoor environments. On the other hand, pervasive technologies have the potential to leverage elders' social interactions by providing novel means to build and manage communities of help and interest among aging adults, their families, and caregivers.

As demonstrated by the several academic and industrial ongoing projects described in the chapter, ubiquitous eldercare has raised great interest in recent years due to both its social and economic relevance. However, despite researchers' ever-increasing energy and enthusiasm, further work is still needed to consolidate the results obtained up to now. Current solutions seem to be more proof-of-concept prototypes of single aspects rather than comprehensive methodological and technical reference guides, and this is motivated by a few relevant reasons. The inherently multidisciplinary nature of ubiquitous eldercare complicates the identification and assessment of a comprehensive well-defined methodology for the design, implementation, and deployment of eldercare systems. It is difficult to orchestrate and merge into a uniform integrated framework the different (and sometimes contrasting) points of view of the various experts involved, ranging from engineers, technicians, and system administrators to social and psychological professionals. Some solutions may be effective from the point of view of the communication and software-engineering requirements, but may fail to properly answer social elder needs, thus resulting in very limited applicability.

In addition to methodological issues, several complex technical aspects still require further investigation. First of all, there is the need to develop eldercare solutions that are dynamically customizable to adapt their assistance and/or care networking support to the elders' individual needs and requirements. In addition, as ubiquitous eldercare systems are becoming more and more compound, with lots of heterogeneous sensors and monitoring devices inter-working, novel management mechanisms and tools are required to replace explicit human control and reactive management with effective, automated, and proactive management. Autonomic computing seems a viable way to enable eldercare services to autonomously configure, reconfigure, heal, and maintain themselves without the need for continuous human intervention. For instance, eldercare services with self-healing properties could autonomously detect, diagnose, and repair localized problems resulting from localized software bugs or hardware failures, for example, through a regression tester (Kephart & Chess, 2003).

Another major technical aspect that needs further analysis and investigation is to identify a reasonable trade-off between the two contrasting needs of (a) elder privacy, and (b) the visibility of monitored, sensitive heath-related information to caregivers. Elder people tend to be reluctant in installing monitoring devices in their living environments if they feel to be spied on against their will. However, a ubiquitous eldercare solution necessitates the provision of relevant data about elder conditions to caregivers. In addition, establishing a proper trade-off between privacy and visibility of health indicators is also difficult due to legal issues: The disclosure of sensitive medical data may involve liabilities. Some partial, country-specific regulations are starting to emerge, for example, the U.S. Health Insurance Portability Accountability Act for the management of medical records. However, no well-established international laws uniformly and appropriately regulate the pervasive eldercare scenario, where the availability of a professional IT staff managing the information infrastructure and the related data privacy aspects is typically unfeasible.

Finally, it is worth noticing that another important obstacle to wider acceptance and diffusion of ubiquitous eldercare solutions is the lack of standards for the acquisition, monitoring, distribution, delivery, and elaboration of health-related and emergency-related information. The extremely wide potential portfolio of ubiquitous eldercare solutions can only be supported by having an appropriate assurance of technology compatibility in place at all levels, from monitoring to networking, and from presentation to data maintenance.

Acknowledgments

This work is supported by the FIRB Wide-Scale Broadband Middleware for Network Distributed Services (WEB-MINDS) project and the Infrastruture Software per Reti Ad-hoc Orientate ad Ambienti Difficili (IS-MANET) project.

References

Asada, H., Shaltis, P., Reisner, A., Rhee, S., & Hutchinson, R. (2003). Mobile monitoring with wearable photoplethysmographic biosensors. *IEEE Engineering in Medicine and Biology Magazine, 22*(3), 28-40.

Balabanovic, M., Chu, L., & Wolff, G. (2000). Storytelling with digital photographs. *CHI Letters, 2*(1), 564-570.

Bardram, J. E. (2005). Activity-based computing: Support for mobility and collaboration in ubiquitous computing. *Personal and Ubiquitous Computing, 9*(5), 312-322.

Bath Institute of Medical Engineering. (2006). Retrieved from http://www.bath.ac.uk/bime/

Bottazzi, D., Corradi, A., & Montanari, R. (2005). Context-awareness for impromptu collaboration in MANETs. *Proceedings of the 2nd Annual Conference on Wireless On-demand Network Systems and Services (WONS'05)* (pp. 16-25).

Canada Mortgage and Housing Corporation. (1988). *The study of emergency response systems for the elderly*. Ottawa, Canada: MacLaren-Plansearch.

Chockler, G. V., Keidar, I., & Vintenberg, R. (2001). Group communication specifications: A comprehensive study. *ACM Computing Surveys, 33*(4), 427-469.

Consolvo, S., Roessler, P., Shelton, B. E., LaMarca, A., Schilit, B., & Bly, S. (2004). Technology for care networks of elders. *IEEE Pervasive Computing Magazine, 3*(2), 22-29.

Dey, A. K. (2001). Understanding and using context. *Personal and Ubiquitous Computing Journal, 5*(1), 4-7.

Dey, A. K., Abowd, G. D., & Salber, D. (1999). A context-based infrastructure for smart environments. *Proceedings of the 1st International Workshop on Managing Interactions in Smart Environments* (MANSE 99), 114-128.

Doughty, K., Lewis, R., & McIntosh, A. (2000). The design of a practical and reliable fall detector for community and institutional telecare. *Journal of Telemedicine and Telecare, 6*(Suppl. 1), 150-154.

Elder Services Network Links. (2006). Retrieved from http://elder-services.net/computer.htm

ENABLE Project. (2006). Retrieved from http://www.enableproject.org/

Haigh, K. Z., Phelps, J., & Geib, C. W. (2002). An open agent architecture for assisting elder independence. *Proceedings of the 1st International Joint Conference on Autonomous Agents and Multiagent Systems* (AAMAS 02), 578-586.

Helal, S., Mann, W., El-Zabadani, H., King, J., Kaddourah, Y., & Jansen, E. (2005). The Gator Tech Smart House: A programmable pervasive space. *IEEE Computer Magazine, 38*(3), 50-60.

Helal, S., Winkler, B., Lee, C., Kaddourah, Y., Ran, L., Giraldo, C., et al. (2003). Enabling location-aware pervasive computing applications for the elderly. *Proceedings of the 1st IEEE International Conference on Pervasive Computing and Communication (PerCom 2003)* (pp. 531-538).

Hirsh, T., Forlizzi, J., Hyder, E., Goetz, J., Kurtz, C., & Stroback, J. (2000). The ELDer Project: Social, emotional, and environmental factors in the design of eldercare technologies. *Proceedings of the ACM Conference on Universal Usability (CUU 2000)* (pp. 72-79).

Keller, I., Van der Hoog, W., & Stappers, P. J. (2004). Gust of me: Reconnecting mother and son. *IEEE Pervasive Computing Magazine, 3*(1), 22-28.

Kephart, J. O., & Chess, D. M. (2003). The vision of autonomic computing. *IEEE Computer Magazine, 36*(1), 41-50.

Koile, K., Tollmar, K., Demirdjian, D., Shrobe, H., & Darrell, T. (2003). Activity zones for context-aware computing. In *Lecture notes in computer science: Vol. 2864. Proceedings of the 5th International Conference on Ubiquitous Computing, UbiComp 2003* (pp. 90-106). Seattle, WA: Springer-Verlag.

Liszka, K. J., Mackin, M. A., Lichter, M. J., York, D. W., Pillai, D., & Rosenbaum, D. S. (2004). Keeping a beat on the Earth. *IEEE Pervasive Computing Magazine, 3*(4), 42-49.

Mann, W. C. (2004). The aging population and its needs. *IEEE Pervasive Computing Magazine, 3*(2), 12-14.

McKenna, S. J., Marquis-Faulkes, F., Gregor, P., & Newell, A. F. (2003). Scenario-based drama as a tool for investigating user requirements with application to home monitoring for elderly people. *Proceedings of the 10th International Conference on Human-Computer Interaction* (HCI 2003).

Mihailidis, A., & Fernie, G. (2002). Context-aware assistive devices for older adults with dementia. *Gerontechnology, 2*(2), 173-189.

Mihailidis, A., Fernie, G. R., & Barbenel, J. C. (2001). The use of artificial intelligence in the design of an intelligent cognitive orthosis for people with dementia. *Assistive Technology, 13*(1), 23-39.

Mihailidis, A., Fernie, G. R., & Cleghorn, W. L. (2000). The development of a computerized cueing device to help people with dementia to be more independent. *Technology & Disability, 13*(1), 23-40.

Miskelly, F. G. (2001). Assistive technology in elderly care. *Age and Ageing, 3*(6), 455-458.

Morris, M., & Lundell, J. (2003). *Ubiquitous computing for cognitive decline: Findings from Intel's proactive health research.* Seattle, WA: Intel Corp.

Munoz, M. A., Rodriguez, M., Favela, J., Martinez-Garcia, A. I., & Gonzales, V. M. (2003). Context-aware mobile communication in hospitals. *IEEE Computer Magazine, 36*(9), 38-46.

Mynatt, E. D., Rowan, J., Craighill, S., & Jacobs, A. (2001). Digital family portraits: Providing peace of mind for extended family members. *Proceedings of the 2001 ACM Conference on Human Factors in Computing Systems (CHI 2001)* (pp. 333-340).

Pirkl, J., & Pulos, A. (1997). *Transgenerational design: Products for an aging population.* New York: John Wiley & Sons.

Pollack, M. E., Brown, L., Colbry, D., McCarthy, C. E., Orosz, C., Peintner, B., et al. (2003). Autominder: An intelligent cognitive orthotic system for people with memory impairment. *Robotics and Autonomous Systems, 44*(3-4), 273-282.

Santana, P. C., Rodríguez, M. D., González, V. M., Castro, L. A., & Andrade, A. G. (2005). Supporting emotional ties among Mexican elders and their families living abroad. *Proceedings of the Conference on Human Factors in Computing Systems (CHI 05)* (pp. 2099-2103).

Seniors-Site.Com. (2006). Retrieved from http://seniors-site.com/

Stanford, V. (2002). Using pervasive computing to deliver Eldercare. *IEEE Pervasive Computing Magazine, 1*(1), 10-13.

Stevens, M., Abowd, G. D., Truong, K. N., & Vollmer, F. (2003). Getting into the living memory box: Family archives & holistic design. *Personal Ubiquitous Computing, 7*(3-4), 210-216.

United Nations Population Division. (2006). *World population prospects: The 2000 revision.* Retrieved from http://www.un.org/esa/population/publications/wpp2000/

Wilson, D., & Atkeson, C. (2003). *The Narrator: A daily activity summarizer using simple sensors in an instrumented environment.* Pittsburgh, PA: Carnegie Mellon University.

Section III

Security, Reliability, and Interoperability

Chapter VII

Bringing Secure Wireless Technology to the Bedside:
A Case Study of Two Canadian Healthcare Organizations

Dawn-Marie Turner, DM Turner Informatics Consulting Inc., Canada

Sunil Hazari, University of West Georgia, USA

Abstract

Wireless technology has broad implications for the healthcare environment. Despite its promise, this new technology has raised questions about security and privacy of sensitive data that is prevalent in healthcare organizations. All healthcare organizations are governed by legislation and regulations, and the implementation of enterprise applications using new technology is comparatively more difficult than in other industries. Using a configuration-idiographic case-study approach, this study investigated challenges faced by two Canadian healthcare organizations. In addition to interviews with management and staff of the organizations, a walk-through was also conducted to observe and collect first-hand data of the implementation of wireless technology in the clinical environment. In the organizations under examination, it was found that wireless technology is being implemented gradually to augment the wired network. Problems associated with implementing wireless technology in these Canadian organizations are also discussed. Because of different standards in this technology, the two organizations are following different upgrade paths. Based on the data collected, best practices for secure wireless access in these organizations are proposed.

Introduction

Technology, the Internet, and healthcare reform are converging to change the healthcare environment and create a seamless integrated healthcare network. This seamless network will facilitate the flow of information from multiple sources to multiple healthcare providers, administrators, patients, and other support services 24 hours a day, seven days a week, among multiple sites (Masys & Baker, 1997). Implementing and managing such a network within the healthcare environment poses unique challenges. First, medical and health information is highly sensitive; therefore security and privacy of the information must be a top priority. Security and privacy in healthcare is governed by legislation and regulation. In Manitoba, this means the Personal Health Information Act (PHIA). PHIA specifies how medical information can be accessed, by whom, and for what purposes. It also states the security and privacy regulations for all health information systems used within the province. Second, unlike other industries, medical care is not delivered in the same place even by the same healthcare professional, necessitating the need for multiple access points (APs) for the same information. For example, a physician on rounds moves from one patient to another, each of whom may reside in a different room, necessitating the need for network access in each room to record and receive data and communicate with other needed services such as pharmacy or nursing.

The challenge in creating a seamless network in healthcare is how to provide information to multiple users at the point in which they will require the information to deliver effective patient care. A wireless network may offer the opportunity to meet this challenge and provide significant benefits to the healthcare system. A wireless local area network (WLAN) offers improved accuracy and efficiency for documenting nursing care, decreased preventable medication error through better point-of-care medication-administration systems, an increase in patient satisfaction, and efficiency in admission and discharge and other health administration processes (Sims, 2004). Additional technical benefits include lower costs, less cabling, availability of the network in locations not accessible with a wired connection, and the ability to adapt to growth easier.

The implementation of a WLAN is not without its challenges. Some challenges such as performance, speed, and accessibility are similar to those of a wired network, but others such as limited battery power of the devices, necessitating the need for an electrical source if the device is required for extended use; higher risk of equipment loss; and interference with medical equipment are unique to the WLAN environment (Karygiannis & Owens, 2002). Multiple standards and the fact that a WLAN does not usually replace a wired network but augments it increases the complexity of the management and compatibility of new systems (Drew, 2003). However, security is the biggest challenge facing a healthcare organization contemplating a wireless network. Wireless networks pose an increased risk of eavesdropping, hackers, and rogue devices (Sims, 2004). Securing a WLAN and the perception of its security may be one of the most limiting factors in the widespread use of WLAN in healthcare today (Campbell & Durigon, 2003).

Objectives of the Study

The objectives of this study were the following:

a.   To gain an understanding of wireless-technology standards and their application within healthcare.
b.   To articulate the security issues of wireless technologies within healthcare.
c.   To identify the potential for best practice in the implementation of wireless technology in healthcare using a case-study methodology.

Wireless Technology in Healthcare

As the development and use of the electronic patient record progresses, mobile devices will become more common. Healthcare providers will begin to use these devices to access information previously only available in the paper-based chart or record. The increased use of wireless will help to stabilize wireless communication standards (Campbell & Durigon, 2003). Currently there are three standards used for wireless networking, wireless fidelity (Wi-Fi), mobile communications (cell phones), and Bluetooth. Wi-Fi was the standard used in the case studies in this research, therefore it is the standard discussed in this literature review.

The Wi-Fi standard encodes the data and then sends them over a selected channel using radio-wave frequencies. The connection is made through the use of a wireless network card within the device and a connection to an access point creating a wireless LAN. If the access point is connected directly to the corporate network or the Internet, the wireless user will have direct access to the corporate network or

Internet (Campbell & Durigon, 2003). The standard for Wi-Fi (802.11) has been established by the Institute of Electrical and Electronics Engineers (IEEE), a professional organization of engineers, students, and scientists. This standard (802.11) was the original standard set for wireless computing and established the protocols to be used between a wireless device and access point, or two wireless devices (Drew, 2003). Revisions and updates to 802.11 have resulted in several versions of the 802.11 standard. Prior to implementing WLAN, a healthcare organization needs to select the variation of the Wi-Fi standard it will use. Choosing which variation will depend on the data-transmission needs, cost, and the number of devices accessing the network. Three variations of the 802.11 standard currently being used by the case studies within this chapter will be discussed (802.11a, 802.11b, 802.11g).

Specification 802.11b, completed in 1999, is probably the most widely used standard. It is a physical-layer standard and operates in the 2.4 GHz frequency, providing users with 11Mbs throughput between the wireless device and the AP, depending on the distance between the device, the number of users, and any other interference. This standard has three available radio channels (Campbell & Durigon, 2003; Drew, 2003). One disadvantage to 802.11b is it operates in the same frequency as most medical devices such as ultrasound, sterilizers, and treatment or diagnostic devices. Therefore using a wireless device at this frequency requires all devices to be tested for potential interference with existing medical equipment sharing the same frequency, special consideration to the placement of the access nodes, and frequent retesting for possible interference.

Specification 802.11a also completed in 1999 provides users with a faster throughput at 54Mbs using the 5 GHz frequency spectrum. In addition to its faster throughput, 802.11a offers two advantages over 802.11b. First, the 5 GHz frequency is not shared by other commonly used devices such as microwaves, cellular phones, and medical monitoring equipment, making interference from these devices less of an issue. Second, 802.11a opens more channels, making the network more available; with eight vs. three channels, it offers better protection against possible interference from neighboring access points (Campbell & Durigon, 2003). However, its disadvantage is the need for more access points because the higher throughput is gained at the cost of a shorter transmission distance, making 802.11a more expensive to implement, something that must be considered when choosing which standard to use. 802.11a is also not backward compatible. This means a healthcare organization that has already implemented a wireless network using 802.11b must replace their access nodes for compatibility with 802.11a. One solution for this is the use of dual-band access points that are certified to work with 802.11b and 802.11a, allowing organizations to leverage existing technology when upgrading to the new standard (Karygiannis & Owens, 2002). The newest standard, 802.11g, was developed in 2001 as a direct result of the compatibility issues between 802.11b and 802.11a. It provides the throughput of 802.11a but is backward compatible with 802.11b (Campbell & Durigon). As such, organizations that have already invested in 802.11b technologies without the use of dual-band access points can upgrade to the new standard without the expense of new hardware.

Authentication is a very important component for healthcare organizations. It refers to the ability to verify the identity of client stations or individuals accessing health data over a network, and deny access to those not providing the correct electronic credentials. The Wired Equivalent Privacy (WEP) protocol defines two types of authentication: open-system and shared-key authentication. Open-key authentication is not a true authentication process because it only requires a one-way channel. Access points using open-system authentication will accept a mobile device on the basis of it having a media access control (MAC) address and does not verify if it is an authenticated address within the network. Shared-key authentication requires a two-way interchange between the access point and the device based on cryptography. In this scenario, the client requesting access to the WLAN sends a message to the access point, and the access point responds with a challenge to the client requesting it to identify itself using its special key. The access point then decrypts the message and if it matches the values allowed, the client is authenticated to the network. The WEP protocol only requires open-system authentication, creating a potential security risk if shared-key authentication is not also implemented within the organization (Newman, 2003). The need for authentication of wireless devices within a healthcare facility is extremely important because of the number of transient population (e.g., patients, visitors) that has the potential to tap into the health information network and have access to sensitive patient records.

Confidentiality or privacy refers to protecting the data from eavesdropping either intentionally or unintentionally through cryptographic techniques. (Privacy issues in healthcare are discussed later.) The WEP protocol uses the Rivest Cipher 4 (RC4) symmetric key, stream cipher algorithm to generate a pseudo-random data sequence supporting a 40-bit encryption for the shared key. This is a weak encryption system and on a busy network could be cracked in a matter of hours (Sims, 2004). Integrity ensures messages sent are not modified during transmission. The service was developed to reject any messages that appeared to have been modified during transmission. The technique used within the WEP is "a simple encrypted cyclic redundancy check (CRC) approach that after sealing the packet encrypts it for transmission where on receipt of the packet, it is decrypted and compared to the original. If they are not equal, an error message is sent. Unfortunately, CRC, unlike a hash code or message authentication code, is not cryptographically secure. Although the WEP provides security services, it is clear they are not sufficient to provide the level of security required for the sensitive information being transmitted within a healthcare institution (Berghel & Uecker, 2005).

It appears clear from this discussion that standards alone will not create a secure network, and controls are also needed. Controls are the mechanisms that reduce or eliminate threats to the organization's computer systems and network (Fitzgerald & Dennis, 2002). There are typically three types of controls. Preventative controls, as the name implies, prevents or mitigates the chance of a security breach such as the use of passwords and locking the computer equipment (such as those located in areas like nursing stations and administrative offices). Detective controls are those strategies and mechanisms used to identify when a security breach has occurred such as identifying when an unknown address has tried to gain entry (such as by implementing an intrusion-detection system on the hospital's internal network). Detection controls usually include reporting and may include an alarm function to alert network personnel of potential threats. Corrective controls correct or fix an unwanted event or threat. Controls should be used to develop specific countermeasures to address the vulnerabilities with using a WLAN. The application of specific countermeasures minimizes the risks to create a more secure network. As with wired networks, risks cannot be completely eliminated, but through the appropriate application of countermeasures and the use of controls, risk can be reduced to a level that is acceptable to the organization. Countermeasures can be divided into three broad areas: management, operational, and software.

Management countermeasures usually focus on the preventative level. They start with a comprehensive security policy outlining such things as who has access to the network, authorization levels, the installation of access points, configuration management, and the reporting of loss or stolen wireless devices. In Manitoba, the security policy must also include a signed confidentiality agreement between the user and the organization outlining what constitutes a breach and the consequences of any breaches.

Operational countermeasures offer both preventative and detection controls. These include the physical security measures taken to ensure only authorized personnel have access to the devices and networks through such things as identification badges, locking the equipment, security guards and video cameras, the use of passwords or biometrics, and the use of site survey tools for mapping access points and ensuring coverage remains within the intended range. It should also include logging and auditing of all accesses and attempted access to the network to identify if unauthorized use has occurred.

Technical countermeasures include the use of software and hardware to protect the network such as proper access-point configurations, software patches and upgrades, authentication, intrusion-detection systems, encryption, the use of a virtual private network (VPN), firewalls, and public-key infrastructure (PKI). The use of a VPN within a wireless network can afford the same level of protection it does within the wired environment. Like in a wired environment, a VPN creates an encrypted secure channel between the user's wireless device and the network, thus hiding the transmission (Kilpatrick, 2003). A firewall is one of the most formidable lines of system defense because it prevents unauthorized users and creates an invisible wall to potential intruders (Campbell & Durigon, 2003; Derba & Siegal, 2003). The goal is to ensure that only authorized individuals are able to view an individual's healthcare record.

Implementing a wireless network in any healthcare environment should be considered carefully with a clear business need. It must also consider the highly sensitive nature of the information being transmitted coupled with the risk of any potential breach of the system. Once the decision has been made to implement wireless, a careful assessment of capabilities must be matched to the goals and objectives for the wireless network. Only then can related decisions about standards, hardware, and software be made.

Privacy Issues in Healthcare

The significance of information privacy will continue to escalate in proportion to the value of information (Rust, Kannan, & Peng, 2002). Information privacy in healthcare organizations is related to information security. It is important to note that an organization may have information security without privacy, but it is not possible to have privacy without having information security controls (preventative, detective, or corrective as discussed earlier). While wireless technology offers convenience and potential that shows promise for improving healthcare delivery, the right to privacy for patients must be protected. In wireless transmission, data is not being confined to a physical medium so that it remains secure when being transmitted from node to node. Therefore, it is necessary that healthcare organizations should develop and implement system security and privacy strategies to protect data and information stored in research and clinical databases. According to Huston (2001), confidentiality and security of a patient's health information has always been important, and with the ease of access afforded electronically, security will likely be more difficult to provide without advanced planning.

As healthcare organizations move toward converting paper-based records and communication processes to digital formats that can be easily stored and manipulated for administrative decision making, intranets and extranets are established within and outside the boundaries of the healthcare organization. There needs to be a strategic aspect to maintaining security, privacy, and standardization of networks that carry data and information within and outside the organization. A common example is the use of a networked decision-support system that is specially developed for supporting decision making related to the solution of a particular healthcare-management problem (Turban, 1993). This type of system ties in with centralized databases so effectively, controls need to be present in these databases to ensure confidentiality, integrity, and availability of data.

Government regulations (e.g., PHIA in Canada, and HIPAA [Health Insurance Portability and Accountability Act] in USA) have provided guidelines to healthcare organizations to maintain privacy and security aspects of the transmission and maintenance of patient records. Healthcare organizations previously outsourced services such as transcriptions, which made it possible to identify patient data. Regulations now hold providers responsible for auditing policies and procedures of contracted firms (Walker & Spencer, 2000). This was done in an effort to safeguard the privacy of patient data. Similarly, policies and procedures for handling patient data that were previously written for the "paper world" are being revised to comply with electronic storage, access, sharing, and transmission of data (Shortell & Kaluzny, 1994).

Methodology

The case study is a widely used method of qualitative research within information systems and is an effective design for understanding the organizational context of information-technology innovations. A case study is defined as "an empirical enquiry that investigates a contemporary phenomenon within its real-life context, especially when the boundaries between phenomenon and context are not clearly evident and it relies on multiple sources of evidence" (Darke, Shanks, & Broadbent, 1998, p. 273). Case-study research is often used to describe, test, or develop theory. This type of case-study research is called the configuration-idiographic study, which is used to describe events and their circumstances to identify relationships but not necessarily generate theoretical interpretations (Smith, 1990). As the purpose of this study was explanatory, case selection was based on the available cases offering the greatest explanatory power.

Two midsize healthcare organizations in Canada using a WLAN were approached and agreed to participate in the case study. Both organizations provide inpatient and outpatient treatment facilities and clinics. One organization's WLAN extended beyond the physical boundaries of the institution, enabling wireless access to the corporate network in two satellite facilities approximately 2 to 5 km away. Interviews were conducted using a semi-structured interview format with one or all of the following personnel: the network manager, and IT security and systems analysts. The interview questions were developed to identify the characteristics of the network and current practices used to secure the organization's wireless network and devices. In addition to the interviews, a walk-through was conducted at one of the sites to view the wireless device within the clinical environment.

The interview questions from each case study were analyzed using a qualitative inductive approach. The goal of this approach was to explore and analyze existing practices of the organizations within the context of recommendations, practices, and standards identified in the literature. There was no effort made to quantify results or compare one organization with the other. The responses to the interview questions were analyzed to answer each of the research questions. Responses to the first research question were assessed against the three primary wireless threats (identified in the literature): malicious hackers, eavesdropping (war driving), and rogue wireless devices. Best practice was assessed using the National Institute of Standards and Technology (NIST) steps for a secure wireless LAN and the fit with current organizational practices to these recommendations. Data was also collected to assess the rationale for selecting wireless technology, the current standard implemented, and the location of wireless within the organization. These areas were not categorized but will be discussed.

Findings

Both organizations in the case study indicated wireless was not a stand-alone network but augmented the existing wired network to provide healthcare professionals with point-of-care access to patients' electronic records housed on the corporate network. Wireless was also implemented to reduce overall costs and ease the management of providing point-of-care access because the wireless technology allowed multiple users to use the same equipment vs. requiring the purchasing of a laptop for each user. Additionally, IEEE 802.11b was the current standard within both organizations. However, future upgrades were split with one organization choosing to upgrade to 802.11a and the other choosing 802.11g. An increased volume of users in one organization was the rationale for the planned upgrade to 802.11a. Although prior implementation of dual-band access nodes meant compatibility with the existing network was not an issue, the increased cost due to the greater number of required access nodes was slowing down the rate of growth. The need for higher transmission rates and compatibility with the existing WLAN was the second organization's rationale for upgrading to 802.11g.

In both organizations, interference with medical equipment was a consideration and required all wireless devices to be tested for compatibility by the biomedical engineering department. Although no issues were found in either organization, one organization had implemented a policy that dictated wireless devices were not permitted in areas with highly critical medical monitoring equipment such as the intensive care units. Each organization identified their first step in securing wireless for healthcare was to enable security within the wireless standard, usually WEP. However, it was identified that this level of security was not enough as one network specialist indicated: "It is not enough to use the security that comes with the system. You need to layer your security; the default settings are not secure enough."

Securing and protecting the wireless network from eavesdropping, malicious hackers, and rogue devices or access points was approached in both organizations from four perspectives: securing the wireless network, securing the device, protecting the data, and protecting the larger corporate network. Securing the wireless network was accomplished through the configuration of the MAC addresses requiring authentication to the network (access nodes will only talk to addresses they know), and hiding the name of the network was also done through the configuration. As one organization indicated, this means "anyone scanning for a network might still find it, but because they don't know its name, it will be inaccessible." In addition to authentication of the MAC address, one organization installed a firewall between the WLAN and the corporate network; the firewall actively scanned the airwaves for unrecognized MAC addresses and when it detected something, it sent a warning that unidentified addresses had attempted access to the system. Additionally, the firewall provided the organization with end-to-end 128-bit encryption and authentication between the WLAN and the corporate network. Furthermore, protection of the network was provided through software monitor switches that looked for and detected unauthorized access nodes. Unauthorized nodes were immediately removed and the organization maintains a strict policy regarding the installation of unapproved access points.

Securing the devices was accomplished in both organizations first through a security policy that specified who could use a wireless device, what authorization level they had, and what their access level was. Second, each device was inventoried and a hardware log was maintained. Finally, the devices themselves were secured. Both organizations in this study use computers on wheels (COWS) as the wireless devices. These are laptop computers secured to a cart or mobile station allowing for easy movement within the organization, but making it difficult to remove from the property. Consideration was given to the use of other wireless devices such as personal digital assistants (PDAs); however, these were considered to pose an increased security risk due to their lack of direct connectivity to the network and the need to store personal health information (even temporarily) on the device. As one organization noted, "When personal health information resides on the device, the loss or theft of a device jeopardizes the confidentiality of the information stored on that device." The storage of personal health information on the device even for a short time also prevented real-time access to information by the health professional. The need to synch the device created a time lag between when the device recorded the information and when it is was entered into the network and available to another healthcare professional.

Protecting health and medical information residing on the corporate network was done through enhanced 128-bit encryption that was centrally controlled and configured for end-to-end data encryption. As one network manager stated, "We made the assumption that people may break in, so we make the data stream unreadable…essentially they get garbage." One organization also indicated the wireless network is treated as a hostile environment, therefore without the proper authentication; even someone that manages to get into the wireless network cannot access the corporate network. As indicated previously, authentication was a two step process, authenticating first to the wireless network and then to the LAN. Finally, access to the corporate network in both organizations was role based as defined by PHIA, giving access to the corporate network only to the level required for the user to deliver safe patient care.

Table 1. Establishing best practice for wireless in healthcare

Wireless Security Steps	Case-Study Findings
1. Maintain a full topology of all wireless connections and access points	Prior to implementation of the network, an access coverage map was drawn to identify the placement of nodes and allow for the straddling of frequencies.
2. Label and maintain an inventory of all wireless devices	All wireless equipment was inventoried. The security policy prohibits the use of personal wireless devices, e.g., PDAs and laptops. An inventory and table of all authorized MAC addresses was retained.
3. Create regular backups	No data were stored on wireless devices; they were all contained within the corporate network. The WLAN serves only as an access mechanism to the corporate network or electronic patient record. Routine back up of servers is done daily.
4. Perform regular security tests of network and devices	All equipment was tested at the time of deployment. Penetration tests were completed routinely (no indication of frequency was provided) and routine vulnerability testing by the third party were completed randomly as outlined in the security policy.
5. Perform random but regularly timed audits	All access to the WLAN, LAN, and applications are logged and available for audit as specified in PHIA. Routine education of staff concerning proper use of WLAN and devices was conducted.

Table 1. Continued

6. Apply patches and security enhancements	Software monitors the switches to detect unauthorized access nodes. Antiviral software is installed and updated regularly on all devices. Current security patches apply in both organizations.
7. Monitor the industry for new standards affecting security and new products to enhance security	802.1x security enhancements were explored.
8. Monitor vigilantly for new threats and vulnerabilities	A firewall was installed for active monitoring of the airwaves. WLAN was treated as a hostile environment and appropriate controls were applied.

Best-Practice Wireless in Healthcare

NIST outlined eight steps for maintaining a secure wireless network (Karygiannis & Owens, 2002). A further analysis of the data collected from the case studies appears to support these eight steps as a foundation on which to build best practice for a secure wireless network in a healthcare organization. A review of each of these steps is presented in Table 1 with an indication of how the case-study organizations addressed each one. Alone, these eight steps were not enough. For the organizations reviewed in this study, best practice also included a strict and well-published security policy, the use of 128-bit encryption end to end, the use of a firewall between the WLAN and the corporate network, strong authentication of both the device and the user, and the use of a virtual private network for wireless transmission.

Limitations

The availability of suitable cases is one difficulty in case-study methodology (Darke et al., 1998). The slower growth of information technology within healthcare (and in particular wireless) coupled with the highly sensitive nature of the topic (security) limited the number of cases available for study. Only two sites were used for this study, and one site limited the amount of information it provided due to concerns about breaching security. Therefore generalizations are difficult. Another limitation of the study was that due to the time constraints of the study period, only the network manager and systems analyst were interviewed, and not the users of the wireless devices. Organizational security policies, although identified, were also not reviewed.

Conclusion

Wireless technology affords many advantages within the healthcare environment, but it also poses a greater risk if not implemented properly. As one participant noted, "You need to assume your system is vulnerable, then make the data stream unreadable through encryption and limit exposure to the rest of the network through the installation of a firewall between the wireless device and the enterprise network."

Is wireless secure enough for healthcare? Although it is difficult to generalize based on only two case studies, it appears a secure WLAN can be implemented to meet the unique requirements of healthcare. However, this research also demonstrated that the implementation of a secure WLAN needs more careful planning, evaluation, and reassessment of the risks and vulnerabilities than a traditional LAN in order to meet the security and privacy challenges within the healthcare environment. The challenge for any organization would be to maintain the confidentiality, integrity, and availability of data and information in a secure environment that makes it possible to provide the most efficient patient care. To achieve this, current technologies (such as wireless networks) offer healthcare practitioners and administrators the opportunity to positively impact the quality of healthcare not only in offices for administrative purposes, but more importantly clinically by the patients' bedside.

References

Berghel, H., & Uecker, J. (2005). Wifi attack vectors. *Communications of the ACM, 48*(8), 21-28.

Campbell, R., & Durigon, L. (2003). Wireless communication in healthcare: Who will win the right to send data boldly where no data has gone before? *Healthcare Manager, 22*(3), 233-240.

Darke, P., Shanks, G., & Broadbent, M. (1998). Successfully completing case study research: Combining rigour, relevance and pragmatism. *Information Systems Journal, 8,* 273-289.

Derba, M., & Siegal, J. (2003). Wireless networks. *The CPA Journal, 73*(7), 18-21.

Drew, W. (2003). Wireless networks: New meaning to ubiquitous computing. *Journal of Academic Librarianship, 29*(2), 102-106.

Fitzgerald, J., & Dennis, A. (2002). *Business data communications and networking* (7th ed.). New York: John Wiley & Sons.

Huston, T. (2001). Security issues for implementation of e-medical records. *Communications of the ACM, 44*(9), 89-94.

Karygiannis, T., & Owens, L. (2002). *Wireless network security* (NIST special publication 800-48). Gaithersburg, MD: National Institute of Standards and Technology.

Kilpatrick, I. (2003, December). Are you indulging in unprotected wireless. *Logistics and Transport Focus, 5*(10), 20-21.

Masys, D. R., & Baker, D. B. (1997). *Patient-centered access to secure systems online: A secure approach to clinical data access via the World Wide Web.* Retrieved February 13, 2006, from http://www.saic.com/healthcare/sysint/pdf/amia1997.pdf

Newman, R. (2003). *Enterprise security.* Columbus, OH: Prentice-Hall.

Rust, R., Kannan, P., & Peng, N. (2002). The customer economics of Internet privacy. *Journal of the Academy of Marketing Science, 30*(1), 455-464.

Shortell, S. M., & Kaluzny, A. D. (1994). Healthcare management: Organization design and behavior (3rd ed.). New York: Delmar.

Sims, B. (2004). Moving from liability to viability. *Health Management Technology, 25*(2), 32-35.

Smith, C. (1990). The case study: A useful research method for information management. *Journal of Information Technology, 5*(3), 123-133.

Turban, E. (1993). *Decision support and expert systems: Management support systems* (3rd ed.). New York: Macmillan.

Walker, J., & Spencer, J. (2000). Ten deadly sins. *Health Management Technology, 21*(7), 10.

Chapter VIII

Reliability and Evaluation of Health Information Online

Elmer V. Bernstam,
University of Texas, Health Science Center at Houston, USA

Funda Meric-Bernstam,
University of Texas, M.D. Anderson Cancer Center, USA

Abstract

This chapter discusses the problem of how to evaluate online health information. The quality and accuracy of online health information is an area of increasing concern for healthcare professionals and the general public. We define relevant concepts including quality, accuracy, utility, and popularity. Most users access online health information via general-purpose search engines, therefore we briefly review Web search-engine fundamentals. We discuss desirable characteristics for quality-assessment tools and the available evidence regarding their effectiveness and usability. We conclude with advice for healthcare consumers as they search for health information online.

Introduction

The healthy, the newly diagnosed, and the chronically ill turn to the Internet for health information. In spite of some controversy regarding the number of individuals that are accessing online health information at any given time, most experts agree that the numbers are enormous. Indeed, some have estimated that on any given day, more people consult the Internet for health information than see a physician (Fox & Rainie, 2002). Furthermore, there is evidence that patients are making treatment decisions based on the information that they encounter online (Helft, Hlubocky, Gordon, Ratain, & Daugherty, 2000). Therefore, researchers, clinicians, and the general public are increasingly concerned about the reliability of health information online that is directed toward consumers of healthcare. In this chapter, we consider the problem of how to enable healthcare consumers to evaluate online health information.

Definitions: Information Quality vs. Information Accuracy

One of the major impediments to research into online information quality is the lack of clear, consistent, and generally accepted definitions. In this section, we define relevant terms to provide a vocabulary for discussion.

The factual correctness (accuracy) of health information online may be difficult to assess. Indeed, even experts often disagree regarding accuracy. Therefore, review of information content by a panel of experts is considered to be the gold standard of accuracy.

Most Internet users are not healthcare experts. Therefore, they cannot judge the accuracy of online health information. Since consumers cannot assess accuracy, surrogate measures that they can assess are appealing. We refer to these surrogate measures as measures of quality and collections of these measures as quality-assessment tools. Measures of information quality are useful to the extent that they (a) can be effectively assessed by healthcare consumers and (b) correlate with outcomes of interest such as whether the information is factually incorrect or whether the information has the potential to harm health (i.e., if the advice were followed).

An example of an information quality measure is authorship (i.e., is an author identified?). The JAMA benchmarks (Silberg, Lundberg, & Musacchio, 1997) are a commonly cited quality-assessment tool. The JAMA benchmarks consist of four quality measures: authorship, attribution, disclosure, and currency. These generally came to be known as the "clarity in publishing" criteria and are similar in spirit to the type of questions one might ask about a textbook or scientific paper.

Preferably, measures of quality should be based on meaning rather than presentation of information. In contrast, some studies tested superficial characteristics such as the claim of copyright (Fallis & Fricke, 2002). In this study, Web pages that claimed copyright were more likely to be accurate than pages that did not claim copyright. The authors point out, however, that it is simple to claim copyright. Even if such superficial measures correlate with accuracy, they are easy to manipulate. Web-site developers could simply claim copyright without modifying the information displayed on the Web site. Therefore, as superficial measures become more widely used, they will become less useful.

Unfortunately, quality measures are difficult to put into practice. A systematic review of studies assessing the quality of online health information determined that although 70% of the studies found quality to be a problem, there were wide differences in the quality measures used, their operational definitions,[1] and methods in which the analyses were carried out (Eysenbach, Powell, Kuss, & Sa, 2002).

Some have argued that a high-quality Web site should display information that is accurate, easy to understand, specifically tailored to the intended audience, and pleasing to the eye. However, in order to enable communication, more precise definitions are useful. Eysenbach et al.'s (2002) review of the literature distinguished the following: (a) technical quality criteria, defined as "general, domain-independent criteria, i.e., criteria referring to how the information was presented or what metadata[2] was provided," (b) design, which "includes visual aspects of the site such as the colors used or layout," (c) readability, meaning whether the language is easy to read and is understandable, (d) accuracy, or the "degree of concordance with the best evidence or with generally accepted medical practice," and (e) completeness, which refers to the portion of relevant material covered on the Web site. These five characteristics may be independent from each other. For example, it is possible for a high-quality Web site based on technical quality criteria to display false information, or a poor-quality Web site to display accurate information.

Quality-Assessment Tools

In response to widespread concern regarding the quality and accuracy of online health information, many organizations published quality-assessment tools. Consumers are encouraged to evaluate online health information using these quality-assessment tools. In spite of this and consumers' stated concern regarding the accuracy of the information that they see on the Internet, recent studies found little evidence that consumers are considering quality when searching for information online (Eysenbach & Kohler, 2002; Meric et al., 2002).

As mentioned above, quality measures often reflect clarity-in-publishing standards adapted from traditional media such as the display of authorship, attribution, disclosures of conflicts of interest, and so forth. Similarly, a television viewer or newspaper reader may want to look for an identified author and sponsorship information. When reading a scientific paper or book chapter, one looks for references to the published literature that support assertions. It is possible that these measures are less appropriate in some domains. Specifically, complementary and alternative medicine (CAM) is an important topic commonly addressed by health-related Web sites. The accuracy of CAM information is difficult if not impossible to define. By definition, CAM information may not reflect expert medical opinion or conventional standards of care. Furthermore, few CAM Web sites reference traditional biomedical literature (Sagaram, Walji, & Bernstam, 2002; Walji et al., 2004). An alternative is to consider the potential harm of the information if the advice were followed. Clinicians assessing the potential harm of online health information are able to do so with high interobserver reliability (see next paragraph; Walji et al.).

A recent review of quality-rating tools provided by the likes of the American Medical Association and Health on the Net (HON) Foundation, among others, found that no instrument reported interobserver reliability as a measure of its validity (Gagliardi & Jadad, 2002). The purpose of quality-rating tools is to allow those looking for health information to filter out poor-quality content. However, the criteria used must also be unambiguous so that different people using the same criteria will agree upon their usage. Therefore, it is important that different reviewers can agree upon and reproduce the results. In other words, quality measures must have high interobserver reliability. Unfortunately, evaluations of interobserver reliability have not been encouraging. Even trained informaticians are not able to consistently agree on the presence or absence of commonly cited quality criteria (Bernstam, Sagaram, Walji, Johnson, & Meric-Bernstam, 2005; Sagaram, Walji, Meric-Bernstam, Johnson, & Bernstam, 2004). It seems unlikely that consumers without informatics training fare any better. Furthermore, many quality-rating tools (composed of one or more quality criteria) are not practically usable by consumers. Many tools are not publicly available, are composed of criteria that cannot be reliably assessed, or have too many criteria (more than 10 questions to be answered per site; Bernstam, Shelton, Walji, & Meric-Bernstam, 2005). Table 1 provides examples of commonly cited quality criteria, their operational definitions, and expected interobserver reliability.

In addition to questionable usability, there is also conflicting evidence regarding the effectiveness of existing quality-assessment tools to identify false or misleading information online. Some studies show a correlation between quality and accuracy (Meric et al., 2002), while others do not (Walji et al., 2004). The reason(s) for this discrepancy is not clear. One possible explanation is that the correlation is domain dependent. Therefore, studies may differ depending on the specific topic tested (e.g., breast cancer vs. complementary and alternative medicine). For a comprehensive listing of commonly cited quality-assessment tools, see Eysenbach et al. (2002).

Is There Documented Harm from Online Health Information?

In spite of widespread concern regarding the potentially harmful effects of online health information, documented evidence of physical harm is scarce. A systematic review of harm from online information found few reported cases. However, the authors caution that this may be a result of underreporting rather than actual absence of harm (Crocco, Villasis-Keever, & Jadad, 2002). Consequently, some argue that concerns have been overemphasized and benefits to users have been underappreciated (D. Hoch & Ferguson, 2005). This issue remains controversial.

Searching for Health Information
(Web Search Engines)

Few consumers are directed to specific Web sites by their healthcare provider (Reents, 1999; Tang & Newcomb, 1998). Therefore, left to their own devices, consumers use general-purpose search engines to find health information on the Internet (Eysenbach & Kohler, 2002; Fox & Rainie, 2002). Although their search strategies are often suboptimal and they rarely look beyond the first page of results, consumers are generally satisfied with their ability to find information on the Internet (Eysenbach & Kohler; Fox & Rainie, 2000). To a large extent, search engines determine what information is seen by consumers.

Search-engine technology is complex, changes quickly, and is often proprietary. Therefore, a detailed discussion of the inner workings of search engines is beyond the scope of this chapter. However, we will present a basic overview of relevant search-engine characteristics with particular emphasis on Google (http://www.google.com), the most popular search engine in the English-speaking world (Sullivan, 2004).

In general, to use a search engine, the user types in a few terms perceived to be related to his or her information need. An information need is an expression, in the user's own language, of the information that she or he desires. For example, "Should I use tamoxifen to keep my breast cancer from coming back?" is an information need. In order to satisfy this information need, a user may issue the following query: "breast cancer tamoxifen." A result set is retrieved in response to the query. The user can click on results that seem to be interesting. If none of the results seem interesting, users generally add or subtract terms from a query (Eysenbach & Kohler, 2002). For example, if "breast cancer tamoxifen" does not lead to a successful search, then "breast cancer tamoxifen recurrence" may be a reasonable query to try.

Before Google, most search engines used text-processing techniques to identify relevant Web sites. In other words, the search engine looked for Web pages that contained the query terms. This approach worked well when the Web was small. However, with many billions of Web sites, common queries now return too many results. Therefore, identifying results that are most likely to be useful became increasingly important. The designers of Google noted that the link structure of the Web was a rich source of information. Specifically, incoming links imply importance. Important sites tend to have many incoming links. Their algorithm, known as PageRank, makes use of links, or references from one Web page to another. This approach has proven so successful that most Web search engines now use some form of link analysis (Cho & Roy, 2004). Therefore, information displayed by popular Web sites is likely to be seen by healthcare consumers. Some studies found a correlation between popularity (Google rank) and accuracy (Fricke, Fallis, Jones, & Luszko, 2005) while others did not (Meric et al., 2002).

Search engines drive traffic on the Web. Therefore, Web sites compete for position on prominent search engines. Being ranked highly by Google, for example, will dramatically increase the number of users accessing a Web site (Cho & Roy, 2004). Many search engines, including Google, provide two types of results in response to a query. Algorithmic results are generated by the general-purpose search algorithm. In contrast, sponsored results are links to Web sites that have paid a fee to the search engine and are essentially advertisements. These are likely to contain commercial offers, often related to complementary and alternative medicine (Walji, Sagaram, Meric-Bernstam, Johnson, & Bernstam, 2005).

Unfortunately, not all search engines clearly identify sponsored results. Consumers are unaware of the existence of sponsored listings and do not differentiate them from nonsponsored results, which are retrieved algorithmically (i.e., using an algorithm that does not factor payment for placement). In fact, two in five (41%) links clicked by consumers are sponsored results (Marable, 2003). The Federal Trade Commission has recently recommended that search engines clearly identify sponsored listings (Hippsley, 2002).

Advice to those Seeking Health Information Online

In this section, we consider the advice that can be offered to users searching for health information online in the absence of usable, reliable quality-assessment tools. Unfortunately, we cannot recommend a quality-assessment tool that can be used to identify problematic health information online. Similarly, we cannot recommend a Web search engine that lists only trustworthy information. In spite of this, we can recommend simple practices that are likely to minimize exposure to inaccurate information online.

Access Sites Known to be Trustworthy

Multiple government agencies such as the National Institutes of Health (NIH, USA) and the National Health Service (NHS) of the United Kingdom maintain Web sites with extensive libraries of health information intended for consumers. MedlinePLUS, maintained by the National Library of Medicine (USA), is a specific example. There are also disease-specific organizations that maintain high-quality content on every imaginable topic. Prominent examples include the National Cancer Institute (USA) and the American Heart Association (USA). Table 2 provides a small list of trustworthy Web sites with stable URLs (uniform resource locators). However, we emphasize that the Web is a dynamic medium and therefore any list of Web sites is dated by the time that it is published in print.

Distinguish Algorithmic from Sponsored Results

As discussed above, sponsored results are basically advertisements for particular Web sites that are often financed by commercial interests. Therefore, many sponsored links lead to Web sites attempting to sell products or services. To their credit, Google clearly identifies and separates sponsored results from algorithmic results. Searchers should be aware of how they encountered the Web site that they are viewing.

Consider the Source

Common sense tells us that information displayed by Web sites with commercial interests (i.e., those that are selling products) may not be entirely unbiased. Therefore, users would do well to consider the source, just as they would when evaluating non-health information. In general, government, organization, or academic Web sites are more likely to provide impartial advice compared to company (commercial) Web sites. However, this is a generalization and there may be exceptions.

Check with a Healthcare Professional

In some ways, online health information is no different from health information provided via any other medium. When in doubt, or when making important decisions, there is still no substitute for professional assessment and judgment. Therefore, the prudent health seeker will use online resources for education, but review important decisions with their healthcare provider.

Future Directions

As of this writing, traditional quality-assessment tools do not sufficiently address the concern regarding the quality and accuracy of health information online. Clearly, self-regulation (the status quo) does not prevent or identify false, misleading, or potentially harmful online health information. Regulation, whether by government(s) or nongovernmental organizations, does not seem viable. Indeed, even voluntary standards such as the HON code of conduct are difficult to enforce (Meric et al., 2002). In this section, we consider promising new approaches with the potential to address the shortfalls of traditional quality-assessment tools.

Distributed Human Annotation

There is too much Web content for any independent organization to evaluate and validate. Since content is constantly changing, evaluation and validation must be repeated to ensure compliance over time. A potential solution is to create a framework and process by which independent evaluators can collaboratively annotate content. Evaluators include the content providers (i.e., the Web-site developers), information specialists, and healthcare professionals. Therefore, the Web-site developer provides metadata such as authorship information. The metadata is checked by a nonmedical information specialist. Finally, the Web-site content is reviewed by a healthcare professional. Once the Web site has been evaluated, it is tagged with descriptors. Therefore, users can search for a Web site that has specific characteristics (e.g., has been evaluated by a healthcare professional). The MedCIRCLE (http://www.med-circle.org) project provides a vocabulary (HIDDEL, Health Information Disclosure, Description and Evaluation Language) to allow independent evaluators to describe a Web site in a consistent manner.

The advantage of this approach is that multiple independent evaluators can effectively collaborate. Consistent self-rating by Web-site developers may be better than no rating at all. However, as discussed above, self-regulation does not always ensure compliance with quality standards. In addition, independent manual Web-site evaluation is laborious. Although a distributed approach is likely to be more effective than one that relies on a single central authority, the Web is vast and constantly changing. Therefore, any validated subset of the Web is likely to be a very small percentage of all health information available online.

Reputation-Based Systems

Reputation-based systems explicitly model trust, which is hearsay evidence of predictable behavior. There are multiple examples of reputation-based systems in common use outside of healthcare. Perhaps the best known reputation-based system is the online auction company eBay (http://www.ebay.com) that brokers transactions (auctions) between buyers and sellers. After each transaction, feedback is solicited from both parties. Users who do not behave honorably acquire a bad reputation that is visible to the community. Similar systems have been developed for e-mail networks (Golbeck & Hendler, 2004). For online health information, one could develop algorithms to identify reliable sites using link analysis given a small set of verified (known to be good) sites such as those shown in Table 2.

Reproducibility

Unlike the traditional print media, Web sites do not exist independently of each other. In other words, a user has access to multiple Web sites from the same terminal. In contrast, when reading a book or journal article on paper, one may not have access to other books or articles on the same topic. It is easy to access multiple Web sites on the same topic to see if there is a consensus regarding specific information. One might hypothesize that if information is repeated on multiple independent Web sites, it is more reliable than information found on only a single Web site. However, this heuristic has not, to our knowledge, been systematically evaluated.

Self-Correction Hypothesis

The self-correction hypothesis refers to the possibility that users can identify and correct problematic health information online without professional guidance or supervision (Fennberg, Licht, Kane, Moran, & Smith, 1996; Ferguson, 1995; D. B. Hoch, Norris, Lester, & Marcus, 1999). In a sense, this is an extension of the approach used by open-source projects such as the online encyclopedia Wikipedia (http://www.wikipedia.org) and the open-source software movement where the user community identifies and corrects errors in content or computer code. Wikipedia allows any Web user to create and modify content. A recent analysis of Wikipedia content found that Wikipedia was generally accurate compared to a traditionally curated online encyclopedia like Britannica (Giles, 2005). In the context of online health information, we do not know whether self-correction really works. However, preliminary data are promising (Esquivel, Meric-Bernstam, & Bernstam, 2006). Therefore, it is possible that, given a sufficiently large and active forum, the Web can police itself.

Conclusion

Many Web users are searching for health information online and express concern regarding the quality and accuracy of online health information. Although there is evidence that online information affects healthcare decisions, there is little evidence that users are considering quality when accessing health information online. Helping users identify problematic health information online remains an open problem. Currently available quality-assessment tools cannot be reliably assessed by Web users and may not be effective at identifying problematic health information online. Fortunately, this is a young and dynamic field. Promising alternatives to currently available quality-assessment tools are subjects of ongoing research.

References

Bernstam, E. V., Sagaram, S., Walji, M., Johnson, C. W., & Meric-Bernstam, F. (2005). Usability of quality measures for online health information: Can commonly used technical quality criteria be reliably assessed? *International Journal of Medical Informatics, 74*(7-8), 675-683.

Bernstam, E. V., Shelton, D. M., Walji, M., & Meric-Bernstam, F. (2005). Instruments to assess the quality of health information on the World Wide Web: What can our patients actually use? *International Journal of Medical Informatics, 74*(1), 13-19.

Cho, J., & Roy, S. (2004). *Impact of search engines on page popularity.* Paper presented at the WWW2004, New York.

Crocco, A. G., Villasis-Keever, M., & Jadad, A. R. (2002). Analysis of cases of harm associated with use of health information on the Internet. *JAMA, 287,* 2869-2871.

Eisenberg, D. M., Kessler, R. C., Rompay, M. I. V., Kaptchuk, T. J., Wilkey, S. A., Appel, S., et al. (2001). Perceptions about complementary therapies relative to conventional therapies among adults who use both: Results from a national survey. *Annals of Internal Medicine, 135*(5), 344-351.

Esquivel, A. E., Meric-Bernstam, F., & Bernstam, E. V. (in press). Accuracy and self correction of information received from an Internet breast cancer list: Analysis of posting content. *BMJ.*

Eysenbach, G., & Kohler, C. (2002). How do consumers search for and appraise health information on the World Wide Web? Qualitative study using focus groups, usability tests, and in-depth interviews. *BMJ, 324*(7337), 573-577.

Eysenbach, G., Powell, J., Kuss, O., & Sa, E. R. (2002). Empirical studies assessing the quality of health information for consumers on the World Wide Web: A systematic review. *JAMA, 287*(20), 2691-2700.

Fallis, D., & Fricke, M. (2002). Indicators of accuracy of consumer health information on the Internet: A study of indicators relating to information for managing fever in children in the home. *J Am Med Inform Assoc, 9*(1), 73-79.

Fennberg, A., Licht, J., Kane, K., Moran, K., & Smith, R. (1996). The online patient meeting. *Journal of Neurological Science, 139*(Suppl.), 129-131.

Ferguson, T. (1995). Consumer health informatics. *Healthcare Forum Journal,* 28-33.

Fox, S., & Rainie, L. (2000). *The online healthcare revolution: How the Web helps Americans take better care of themselves.* Washington, DC: Pew Internet and American Life Project.

Fox, S., & Rainie, L. (2002). *Vital decisions: How Internet users decide what information to trust when they or their loved ones are sick.* Washington, DC: Pew Internet & American Life Project.

Fricke, M., Fallis, D., Jones, M., & Luszko, G. M. (2005). Consumer health information on the Internet about carpal tunnel syndrome: Indicators of accuracy. *Am J Med, 118*(2), 168-174.

Gagliardi, A., & Jadad, A. R. (2002). Examination of instruments used to rate quality of health information on the Internet: Chronicle of a voyage with an unclear destination. *BMJ, 324*(7337), 569-573.

Giles, J. (2005). Internet encyclopaedias go head to head. *Nature, 438*(7070), 900-901.

Golbeck, J., & Hendler, J. (2004, October). *Accuracy of metrics for inferring trust and reputation.* Paper presented at the 14th International Conference on Knowledge Engineering and Knowledge Management, Northamptonshire, United Kingdom.

Helft, P. R., Hlubocky, F. J., Gordon, E. J., Ratain, M. J., & Daugherty, C. (2000). *Hope and the media in advanced cancer patients.* Paper presented at the American Society of Clinical Oncology 36th Annual Meeting, New Orleans, LA.

Hippsley, H. (2002). *Letter to Gary Ruskin: Executive director of Commercial Alert.* Retrieved from http://www.ftc.gov/os/closings/staff/commercialalertletter.htm

Hoch, D., & Ferguson, T. (2005). What I've learned from e-patients. *PLoS Med, 2*(8), e206.

Hoch, D. B., Norris, D., Lester, J. E., & Marcus, A. D. (1999). Information exchange in an epilepsy forum on the World Wide Web. *Seizure, 8*(1), 30-34.

Marable, L. (2003). False oracles: Consumer reaction to learning the truth about how search engines work: Results of an ethnographic study. *Consumer Reports WebWatch*. Retrieved February 27, 2006, from http://64.78.25.46/view-article. cfm?id=10171&at=510

Meric, F., Bernstam, E. V., Mirza, N. Q., Hunt, K. K., Ames, F. C., Ross, M. I., et al. (2002). Breast cancer on the World Wide Web: Cross sectional survey of quality of information and popularity of Websites. *BMJ, 324*(7337), 577-581.

Reents, S. (1999). *Impacts of the Internet on the doctor-patient relationship: The rise of the Internet health consumer.* Cyber Dialogue, Inc.

Sagaram, S., Walji, M., & Bernstam, E. (2002). Evaluating the prevalence, content and readability of complementary and alternative medicine (CAM) Web pages on the Internet. *Proceedings of the AMIA Symposium*, 672-676.

Sagaram, S., Walji, M., Meric-Bernstam, F., Johnson, C., & Bernstam, E. (2004). Inter-observer agreement for quality measures applied to online health information. In *Medinfo* (Vol. 2004, pp. 1308-1312).

Silberg, W. M., Lundberg, G. D., & Musacchio, R. A. (1997). Assessing, controlling, and assuring the quality of medical information on the Internet: Caveant lector et viewor—Let the reader and viewer beware. *JAMA, 277*(15), 1244-1245.

Sullivan, D. (2004). *comScore Media Metrix Search Engine ratings*. Retrieved December 4, 2004, from http://searchenginewatch.com/reports/article.php/2156431

Tang, P. C., & Newcomb, C. (1998). Informing patients: A guide for providing patient health information. *JAMIA, 5*(6), 563-570.

Walji, M., Sagaram, S., Meric-Bernstam, F., Johnson, C. W., & Bernstam, E. V. (2005). Searching for cancer-related information online: Unintended retrieval of complementary and alternative medicine information. *International Journal of Medical Informatics, 74*(7-8), 685-693.

Walji, M., Sagaram, S., Sagaram, D., Meric-Bernstam, F., Johnson, C., Mirza, N. Q., et al. (2004). Efficacy of quality criteria to identify potentially harmful information: A cross-sectional survey of complementary and alternative medicine Web sites. *J Med Internet Res, 6*(2), e21.

Appendix: Tables

Table 1. Commonly cited quality criteria

Quality Criterion	Operational Definition	Page (P) or Site (S) Review	Allowed Options	Interobserver Reliability*	
				% Agreement	K
Disclosure of authorship	Name of the person(s) or organization(s) present that is attributed as the creator or producer of the presented information	P	Yes No	0.857	-
Disclosure of ownership	Indication of the entity that owns the information presented on the Web site (e.g., organization logo)	S	Yes No	0.571	0.121
Sources clear	Claims backed up with a source (e.g., reference, expert opinion, or bibliography)	P	Yes No	0.905	0.800
Disclosure of sponsorship	Funding source or commercial aim/intent of the Web site or organization disclosed (e.g., organization mission on "About Us" page)	S	Yes No	0.238	-
Disclosure of advertising	Clear distinction (visual or by text) between advertising (paid piece of information or banner ad) and content (does not include product listings or a "buy now" link)	P	Yes No	0.143	-
Statement of purpose	General purpose or aims behind the Web site or organization may be found on front page, or under "About Us" or "Contact Us"	S	Yes No	0.524	0.186
General disclosures	Disclosure of either authorship, ownership, sponsorship, or currency of information	P	Yes No	0.667	0.310

Table 1. continued

Date of creation disclosed	Date disclosed when information was produced or reported (without the phrase "date created," the given date may be assumed to be the date created, such as with news organizations)	P	Yes No	0.857	0.504
Date of last update disclosed	Date disclosed of any revision or update	P	Yes No	0.905	-
Date of creation or update disclosed	Either creation or last-update date disclosed	P	Yes No	0.905	0.696
Authors' credentials disclosed	Disclosure of authority and qualification (MD, PhD, ND, etc.) of author. Disclosure does not include "Dr." or "Professor" (NA if no author)	P	Yes No NA	0.952	-
Credentials of physicians disclosed	Disclosure of credentials of physician (MD or including area of specialization) (NA if no author, author not physician, or not known if physician)	P	Yes No NA	0.857	-
Authors' affiliation disclosed	Disclosure of author's affiliations or relationships with relevant entity (NA if no author)	P	Yes No NA	0.905	0.738
Internal search engine present	Presence or absence of a search engine (can be any type of search engine, including a product search engine)	S	Yes No	0.857	0.715
Links provided	Presence of links (internal or external but not within the page; anchor related to topic for further information)	P	Yes No	0.762	0.516

Table 1. continued

References provided	Presence of conventional references or citations relevant to information on the page. Link to reference also acceptable	P	Yes No	0.952	0.897
Feedback mechanism provided	Author, editor, webmaster, or other official can be contacted. Presence of e-mail address, telephone, fax, or online form	S	Yes No	0.762	0.348
Fax number provided	Presence of a fax number for contact purposes	S	Yes No	0.762	0.516
E-mail address provided	Presence of an e-mail address for contact purposes. E-mail address does not include an e-form	S	Yes No	0.857	0.488
General disclaimers provided	Presence of a general disclaimer such as "Not a substitute for professional care" or "For educational purposes only," or a link to a disclaimer	S	Yes No	0.857	0.717
Copyright notice	Presence of a copyright notice	S	Yes No	0.857	0.710
Editorial review process	Presence of claim of use of an editorial review process or the listing of an editorial review committee or medical advisory board	S	Yes No	0.952	0.644

* *Interobserver reliability from Bernstam, Sagaram, et al. (2005) using precalibration values.*

K = Cohen's kappa, which cannot be calculated for some data.

Table 2. Trustworthy Web sites

	Organization	URL	Topic
General Health Information	MedlinePLUS (National Library of Medicine, NIH, USA)	http://www.medlineplus.gov	Consumer-oriented health information
	Office of Disease Prevention and Health Promotion, Department of Health & Human Services (USA)	http://www.healthfinder.gov	Consumer-oriented health information (not cancer specific)
	National Institute of Health (USA)	http://clinicaltrials.gov	Information about government- and privately funded clinical research
	NHS Direct Online (UK)	http://www.nhsdirect.nhs.uk	General health information from the British government
Disease- or Treatment-Specific Health Information	National Center for Complementary and Alternative Medicine (NIH, USA)	http://nccam.nih.gov/	Complementary and alternative medicine information
	American Heart Association (USA)	http://www.americanheart.org	Information about heart-related illnesses
	National Cancer Institute (NIH, USA)	http://www.nci.nih.gov	Information about cancer

Chapter IX

Integrating Mobile-Based Systems with Healthcare Databases

Yu Jiao, Oak Ridge National Laboratory, USA

Ali R. Hurson, Pennsylvania State University, USA

Thomas E. Potok, Oak Ridge National Laboratory, USA

Barbara G. Beckerman, Oak Ridge National Laboratory, USA

Abstract

In this chapter, we discuss issues related to e-health and focus on two major challenges in distributed healthcare database management: database heterogeneity and user mobility. We designed and prototyped a mobile-agent-based mobile data-access system framework that can address these challenges. It applies a thesaurus-based hierarchical database federation to cope with database heterogeneity and utilizes the mobile-agent technology to respond to the complications introduced by user mobility and wireless networks. The functions provided by this system are described in detail and a performance evaluation is also provided.

Introduction

The integration of healthcare management and advances in computer science, especially those in the areas of information-system research, has begotten a new branch of science: e-health. E-health is becoming more and more widely recognized as an essential part for the future of both healthcare management and the health of our children. The 2001 President's Information Technology Advisory Committee, in its report "Transforming Healthcare through Information Technology," noted that information technology "offers the potential to expand access to healthcare significantly, to improve its quality, to reduce its costs, and to transform the conduct of biomedical research"(p. 1). Although much has been done, reality has proven to us that there are still a great number of problems remaining to be taken care of. Health and human-services secretary Mike Leavitt told the Associated Press (2005) in an interview after hurricane Katrina, "There may not have been an experience that demonstrates, for me or the country, more powerfully the need for electronic health records…than Katrina" (p. 1). The article also pointed out that the "federal government's goal is to give most Americans computerized medical records within 10 years"(p. 1).

E-health embraces a broad range of topics, such as telemedicine, medical-record databases, health information systems, genomics, biotechnology, drug-treatment technologies, decision-support systems, and diagnosis aids, just to name a few. In this chapter, we focus on the topic of technologies that deal with integrating mobile-based systems with healthcare databases.

One of the major challenges in healthcare database integration is the fact that the lack of guidance from central authorities has, in many instances, led to incompatible healthcare database systems. Such circumstances have caused problems to arise in the smooth processing of patients between health service units, even within the same health authority (Svensson, 2002). For instance, electronic health record (EHR) systems have been used in practice for many years. However, they are often designed and deployed by different vendors and, thus, patients' information is collected and stored in disparate databases. Due to the lack of uniformity, these systems have very poor interoperability. Even though the wide deployment of networks has enabled us to connect these databases, a large amount of work still needs to be handled manually in order to exchange information between the databases.F

There are two potential solutions to the problems of interoperability and automated information processing: redesigning and reimplementing the existing databases or using a database federation. Redesigning and reimplementing existing databases require large capital investments, and are difficult to achieve. An alternative solution is to build a database federation in which problems caused by database heterogeneity are remedied by the use of a mediator: metadata. This approach is often referred to as the multidatabase solution (Bright, Hurson, & Pakzad, 1994).

The Internet and the client-server-based computing paradigm have enabled us to access distributed information remotely, where the data servers act primarily as an information repository, the user's workstation bears the brunt of the processing responsibility, and the client and server communicate through a well-formulated network infrastructure. Recently, the surge of portable devices and the wide deployment of wireless networks have ushered a new era of mobile computing. Users access information via wireless media and from lightweight and less powerful portable devices. This paradigm shift permits the exchange of information in real time without barriers of physical locations. This is particularly helpful in situations where emergency medical teams need to access patients' information as soon as possible at a disaster site (Potok, Phillips, Pollock, & Loebl, 2003). However, mobile computing has also brought upon several technical challenges. First, unlike workstations, portable devices usually have limited CPU (central processing unit) processing capability and limited battery capacity. Second, low bandwidth and intermittent wireless network connections are often debilitating to client-server applications that depend on reliable network communications.

The mobile-agent-based distributed system design paradigm can address the aforementioned limitations. Unlike the client-server-based computational model, which moves data to computation, mobile agents move computation to data. This allows mobile users to take advantage of the more powerful servers on the wired networks. In addition, mobile agents are intelligent and independent entities that posses decision-making capabilities. Once dispatched, they are able to fulfill tasks without the intervention of the agent owner. Network connectivity is only required at the time of an agent's submission and retraction. Therefore, the use of mobile agents alleviates constraints such as connectivity, bandwidth, energy, and so forth.

We proposed and developed a prototype of a novel mobile-agent-based mobile data-access system (MAMDAS) for heterogeneous healthcare database integration and information retrieval (Jiao & Hurson, 2004). The system adopts the summary-schemas model (SSM; Bright et al., 1994) as its underlying multidatabase organization model. Queries are carried out by mobile agents on behalf of users. Via a medical thesaurus, created by combining the Medical Subject Headings (MeSH) thesaurus (Chevy, 2000) and an English-language thesaurus WordNet (Miller, Beckwith, Fellbaum, Gross, & Miller, 1990), MAMDAS supports imprecise queries and provides functions for user education.

The purpose of this chapter is to provide details about the tools we developed for disparate healthcare database management and their potential applications. The rest of the chapter is organized as follows. First it provides the background, and then presents the design, functions, application, and performance evaluation of MAMDAS and a medical thesaurus *MEDTHES*. Finally, we summarize this chapter and discuss future trends.

Background

In this section, we provide an overview of the current solutions to healthcare database management and introduce the two technologies on which we built our system: SSM and mobile-agent technology.

Healthcare Database Systems

The Veterans Health Administration (VHA) clinical information system began in 1982 as the Decentralized Hospital Computer Program (DHCP) and is now known as VistA (Veterans Health Information Systems and Technology Architecture; Hynes, Perrin, Rappaport, Stevens, & Demakis, 2004). VistA has evolved into a very rich healthcare information system that provides the information-technology framework for VHA's approximately 1,300 sites of care. VistA is built on a client-server architecture that ties together workstations and personal computers with nationally mandated and locally adapted software access methods. More specifically, VistA comprises more than 100 applications that clinicians access via the Computerized Patient Record System (CPRS) GUI (graphical user interface) to pull all the clinical information together from the underlying facility-based programming environment. CPRS provides a single interface for healthcare providers to review and update a patient's medical record. More than 100,000 VHA healthcare providers nationwide currently use CPRS. One important reason for VistA's success is the existence of a central authority. All VHA facilities are mandated to apply the same database-management system and unified access methods, which significantly eases the problem of interoperability among systems at different sites. Unfortunately, this uniformity is not a norm in today's healthcare databases. More often, we have to deal with heterogeneous databases that are designed and developed by different vendors.

IBM's DiscoverLink targets applications from the life-sciences industry (Hass et al., 2001). It is a fusion of two major components: Garlic (Carey et al., 1995) and DataJoiner (Cahmberlin, 1998). Garlic is a federated database-management system prototype developed by IBM Research to integrate heterogeneous data. DataJoiner is an IBM federated database-management product for relational data sources based on DATABASE 2 (Cahmberlin). It is a mediator system that limits itself to metadata exchange and leaves the data in their original databases and format. When an application submits a query to the DiscoveryLink server, the server identifies the relevant data sources and develops a query execution plan for obtaining the requested data. The server communicates with a data source by means of a wrapper, a software module tailored to a particular family of data sources. The wrapper is responsible for mapping the information stored by the data source into DiscoveryLink's relational data model, informing the server about the data source's query-processing capability, mapping the query fragments submitted to the wrapper into requests that can be processed using the native query language of the data source, and issuing query requests and returning results. Since data sources may take one of the many formats—relational database, object-oriented database, or flat files such as XML (extensible markup language) files and text files—a wrapper is needed for each format. Thus, wrapper development is the key to the extensibility in DiscoveryLink.

The TAMBIS (Transparent Access to Multiple Bioinformatics Information Sources) project (Stevens et al., 2000), as its name suggests, aims to provide transparent access to disparate biological databases. TAMBIS includes a knowledge base of biological terminology (the biological concept model), a model of the underlying data sources (the source model), and a knowledge-driven user interface. The concept model provides the user with the concepts necessary to construct multiple-source queries, and the source model provides a description of the underlying sources and mappings between the terms used in the sources and the terms defined in the concept model. In other words, TAMBIS utilizes a domain-specific ontology for heterogeneous data-source integration. It is a novel and valid approach. However, the depth and quality of the TAMBIS ontology are difficult to evaluate because the ontology contents are not publicly available.

The PQL query language proposed by Mork, Shaker, Halevy, and Tarczy-Homoch (2002) intends to integrate genetic data distributed across the Internet. It is essentially a query language for semistructured data. It relies on metadata describing the entities and the relationships between entities in a federated schema. These metadata appear to be created manually. While providing a new query language, this approach also raises questions about the accuracy of the metadata and extensibility of the system.

The Query Integration System (QIS) of Marenco, Wang, Shepherd, Miller, and Nadkami (2004) is a database mediator framework that addresses robust data integration from continuously changing heterogeneous data sources in the biosciences. The QIS architecture is based on a set of distributed network-based servers, data-source servers, integration servers, and ontology servers that exchange metadata as well as mappings of both metadata and data elements to elements in an ontology. Metadata version difference determination coupled with the decomposition of stored queries is used as the basis for partial query recovery when the schema of data sources alters. The principal theme of this research is handling schema evolution.

We developed a prototype of a mobile-agent-based mobile data-access system that deals with heterogeneous healthcare data-source integration and information retrieval (Jiao & Hurson, 2004). Our work differs from the previously mentioned research in several ways. First, MAMDAS utilizes the summary-schemas model for multidatabase organization (Bright et al., 1994). The hierarchical structure of SSM enables automated metadata population and improves search efficiency. Second, supporting user mobility is an emerging demand and it has not yet received enough attention in healthcare information-system research. We proposed to apply the mobile-agent technology to cope with this issue. Third, existing biomedical thesauri often demonstrate poor interoperability and reusability due to their nonstandard designs. We modified the MeSH thesaurus (Chevy, 2000) so that it complies with the ANSI/NISO (American National Standard Institute/National Information Standards Organization) Z39.19 monolingual thesaurus-creation standard (NISO, 1994). In addition, most biomedical thesauri and ontologies are tailored to the needs of medical professionals and, thus, nonprofessionals often find them hard to use due to the lack of precise knowledge. We addressed this problem by augmenting MeSH terms with synonyms defined by a general English-lexicon thesaurus WordNet (Miller et al., 1990). Finally, MAMDAS can be coupled with thesauri or ontologies of different domains to provide an information-system infrastructure for various applications with minimal modification. In the following subsections, we briefly discuss the background information pertinent to the development of MAMDAS.

The Summary-Schemas Model

The SSM consists of three major components: a thesaurus, local nodes, and summary-schemas nodes. Figure 1 depicts the structure of the SSM. The thesaurus defines a set of standard terms that can be recognized by the system, namely, global terms, and the categories they belong to. Each physical database (local nodes) may have its own dialect of those terms, called local terms. In order to share information among databases that speak in different dialects, each physical database maintains local-global schema metadata that map each local term into a global term in the format of "local term: global term." Global terms are related through synonym, hypernym, and hyponym links. The thesaurus also uses a semantic-distance metric (SDM) to provide a quantitative measurement of semantic similarity between terms.

Figure 1. A summary-schemas model with M local nodes and N levels

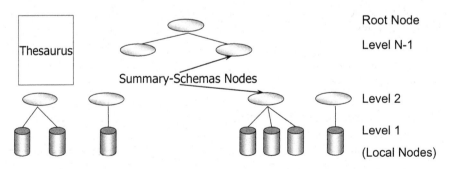

Figure 2. An example of the schema-summarization process

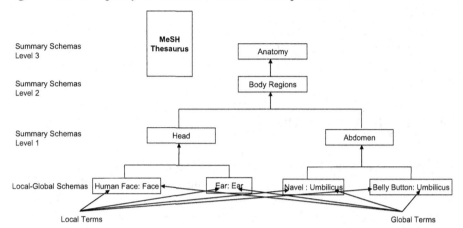

This feature allows for fine-grained semantic-based information retrieval.

The cylinders and the ovals in Figure 1 represent local nodes and summary-schemas nodes, respectively. A local node is a physical database containing real data. A summary-schemas node is a logical database that contains metadata called summary schema, which store global terms and lists of locations where each global term can be found. The summary schema represents the schemas of the summary-schema node's children in a more abstract manner; it contains the hypernyms of the input data. As a result, fewer terms are used to describe the information than the union of the terms in the input schemas.

Figure 2 shows an example of the automated schema-abstraction process of four local terms, *human face, ear, navel,* and *belly button,* under the guidance of the MeSH thesaurus in a bottom-up fashion. First, all local terms are mapped into global terms that are terms defined in MeSH. In the current prototype, this step is done by local database administrators manually. At SSM Level 1, the least common ancestors (immediate hypernyms) of the global terms are automatically identified by searching through the MeSH hierarchy: *Head* is the immediate hypernym of *face* and *ear.* Similarly, *abdomen* is the hypernym of *umbilicus.* At Summary Schemas Level 2, *head* and *abdomen* are further abstracted into *body regions.* Finally, at Level 3, *body region* is found to be a hyponym of 1 of the 15 categories defined in MeSH: *anatomy.*

The SSM is a tightly coupled federated database solution and the administrator is responsible for determining the logical structure of it. In other words, when a node joins or leaves the system, the administrator is notified and changes to the SSM are made accordingly. Note that once the logical structure is determined, the schema-population process is automated and does not require the administrator's attention.

The major contributions of the SSM include preservation of the local autonomy, high expandability and scalability, short response time, and the resolution of imprecise queries. Because of the unique advantages of the SSM, we chose it as our underlying multidatabase organization model.

The Mobile-Agent Technology

An agent is a computer program that acts autonomously on behalf of a person or organization (Lange & Oshima, 1998). A mobile agent is an agent that can move through the heterogeneous network autonomously, migrate from host to host, and interact with other agents (Gray, Kotz, Cybenko, & Rus, 2000). Agent-based distributed application design is gaining prevalence, not because it is an application-specific solution—any application can be realized as efficiently using a combination of traditional techniques. It is more because of the fact that it provides a single framework that allows a wide range of distributed applications to be implemented easily, efficiently, and robustly. Mobile agents have many advantageous properties (Lange & Oshima) and we only highlight some of them here:

- **Support disconnected operations:** Mobile agents can roam the network and fulfill their tasks without the owner's intervention. Thus, the owner only needs to maintain the physical connection during submission and retraction of the agent. This asset makes mobile agents desirable in the mobile computing environment where intermittent network connection is often inevitable.

- **Balance workload:** By migrating from the mobile device to the core network, the agents can take full advantage of the high bandwidth of the wired portion of the network and the high computation capability of servers and workstations. This feature enables mobile devices that have limited resources to provide functions beyond their original capability.

- **Reduce network traffic:** Mobile agents' migration capability allows them to handle tasks locally instead of passing messages between the involved databases. Therefore, fewer messages are needed in accomplishing a task. Consequently, this reduces the chance of message losses and the overhead of retransmission.

Contemporary mobile-agent system implementations fall into two main groups: Java-based and non-Java-based. We argue that Java-based agent systems are better in that the Java language's platform-independent features make it ideal for distributed application design. We chose the IBM Aglet Workbench SDK 2.0 (*IBM Aglets Workbench*, 1996) as the MAMDAS' implementation tool.

Design, Functions, Application, and Performance Evaluation of Mamdas and Medthes

Mobile-Agent-Based Mobile Data-Access System

MAMDAS consists of four major logical components: the host, the administrator, the thesaurus, and the user (Jiao & Hurson, 2004). Figure 3 illustrates the overall architecture of MAMDAS.

The MAMDAS can accommodate an arbitrary number of hosts. A HostMaster agent resides on each host. A host can maintain any number and any type of nodes (local nodes or summary-schemas nodes) based on its resource availability. Each NodeManager agent monitors and manipulates a node. The HostMaster agent is in charge of all the NodeManager agents on that host. Nodes are logically organized into a summary-schemas hierarchy. The system administrators have full control over the structure of the hierarchy. They can construct the structure by using the graphical tools provided by the AdminMaster agent.

In Figure 3, the solid lines depict a possible summary-schemas hierarchy with the darkened node as the root and the arrows indicating the hierarchical relation. The ThesMaster agent acts as a mediator between the thesaurus server and other agents. The dashed lines with arrows indicate the communication between the agents. The DataSearchMaster agent provides a query interface, the data-search window, to the user. It generates a DataSearchWorker agent for each query. The three dash-dot-dot lines depict the scenario that three DataSearchWorker agents are dispatched to different hosts and work concurrently.

Once the administrator decides the summary-schemas hierarchy, commands will be sent out to each involved NodeManager agent to build the structure. NodeManagers at the lower levels export their schemas to their parents. Parent nodes contact the thesaurus and generate an abstract version of their children's schemas. When this process reaches the root, the MAMDAS is ready to accept queries.

The user can start querying by launching the DataSearchMaster on his or her own device, which can be a computer attached to the network or a mobile device. The DataSearchMaster sends out two UserMessengers (not shown in the figure): one to the AdminMaster and one to the ThesMaster. The UserMessengers will return to the DataSearchMaster with the summary-schemas hierarchy and the category information. The DataSearchMaster then creates a data-search window that shows the user the summary-schemas hierarchy and the tree structure of the category. The user can enter the keyword, specify the preferred semantic distance, choose a category, and select a node to start the search. After the user clicks on the "Submit" button, the DataSearchMaster packs the inputs, creates a DataSearchWorker, and passes the

Figure 3. An overview of the MAMDAS system architecture

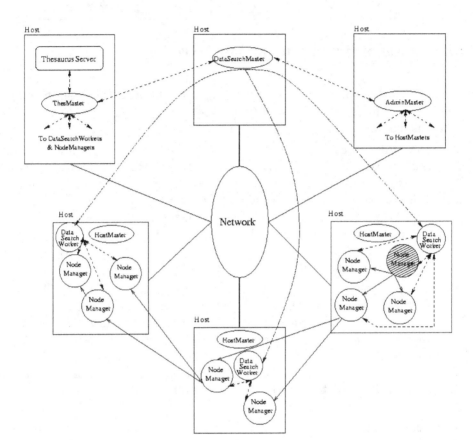

inputs to it as parameters. Since the DataSearchMaster creates a DataSearchWorker to handle each query, the user can submit multiple queries concurrently.

Once dispatched, the DataSearchWorker can intelligently and independently accomplish the search task by making local decisions without the owner's interference. During the query execution, the DataSearchWorker may generate DataSearchSlaves by cloning itself. The slaves can then work in parallel and report results to their creator. Figure 4 describes the search algorithm.

One of the major advantages of the MAMDAS framework is that it supports database heterogeneity and geographical distribution transparency. It provides the users with a uniform access interface. This property of MAMDAS significantly

Figure 4. The search algorithm

```
1      Set all child nodes to be unmarked;
2      WHILE (NOT (all term(s) are examined OR all
       child node(s) are marked))
           IF (term is of interest)
3              Mark all the child nodes that contain
       this term;
4          ELSE
5              CONTINUE;
6          END IF
7      END WHILE
8      IF (no marked child node)
           Go   to   the   parent   node   of
       the  current  node  and  repeat  the  search
           algorithm  (if  a  summary  schema  term
       of  the  parent  node  only  exists  on  the
       current node, we can skip this term);
9      ELSE
10         Create a DataSearchSlave for each marked
       child node;
11         Dispatch the slaves to the destinations
       and repeat the search algorithm;
12     END IF
13
```

eases the use of the system and makes it possible for users with limited computer skills to benefit from it.

A Medical Thesaurus: MEDTHES

The quality of the thesaurus is critical to the effectiveness of MAMDAS because it provides semantic-similarity measures to assist users in performing imprecise queries in which the query term is different than the indexing term of a document. The proliferation of biomedical research and the public demand of e-healthcare systems have stimulated the development of biomedical thesauri. Several examples include MeSH (Chevy, 2000), the Unified Medical Language System (UMLS; McCray & Nelson, 1995), and the Systematized Nomenclature of Medicine (SNOMED; Spackman, Campbell, & Cote, 1997). While the existing medical thesauri have helped immensely in information categorization, indexing, and retrieval, two major problems remain:

- Their designs do not follow any international or national thesaurus standard and therefore they could result in poor interoperability and reusability.
- They do not provide information regarding the semantic similarities among terms and, thus, the users are required to possess precise knowledge of the controlled vocabulary in order to make effective use of the thesaurus.

In order to alleviate these problems, we implemented a new medical thesaurus MEDTHES based on the medical thesaurus MeSH (Chevy, 2000) and the English-language thesaurus WordNet (Miller et al., 1990). It can be used as either a stand-alone thesaurus or an integral part of MAMDAS. In this subsection, we (a) briefly outline the ANSI/NISO standard for thesauri construction, (b) describe the two thesauri that have served as the foundation of MEDTHES, MeSH, and WordNet, (c) explain the concept of semantic similarity, (d) present the implementation of MEDTHES, (e) demonstrate the functions provided by MEDTHES as a stand-alone thesaurus, and (f) show the integration of MEDTHES with MAMDAS.

The ANSI/NISO Z39.19 Standard

The ANSI/NISO Z39.19 standard (NISO, 1994), entitled *American National Standard Guidelines for the Construction, Format, and Management of Monolingual Thesauri*, was developed by NISO and approved by ANSI. It provides guidelines for the design and use of thesauri, including rules for term selection, thesaurus structure, relation definitions, and thesaurus maintenance. Three types of semantic relationships between terms are distinguished in this standard: equivalence, hierarchical, and related. The equivalence relation establishes the link between synonyms, the hierarchical relationship provides links between terms that reflect general concepts (broader terms) and those that represent more specific information (narrower terms), and the related relationship exists among terms that have similar meanings or are often used in the same context but do not have hierarchical relationships. The design of MEDTHES follows this standard.

MeSH

The MeSH (Chevy, 2000) thesaurus is the standardized vocabulary developed by the National Library of Medicine for indexing, cataloging, and searching the medical literature. Currently, it contains approximately 22,000 terms (called descriptors) that describe the biomedical concepts used in health-related databases such as MED-LINE (*MEDLINE*, 2005), which is an online bibliographic database of medicine, nursing, health services, and so forth. All descriptors in MeSH are organized into 15 categories. Each category is then further divided into more specific subcategories. Within each category, descriptors are organized in a hierarchical fashion of up to 11 levels. In addition to the hierarchical structure, MeSH uses "Entry Term" or "See" references to indicate semantic relations such as synonyms, near synonyms, and related concepts of some terms.

Although MeSH is comprehensive and well maintained, it has several drawbacks. First, the synonymous relationship is not clearly listed and not differentiated from the related-term relation in MeSH. Second, the design of MeSH does not follow the ANSI thesaurus standard, which may result in poor interoperability and reusability. Third, MeSH is tailored to the needs of medical professionals. Nonprofessionals often find it hard to perform queries due to the lack of precise knowledge. For instance, a nonprofessional would use search terms such as *navel* and *belly button* instead of the official term *umbilicus* when submitting a query. Unfortunately, the query will fail because these terms are not defined in MeSH. We addressed this problem by augmenting MeSH with the well-defined synonyms found in WordNet, which we will discuss next.

WordNet

WordNet is an online thesaurus that models the lexical knowledge of the English language (Miller et al., 1990). It organizes English nouns, verbs, adjectives, and adverbs into synonym sets, called synsets. In other words, a synset is a list of synonymous terms. Each term in WordNet may have one or more meanings, and each meaning has a synset. Different synsets are connected through hierarchical relationships.

In summary, WordNet is comprehensive and designed with the goal to include every English word; it makes a number of fine-grained distinctions among word meanings. Thus, we decided to take advantage of the well-defined synonyms of WordNet and use them to complement the MeSH thesaurus.

Semantic Similarity

Synonyms and related terms obtained from a thesaurus are often used in query expansion for the purpose of improving the effectiveness of information retrieval (Shiri, Revie, & Chowdhury, 2002). However, in order to improve the quality of document ranking, a more fine-grained measure is needed to describe the degree of semantic similarity, or more generally, the relatedness between two lexically expressed concepts (Budanitsky & Hirst, 2001). Naturally, semantic distance is the inverse of semantic similarity. For example, the semantic distance between synonyms can be defined as zero, and that between antonyms can be defined as infinite.

If a thesaurus provides functions that calculate the semantic similarity between terms, the users can perform fine-tuned queries by limiting the scope of the search via the constraint of semantic distance between the keyword and the search results. The user can indicate how closely the returned terms should be related to the keyword (searched term) by selecting preferred semantic-distance values.

Two main categories of algorithms for computing the semantic distance between terms organized in a hierarchical structure (e.g., WordNet) have been proposed in the literature: distance-based approaches and information-content-based approaches. The general idea behind the distance-based algorithms (Leacock & Chodorow, 1998; Rada, Mili, Bicknell, & Blettner, 1989; Wu & Palmer, 1994) is to find the shortest path between two terms based on the number of edges, and then translate this distance into semantic distance. Information-content-based approaches (Jiang & Conrath, 1997; Rada et al.) are inspired by the perception that pairs of words that share many common contexts are semantically related. Thus, the basic idea of these methods is to quantify the frequency of the co-occurrences of words within various contexts.

In order to avoid the potential bias introduced by context selection, we chose to implement three distance-based algorithms in the MEDTHES prototype: the edge-counting algorithm (Rada et al., 1989), the Leacock and Chodorow (1998) algorithm, and the Wu and Palmer (1994) algorithm.

The Edge-Counting Algorithm

In the edge-counting algorithm, the semantic distance is defined as the number of edges (nodes) along the shortest path between any two terms.

The Leacock and Chodorow Algorithm

The relatedness measure proposed by Leacock and Chodorow (1998) also relies on the shortest path between two terms, t_1 and t_2. The relatedness between two terms, t_1 and t_2, is calculated as follows.

$$relatedness(t_1, t_2) = -\log \frac{len(t_1, t_2)}{2D} \tag{1}$$

where relatedness (t_1, t_2) is the similarity of terms t_1 and t_2, $len(t_1, t_2)$ is the length of the shortest path between two terms (using edge counting), and D is the maximum depth of the structure. Semantic distance is the inverse of relatedness (t_1, t_2), that is,

$$\frac{1}{relatedness(t_1, t_2)}.$$

The Wu and Palmer Algorithm

The Wu and Palmer (1994) algorithm uses the term *score* to define how two terms are related to each other. It measures the score by considering the depth of the two terms t_1 and t_2 in the tree structure, along with the depth of the LCA (least common ancestor). The formula used to calculate the score is shown in Equation 2.

$$score(t_1,t_2) = \frac{2*N_3}{N_1 + N_2 + 2*N_3} \tag{2}$$

where N_1 is the length of the shortest path from t_1 to the LCA, N_2 is the length of the shortest path from t_2 to the LCA, and N_3 is the length of the shortest path from the LCA to the root. The range of relatedness is $0 < score(t_1,t_2) <= 1$. The $score(t_1,t_2)$ is 1 if t_1 and t_2 are the same. Semantic distance is the inverse of $score(t_1,t_2)$, that is,

$$\frac{1}{score(t_1,t_2)}.$$

MEDTHES Design

The taxonomy defined in MeSH is the foundation of MEDTHES. However, several major changes to MeSH have been made: (a) the semantic relations of MeSH were reconstructed according to the ANSI standard, (b) the synonym set of each entry in MeSH was enriched by synonyms extracted from WordNet, and (c) three algorithms of semantic-distance calculation were implemented in order to provide users with fine-grained control over the query results.

MEDTHES adopts the three standard relationships suggested by the ANSI/NISO standard for thesaurus construction: the equivalence relationship, hierarchical relationship, and associative relationship. Terms in MeSH are arranged hierarchically in a tree structure—top down from general to more specific. The broader term (BT) and narrower term (NT) relations can be easily extracted from this hierarchical structure. A program, MeSHFileParser, was developed to automatically parse such information.

In MeSH, synonyms and related terms (RTs) are not clearly differentiated. The definitions of synonyms are neither accurate nor complete. As a result, MeSH is not suitable to be used directly to obtain synonyms. Since the well-defined synonyms are one of the major strengths of WordNet, it was used as a reference when adding synonyms to MEDTHES. A term is selected from the synonym set as the preferred term, which means that term is used for (UF) indexing other terms in the same set. The reverse relation is use, which means that if a keyword is a nonpreferred term, it is substituted with the preferred term before searching.

A term may exist in one or more categories in MeSH. In order to establish a link between a term and the category it belongs to, an additional relationship, subject categories (SCs), was also defined. Table 1 summarizes the relationships used in MEDTHES.

Table 1. Relationship definitions in MEDTHES

ANSI/NISO Relationship	MEDTHES Representation	Abbreviation
Equivalence	Use	USE
	Used For	UF
Hierarchical	Broader Term	BT
	Narrower Term	NT
Associative	Related Term	RT
	Subject Category	SC

Figure 5. Main menu

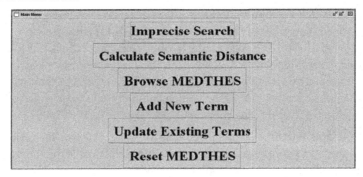

Main Functions of MEDTHES

As shown in Figure 5, functions featured in MEDTHES include carrying out an imprecise search, calculating semantic distance, browsing MEDTHES, adding new terms, updating existing terms, and resetting MEDTHES. In this subsection, we demonstrate the usage of each of these functions.

The imprecise-search function returns terms that are within a certain semantic distance to the keyword under a specific category. In the example shown in Figure 6, the category is Anatomy, the keyword is Brain, the semantic distance is 2.0, and the semantic-distance method selected is edge counting. Figure 7 shows the results and lists all the terms that have a semantic distance, with respect to the term Brain,

Figure 6. Imprecise search

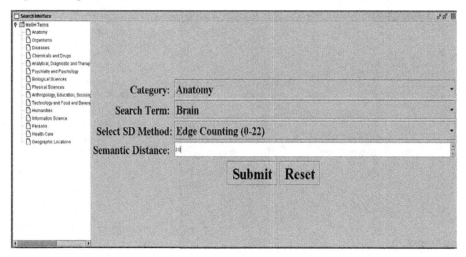

Figure 7. Result of the imprecise search

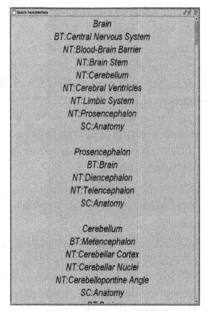

Figure 8. Calculate semantic distance

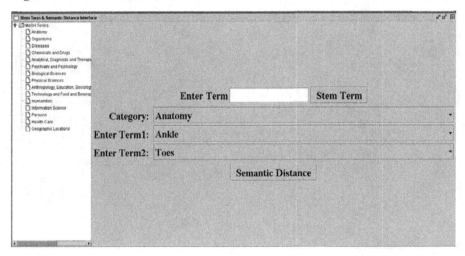

as less than or equal to 2.0 according to the edge-counting algorithm. The BT, NT, UF, and SC of these terms are also listed.

Calculating the semantic distance is the function that can be used to find the semantic distance between any two terms. It also provides the ability to compare the results generated by three different semantic-distance methods. Figure 8 shows an example demonstrating the semantic distances between two terms, Ankle and Toes, calculated by using different algorithms. The results are shown in Figure 9. This function endows the users with the opportunity of becoming familiar with the typical range of values of the various algorithms to aid in better selecting a good value when defining imprecise queries.

Browsing MEDTHES is the function that enables users to learn the content of MEDTHES including the tree structure and the alphabetical list of terms. Figure 10 shows a browser window.

Updating existing terms and adding new terms are functions that give users the freedom of customizing MEDTHES. Updating existing terms is a function that enables users to designate relationships among terms with their own knowledge instead of using the predefined interterm relationships. It helps users to modify their copies of MEDTHES according to their needs, such as changing the relationships among the terms and moving a term from one category to another.

Adding a new term is a function that can be used to add new terms to MEDTHES and establish relationships between the added terms and existing terms. A new term can be entered as the synonym of an existing term, in which case the new term is added

Figure 9. Results of semantic distances

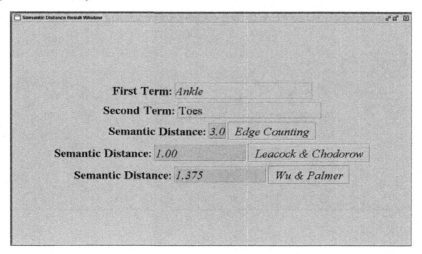

Figure 10. MEDTHES browser

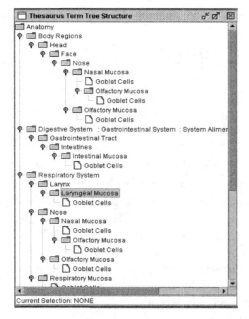

Figure 11. Add new term

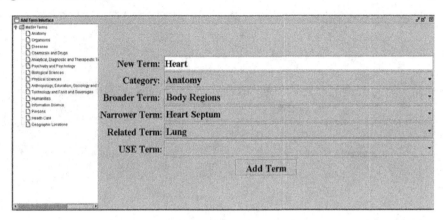

to the synonym set of the existing term, and all other relations, for example, BT, NT, and RT are automatically established. A new term and its relations can also be manually added, if desired. Figure 11 illustrates how to manually add a new term.

Integrate MEDTHES with MAMDAS

When MAMDAS is used in different application domains, the only modification required is to change the thesaurus in a plug-and-play fashion. In other words, we should choose a thesaurus that is the most appropriate for that domain. MAMDAS can work with any thesauri that follow the ANSI/NISO Z39.19 standard. In our study, we integrated MEDTHES with MAMDAS in order to resolve queries in the biomedical domain. Figure 12 shows an example of the MAMDAS data-search GUI.

The top-left window shows the content of MEDTHES, and the lower left window contains the current multidatabase hierarchy. The tree structure shown on the right hand side is an equivalent representation of the multidatabase hierarchy shown on the left. A query can be sent to any node in this hierarchy. In this example, the search keyword is umbilicus in the anatomy category, and the search starts at the root of the multidatabase hierarchy. In other words, every node (data source) in the hierarchy should be searched. The user wishes the search engine to return all terms that have a semantic distance of less than or equal to 3 with respect to the keyword according to the edge-counting algorithm. The search results are shown in Figure 13: Three terms, navel, umbilicus, and belly button, which satisfy the user-specified semantic distance, were found from three different data sources. Although the terms navel and belly button are not the exact lexical match of umbilicus, they are returned because they are defined as synonyms of umbilicus in MEDTHES. This example demonstrates that MEDTHES can be easily and successfully incorporated into MAMDAS and provides semantic-related information.

Performance Evaluation

The effectiveness of information retrieval is assessed by the ability of the system to retrieve relevant documents while at the same time suppressing the retrieval of irrelevant documents. Thesauri-aided query expansion is one frequently used technique in information retrieval that improves the effectiveness of retrieval by adding terms related to the search keyword. We evaluated the practicality and effectiveness of MEDTHES by using it as the reference thesaurus in several query-expansion experiments.

Experiment Setup

We chose to use the OHSUMED(Oregon Health Sciences University's MEDLINE data collection) test collection, a large interactive test collection for information-retrieval evaluation, created by Hersh, Buckley, Leone, and Hickman (1994) at the Oregon Health Sciences University. It is a subset of the MEDLINE medical-domain

Figure 12. MAMDAS data-search GUI

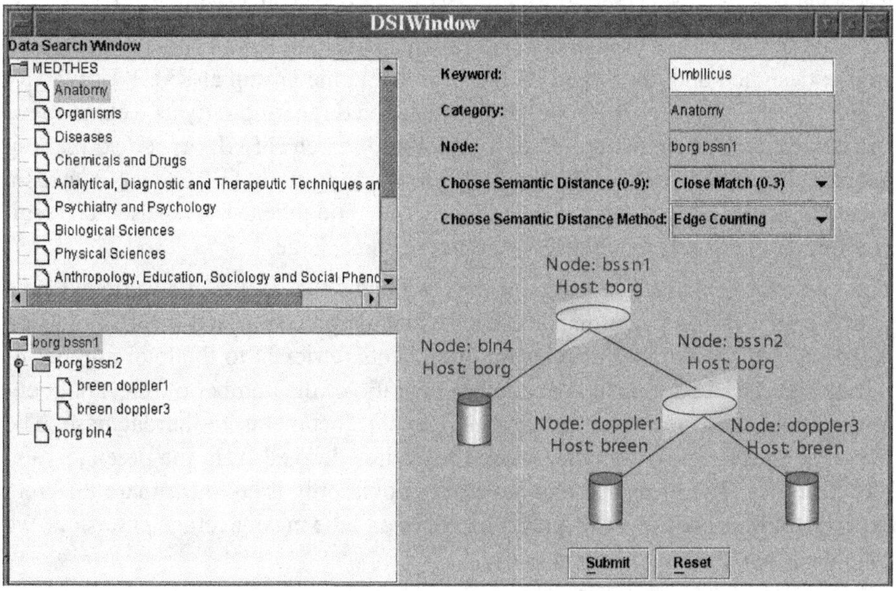

Figure 13. Search results

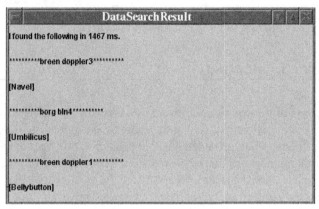

abstracts and consists of 348,566 articles derived from a subset of 270 medical journals over the period from 1987 to 1991. The documents in the test collection are manually indexed by professional indexers by using the MeSH thesaurus. A set of 106 queries and corresponding relevance judgments are provided. The queries were generated by actual physicians in the field of patient care and have at least one definitely relevant document. Each query contains a brief statement about the patient, followed by the actual information needed. OHSUMED has been widely used with many information-retrieval systems to evaluate the performance improvement of query expansion in the medical domain. It was also included in the TREC9 (Text Retrieval Conference, Track 9) Filtering Track (Robertson & Hull, 2001).

The data search engine used in our experiments is Zettair. It is a compact text search engine designed and developed by the Search Engine Group at RMIT University (*Zettair Search Engine*, 2003). It has been designed for simplicity as well as speed and flexibility. Users can use Zettair to index and search HTML (hypertext markup language) or TREC data collections. While working with TREC data, it takes a TREC topic file and an indexed data file as input, and it generates the search result in a format that can be analyzed by the trec-eval package.

The trec-eval package is a program that evaluates the documents retrieved by a search engine using performance metrics such as precision and recall. Precision is the ratio of the number of relevant documents retrieved to the total number of documents retrieved. Recall is defined as the ratio of the number of relevant documents retrieved to the total number of relevant documents in the database. The trec-eval package also contains several measures derived from the precision and recall metrics. The measures that have been commonly used to compare different experiment runs are the recall-precision curve and the mean average precision. We used these two measures in our study.

Figure 14. Experiment setup

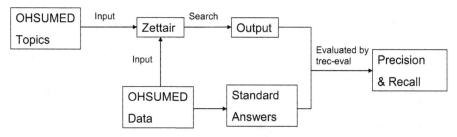

Four main components are included in the experiments (Figure 14): (a) OHSUMED test topics and query-expansion techniques, (b) the OHSUMED test data collection that provides a set of documents, a set of topics, and standard answers, (c) the Zettair search engine that serves as the full-text search and retrieval engine, and (d) the trec-eval package that assesses query results and calculates the precision and recall.

The Zettair search engine takes OHSUMED topics (queries), either expanded or without expansion, and documents from the OHSUMED test collection as input. There are four types of query expansion: synonym expansion (Run 2), narrower term expansion (Run 3), synonym and narrower term expansion (Run 4), and broader term expansion (Run 5). In our experiments, we used the synonyms, narrower terms, and broader terms defined in MEDTHES. The output from Zettair is compared with the standard answers provided by OHSUMED test collections by using the trec-eval package. Precision and recall are then generated automatically. Since the title portion of each topic resembles a typical user query, we used the

Table 2. Five experiment runs

Mode	Name	Description
Run 1	Baseline	Use initial queries without expansion.
Run 2	Synonym Expansion	Initial queries are extended with synonyms.
Run 3	Narrower Term Expansion	Initial queries are extended with narrower terms (one level).
Run 4	Synonym and Narrower Term Expansion	Initial queries are extended with a combination of synonyms and narrower terms (one level).
Run 5	Broader Term Expansion	Initial queries are extended with broader terms (one level).

Figure 15. Eleven-point interpolated precision and recall

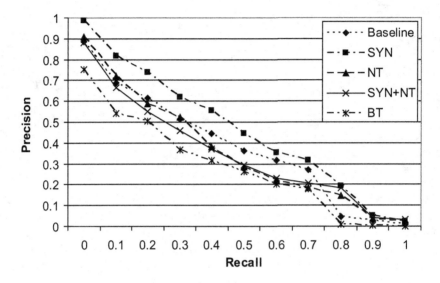

Figure 16. Average precision

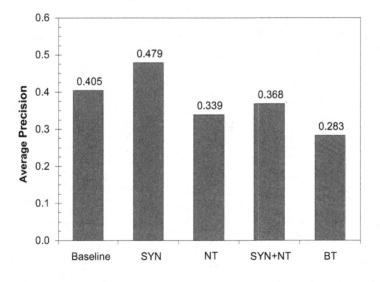

titles in all the experiments. Our experiments included five distinct modes of tests. Table 2 describes each of them.

Experimental Results and Analysis

Figure 15 plots the 11-point interpolated precision and recall curves and Figure 16 shows the average precision of each run. The results show that only synonym expansion (Run 2) can improve the performance. However, narrower term expansion (Run 3), synonym and narrower term expansion (Run 4), and broader term expansion (Run 5) worsens the performance compared to the baseline due to the introduced noise. The general trends demonstrated by our findings are consistent with the results reported by Hersh, Price, and Donohoe (2000), who used the same data collection. However, we observe two major differences. First, the synonym expansion was done manually in the research of Hersh et al., whereas it was automatically added by using WordNet as a reference in our experiments. Second, synonym expansion achieved good average precision: In our work, synonym expansion increases the average precision by 7.4%. In contrast, the results reported by Hersh et al. show that adding synonyms to the query actually degrades the average precision. This evidence has led us to conclude that by combining the strength of MeSH and WordNet, we can enhance the effectiveness of information retrieval without requiring experts' involvement in MeSH modifications.

Conclusion and Future Trends

This study provides an innovative solution, MAMDAS, to integrating a mobile-based system with healthcare databases. It utilizes the summary-schemas model to address the difficulties of heterogeneous data-source integration and exploits the unique characteristics of the mobile-agent paradigm to handle problems in a wireless computing environment.

In order to promote interoperability and reusability, improve the effectiveness of information retrieval, and accommodate clients who are not medical professionals, we redesigned the MeSH thesaurus in accordance to the ANSI standard and augmented it with synonyms from a well-known English-language thesaurus, WordNet. We named our implementation MEDTHES. In addition, we incorporated three semantic-distance calculation algorithms into MEDTHES in order to support imprecise queries. Other than providing the basic search function, like many other thesauri, MEDTHES also empowers the users with functionalities such as adding new terms and updating existing terms for thesaurus customization.

MEDTHES can be used as a stand-alone thesaurus or as an integral part of MAM-DAS. We demonstrate the integration of MEDTHES and MAMDAS via an example. We further quantitatively evaluated the performance of MEDTHES with regard to improving the effectiveness of information retrieval by using it as the reference for query expansion. Experimental results obtained from the standard biomedical test data collection, OHSUMED, show that by combining the strength of MeSH and WordNet, MEDTHES improves the precision of information retrieval by using query expansion with synonyms. However, the results also indicate that query expansion with broader or narrower terms may worsen the performance.

The MAMDAS framework can serve as the core infrastructure of many large, distributed healthcare applications where real-time access to heterogeneous data sources is required. However, we must note that high speed and high precision are not the sole requirements of e-health systems. Information security and user privacy also play an important role in the future of healthcare system development. For example, computer-based population or community health records usually use patient records anonymously. These systems are particularly valuable in public health where one is trying to trace different types of health hazards, linked either to medical, environmental, or social agents. There is certainly important ethical concern in relation to the composition of records and access to them.

The administrative simplification provisions of the Health Insurance Portability and Accountability Act of 1996 (HIPAA) require the Department of Health and Human Services (HHS) to establish national standards for electronic healthcare transactions and a national identifier for providers, health plans, and employers. It also addresses the security and privacy of health data. Adopting these standards will improve the efficiency and effectiveness of the national healthcare system by encouraging the widespread use of electronic data interchange in healthcare.

Traditionally, the AAA techniques are used to secure information systems: authentication, access control, and auditing. A user is authenticated before he or she is allowed to gain entrance to the system. The access-control unit of the system determines and enforces the access privileges associated with each user. System-access information is often kept in a log for further analysis, and this process is often referred to as auditing. As the information systems grow more complex, however, new problems that cannot be addressed by the traditional methods frequently emerge. Many of them are still open-ended questions. Breakthroughs in the security and privacy research field are crucial to the success of e-health systems and appropriate steps must be taken in search for solutions.

References

Associated Press. (2005). *Hurricane highlights need for digital records.* Retrieved September 13, 2005, from http://www.msnbc.msn.com/id/9316246/#storyContinued

Bright, M. W., Hurson, A. R., & Pakzad, S. H. (1994). Automated resolution of semantic heterogeneity in multidatabases. *ACM Transactions on Databases Systems, 19*(2), 212-253.

Budanitsky, A., & Hirst, G. (2001). Semantic distance in WordNet: An experimental, application-oriented evaluation of five measures. *Proceedings of Workshop on WordNet and Other Lexical Resources, Second Meeting of the North American Chapter of the Association for Computational Linguistics*, 29-34.

Carey, M. J., Haas, L. M., Schwarz, P. M., Arya, M., Cody, W. F., Fagin, R., et al. (1995). Towards heterogeneous multimedia information systems: The garlic appraoch. *In Proceedings of the Fifth International Workshop on research Issues in Data Engineering: Distributed Object Management*, (pp. 124-131).

Chamberlin, D. (1998). *A complete guide to DB2 universal database.* San Francisco: Morgan Kaufmann Publishers.

Chevy, C. (2000). Historical notes: Medical subject headings. *Bull Med Library Association, 88*(3), 265-266.

Fedyukin, I. V., Reviakin, Y. G., Orlov, O., Doarn, C. R., Harnett, B. M., & Merrell, R. C. (2002). Experience in the application of Java technologies in telemedicine. *eHealth International, 1*, 3.

Gray, R. S., Kotz, D., Cybenko, G., & Rus, D. (2002). Mobile agents: Motivations and state-of-the-art systems. Technical report: TR2000-365, Darthmouth University.

Hass, L. M., Schwarz, P. M., Kodali, P., Kotlar, E., Rice, J. E., & Swope, W. C. (2001). DiscoveryLink: A system for integrated access to life science data sources. *IBM Systems Journal, 40*(2), 489-511.

Hersh, W., Buckley, C., Leone, T., & Hickman, D. (1994). OHSUMED: An interactive retrieval evaluation and new large text collection for research. *Proceedings of SIGIR '94* (pp. 192-201).

Hersh, W., Price, S., & Donohoe, L. (2000). Assessing thesaurus-based query expansion using the UMLS metathesaurus. *Proceedings of the AMIA Symposium* (pp. 344-348).

Hynes, D. M., Perrin, R. A., Rappaport, S., Stevens, J. M., & Demakis, J. G. (2004). Information resources to support healthcare quality improvement in the veterans health administration. *Journal of American Medical Informatics Association, 11*(5), 344-350.

IBM Aglets Workbench. (1996). Retrieved January 5th, 2005, from http://www.trl. ibm.co.jp/aglets/index.html

Jiang, J., & Conrath, D. (1997). Semantic similarity based on corpus statistics and lexical taxonomy. *Proceedings of International Conference on Research in Computational Linguistics* (pp. 19-33).

Jiao, Y., & Hurson, A. R. (2004). Application of mobile agents in mobile data access systems: A prototype. *Journal of Database Management, 15*(4), 1-24.

Lange, D., & Oshima, M. (1998). *Programming and developing Java mobile agents with aglets.* Reading, MA: Addison Wesley Longman, Inc.

Leacock, C., & Chodorow, M. (1998). Combining local context and WordNet similarity for word sense identification. In C. Fellbaum (Ed.), *WordNet: A lexical reference system and its application.* Cambridge, MA: MIT Press.

Marenco, L., Wang, T. Y., Shepherd G., Miller, P. L., & Nadkarni, P. (2004). QIS: A framework for biomedical database federation. *Journal of American Medical Informatics Association, 11*(6), 523-534.

McCray, A. T., & Nelson, S. J. (1995). The representation of meaning in the UMLS. *Methods of Information in Medicine, 34*, 193-201.

MEDLINE. (2005). Retrieved January 5, 2005, from http://www.nlm.nih.gov/pubs/ factsheets/medline.html

Miller, G. A., Beckwith, R. T., Fellbaum, C. D., Gross, D., & Miller, K. J. (1990). WordNet: An on-line lexical database. *International Journal of Lexicography, 3*(4), 235-244.

Mork, P., Shaker, R., Halevy, A., & Tarczy-Hornoch, P. (2002). PQL: A declarative query language over dynamic biological schemata. *Proceedings of the AMIA Fall Symposium* (pp. 533-537).

National Information Standards Organization (NISO). (1994). *National Information Standards Institute American national standard guidelines for the construction, format, and management of monolingual thesauri.* Bethesda, MD: NISO Press.

Potok, T., Phillips, L., Pollock, R., & Loebl, A. (2003). Suitability of agent technology for military command and control in the future combat system environment. *Proceedings of the 8th International Command and Control Research and Technology Symposium* (pp. 1-20).

President's Information Technology Advisory Committee. (2001). Transforming healthcare through information technology. In *Panel on transforming healthcare.* Arlington, VA: National Coordinating Office for Information Technology Research and Development (pp. 1-17).

Rada, R., Mili, H., Bicknell, E., & Blettner, M. (1989). Development and application of a metric on semantic nets. *IEEE Transactions on Systems, Man, and Cybernetics, 19*(1), 17-30.

Robertson, S., & Hull, D. A. (2001). The TREC-9 filtering track final report. *The Ninth Text Retrieval Conference (TREC-9)*, 25-40.

Shiri, A. A., Revie, C., & Chowdhury, G. (2002). Thesaurus-assisted search term selection and query expansion: A review of user-centered studies. *Knowledge Organization, 29*(1), 1-19.

Spackman, K. A., Campbell, K. E., & Cote, R. A. (1997). SNOMED RT: A reference terminology for healthcare. *Proceedings of AMIA Annual Fall Symposium* (pp. 640-644).

Stevens, R., Baker, P., Bechhofer, S., Ng, G., Jacoby, A., Paton, N. W., et al. (2000). TAMBIS: Transparent access to multiple bioinformatics information sources. *Bioinformatic, 16*(2), 184-186.

Svensson, P. (2002). eHealth application in healthcare management. *eHealth International, 1*, 1.

Wu, Z., & Palmer, M. (1994). Verb semantics and lexical selection. *Proceedings of the 32nd Annual Meeting of the Association for Computational Linguistics* (pp. 133-138).

Zettair Search Engine. (2003). Retrieved January 5, 2005, from http://www.seg.rmit.edu.au/zettair

Section IV

Patient Empowerment

Chapter X

Utilizing Mobile Phones as Patient Terminal in Managing Chronic Diseases

Alexander Kollmann, ARC Seibersdorf Research GmbH, Austria

Peter Kastner, ARC Seibersdorf Research GmbH, Austria

Guenter Schreier, ARC Seibersdorf Research GmbH, Austria

Abstract

Mobile information and communication technologies are advancing rapidly and provide great opportunities for home monitoring applications in particular for outpatients and patients suffering from chronic diseases. Because of the ubiquitous availability of mobile phones, these devices can be considered as patient terminals of choice to provide a telemedical interaction between patients and caregivers. The most challenging part still is the patient terminal, that is, to offer the user a method to enter measured data into a system as well as to receive feedback in a comfortable way. The objective of this chapter is to present and compare solutions for mobile-phone-based patient terminals as developed by us and other authors.

Introduction

"Telemonitoring is defined as the use of audio, video, and other telecommunications and electronic information processing technologies to monitor patient status at a distance" (Field, 1996, p. 271). This concept may be particularly suitable in the management of chronic diseases where a close partnership as well as collaboration between patient and healthcare provider are essential. New paradigms such as prevention and patient empowerment promote the development of novel care approaches in which outpatient monitoring is a basic aspect.

Rapid advancements of information and communication technologies and the increasing availability of mobile phones open new perspectives in using these devices for tele-monitoring applications to deliver healthcare to people geographically remote from physicians or medical centers. The possibility to use the mobile phone for standard voice communication as well as for the transmission of a variety of multimedia information like text, audio, images, and videos makes it the communication interface of choice for patient-centered tele-monitoring applications.

The basic idea is to track patients' personal health status using the mobile phone as a patient terminal and to send the data to a remote monitoring centre. An automated monitoring process checks the values and gives feedback in order to guide the patient through the self-managing process and to turn the doctor's or other caregiver's attention to the patient when necessary by means of notifications and alerts. The most challenging part in this scenario still is the patient terminal, that is, to offer the user an easy method to enter measured data into a system as well as to get feedback of the current health status in a comfortable way.

This chapter will focus on the usage of mobile phones in the management of chronic diseases and gives an overview of available technologies. Furthermore, it will present and compare already implemented mobile-phone-based home monitoring concepts as developed by us and other authors.

Brief Historical Outline

Basically, tele-monitoring combines topics from the fields of medicine, information and communication technology, and computer science. Particularly, information and communication technologies have undergone rapid advancements over the past decades, driven by the needs of modern information society. Communication devices such as mobile phones or personal digital assistants (PDAs) became smaller and more powerful, and advanced from single-purpose stand-alone devices to multipurpose networked devices that make them usable for tele-monitoring applications indeed.

However, reviewing the literature, the exact date when tele-monitoring was first mentioned in healthcare is still unknown (Brown, 1995). Starting with the first words transmitted by telephone in 1876 by Alexander Graham Bell, communication technology was ready to be used to facilitate healthcare services. For example, William Einthoven, the father of electrocardiography (ECG), transmitted ECG signals over wired telephone lines in 1906 (Barold, 2003).

In the 1930s, when the telephone became standard equipment in households, it also became the mainstay of medical communication and remained a major element until today. Wireless communication technologies were invented at the same time. Around the time of World War I, radio communication was established in a wider area, and, around 1930, it was used in remote areas such as Alaska and Australia to transfer medical information (Zundel, 1996).

Besides pioneering efforts of a few physicians using off-the-shelf commercial equipment to overcome time and distance barriers, current tele-monitoring concepts originated from developments in the manned space-flight program introduced by the National Aeronautics and Space Administration (NASA) in the early 1960s (Brown, 1995). The main intention was to monitor physiological parameters like heart rate, body temperature, ECG, and oxygen and carbon-dioxide concentration of astronauts in space and transmit the data to earth in order to establish an understanding of the health and well-being of the astronauts while they were in orbit.

Nowadays, tele-monitoring is adjudicated an important role in health systems since for a number of indications the cost effectiveness and the medical benefits have been approved (Meystre, 2005). Moreover, there is still a driving force to improve outpatient care by shifting tasks from hospitals to patients' homes, particularly in the management of chronic conditions, which will be discussed in detail in this chapter.

Managing Chronic Diseases

Chronic diseases are cited in literature as diseases that have one or more of the following characteristics: They are permanent, leave residual disability, are caused by nonreversible pathological alteration, require special training of the patient for rehabilitation, or may be expected to require a long period of supervision, observation, or care (National Library of Medicine, 2005).

Chronic diseases such as heart failure, hypertension, cancer, diabetes, and asthma were the major cause of death (59% of the 57 million deaths annually) and global burden of disease (46%) in 2004 (World Health Organization [WHO], 2005).

It is also obvious that chronic diseases tend to become more common with age. WHO reported that in 1998, 88% of the population in developed countries over

Figure 1. Three phases of monitoring (pretreatment, adoption, long-term treatment)

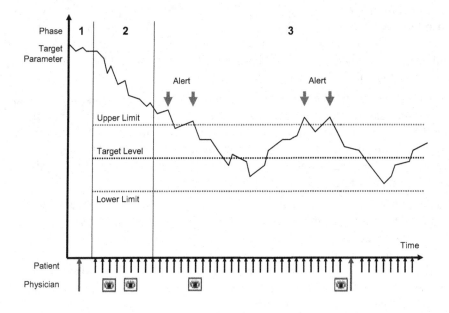

65 years old suffered from at least one chronic health condition. Since populations are aging worldwide and chronic diseases have a significant impact on healthcare systems, new strategies in prevention, early detection of illness pattern, and long-term treatment are needed.

To address the quality and effectiveness of healthcare services for chronic-disease management, the chronic care model was developed at the MacColl Institute for Healthcare Innovation in 1998 (Wagner, 1998). The model describes the transformation of healthcare from a system that is essentially reactive, responding mainly when a person is sick, to one that is proactive and focused on keeping a person as healthy as possible. Besides strategic and lasting changes in health systems, the model suggests enforcing patients' central role in their care and self-management (Epping-Jordan, Pruitt, Bengoa, & Wagner, 2004).

Effective self-management support means more than just telling patients what to do. Because of the fact that neither the chronic condition nor its consequences are static, the process of self-management might be complex and patients are often overstrained. Thus, a close partnership between patients and caregivers is essential in order to guide the patient through the self-management process and to the best possible health status.

It is widely recognized that the monitoring of health-related parameters like blood pressure, blood glucose, or well-being in regular intervals is the central element in the strategy of an effective self-management of chronic conditions. Basically, monitoring is defined as periodic measurements that guide the management of a chronic or recurrent condition. It can have an impact on the improvement of therapy adherence, the better selection of treatments based on individual response, or the better titration of medication (Glasziou, Irwig, & Mant, 2005).

Depending on the target parameter and the phase of treatment, monitoring can be done by clinicians, patients, or both. The strategy of monitoring can be divided mainly into three phases shown in Figure 1. Blue arrows indicate visits at the physician's office. The eye icon indicates tele-monitoring activity. Short, black arrows show the patient's self-monitoring activity.

- **Pretreatment.** Monitoring before treatment is mostly performed at the physician's office by the physician. The main goal is to access the patient's current health status and to verify the need for a medical intervention. Beside standard measurements like the assessment of blood pressure, more complex diagnostic tests like blood tests or urine tests are performed to confirm decisions. At that stage, a person has to be accustomed to his or her new role as a patient in the patient-physician relation. Moreover, objectives of treatment, monitoring parameters, target levels, and their upper and lower limits should be well defined.

- **Adoption.** After establishing the target level, application of medication will be started in order to reach the objectives of treatment. Monitoring disease-related parameters, for example, blood pressure in case of hypertension, will show the effect of treatment in a representative way. In case of the necessity to titrate medication, monitoring has to be done very carefully and in shorter intervals to avoid abnormalities or the worsening of the health status.

- **Long-term treatment.** Although the target level has been reached periodically, monitoring is essential because the course of a chronic condition is rarely static. Overstepping or undershooting the predefined target level should be observed carefully. Additionally, alerts could be generated in order to turn the patient's and physician's attention to this special situation. Periods of problems can easily be identified via graphical representation. If noticed problems cannot be solved via tele-medical intervention, the patient has to be ordered to the physician's office for further examination. However, the main objective of tele-monitoring is to detect illness patterns at the earliest possible stage in order to avoid emergency situations and hospitalization.

Today, patients are asked to track their key measures like blood pressure, heart rate, diabetes-relevant data, well-being, or side effects of medication by daily taking notes on a piece of paper, called a health-data diary. The captured data are expected to show trends in the illness patterns and to help the doctor to guide the patient to the best possible health status.

However, patients' motivation for using the conventional method in self-management is often poor. This is hardly surprising since patients are often confronted with complex documents and instructions. On the other hand, paper-based diaries lack proper data representation, feedback, and timely delivery of data. Therefore, an easy-to-use and patient-centered data-acquisition system is essential to guide the patient through data capturing and the complex process of self-management.

Using Information and Communication Technologies in Home Monitoring Environment

The effective management of chronic diseases requires a close partnership between the patient and healthcare provider, which can be supported by contemporary information and communication technologies (Celler, Lovell, & Basilakis, 2003). The timely delivery of data is indispensable to detect an aggravation in illness patterns and to ensure appropriate medical decisions at the earliest possible stage.

The basic idea is to track the patient's personal health status using a patient terminal, and to transmit the data to a remote monitoring centre where the data are processed and trends, statistics, and graphical representations are generated. An automated monitoring process checks the values and gives feedback. When necessary, the

Figure 2. Main components of a home monitoring platform

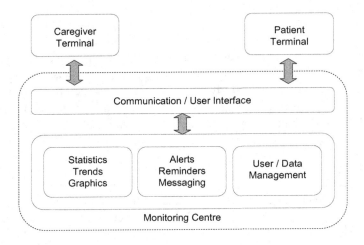

doctor's or other caregivers' attention will be turned to the patient by the means of notifications or alerts.

In our research, we developed a home monitoring platform for regular and home-based measurement and transmission of health parameters like blood pressure, body weight, symptoms, and medication. The system, shown in Figure 2, has been built using mostly standard components and state-of-the-art Internet technology, and it comprises the following.

Monitoring Centre

A 24/7-accessible server system receives, stores, and processes the data. The results are forwarded to the caregivers in a standardized and easily comprehensible format. Central components are user and data management to ensure the security, integrity, and traceability of data. Role-based, hierarchical user management guarantees that only authorized users are able to view, edit, or enter data in the system.

Patient Terminal

 The patient terminal provides the user with an adequate communication interface for entering data into the system as well as to receive feedback or medical advice. Although the availability of PCs (personal computers) with Internet access is already high in developed countries, the use of computers still presents a barrier to the adoption of Web-based solutions, especially by the elderly and technically unskilled people. Hence, different solutions have been developed during the last years to give such users an easy-to-use and intuitive way for entering their health data comfortably using manual, automatic, or semiautomatic methods in conjunction with standard medical measurement devices.

Caregiver Terminal

The Web-based caregiver terminal gives authorized people access to assigned patients and their data via PC or mobile Internet-enabled devices like PDAs or tablet PCs. Specially designed GUI (graphical user interface) components give the physician a quick overview and support straightforward navigation. Transmitted values as well as statistics, trends, and graphical representations can be accessed easily.

Statistics, Trends, and Graphics

An automated process analyses incoming data and generates statistics, trends, and graphical representation. Additionally, the process is able to check for alert conditions (limits given by the responsible physician) or to correlate the values with predefined patterns to identify abnormalities in the course of treatment.

Messaging System

In case of detected aberrance, notification messages will be sent to the physician or the patient via SMS (short-message service) or e-mail immediately in order to initiate further action. Additionally, it turned out that reminder messages are useful for patients and caregivers. The possibility to send interactive messages in several ways enforces communication between patients and caregivers.

Communication and User Interface

Besides the possibility to access or enter data via a Web interface, interfaces to several wireless communication technologies such as WAP (wireless application protocol), SMS, and MMS (multimedia messaging service) are provided in order to send or receive data.

Going Wireless

A current trend in information and communication technologies is the convergence of wireless communication and computer networks as well as a moving from stand–alone, single-purpose devices to multipurpose network devices. Up to now, several studies have demonstrated that tele-monitoring using information and communication technologies on different integration levels - reaching from a simple telephone call to wearable or implantable sensor systems - can effectively assist patients in the management of care. Although the use of older approaches (telephone, fax) is still common, latest innovations in computer and network technologies can be considered for tele-monitoring applications to support the patient with a method to enter and transmit data in a comfortable way.

The Telephone as Communication Interface

Telephone care services represent the oldest method to deliver medical advice to the patients' homes. Nowadays, standard telephone lines are available in almost every household and even technically unskilled people are able to handle the telephone. For monitoring purposes, patients are contacted by a call centre to hand over their key measurements in predefined follow-up intervals. Specially trained personnel check the values against the patient's history and predefined limits and give feedback in order to assist the patient in self-management.

The usage of telephone interventions in the management of outpatients suffering from chronic heart failure was demonstrated in the DIAL(randomised trial of telephone intervention in chronic heart failure) trial (GESICA, 2005): 1,518 patients with stable chronic heart failure and optimal treatment were enrolled in 51 centers in Argentina. The DIAL trial intervention strategy was based on frequent telephone follow-ups provided by nurses trained in heart failure. The purpose of interventions was mainly to educate and monitor the patients as well as to increase the patients' adherence to diet and drug treatments. The results indicate that patients in the usual-care group were more likely to be admitted to the hospital for reasons of worsening and they were also more likely to die than patients in the tele-monitoring group who received telephone intervention. Moreover, patients in the intervention group showed a better quality-of-life score than patients randomized to the usual-care group.

Besides the possibility to use plain old telephone systems (POTS) for voice communication, they can also be used to access the Internet and network-based data services. Medical measurement devices like blood-pressure meters, blood-glucose meters, or scales equipped with modems can be connected to the POTS directly or by the means of a home terminal.

The Trans-European Network Home-Care Management Systems study (TEN-HMS) was the first large-scale, randomized, prospective clinical trial to decide if home-based tele-monitoring services for heart-failure patients are able to reduce hospitalizations and to improve patient well-being while reducing the overall costs of care (Cleland, Louis, Rigby, Janssens, Balk, & TEN-HMS, 2005). In total, 426 patients were randomized to the control group, nurse telephone-support group, or tele-monitoring group. The results indicate that patients randomized to the tele-monitoring group faced a reduced number of days spent in the hospital (minus 26%). Furthermore, it led to an overall cost saving compared to the nurse telephone-support group (minus 10%). Tele-monitoring also significantly improved survival rates relative to the usual-care group and led to high levels of patient satisfaction.

Mobile Phone as Communication Interface

Recent statistics indicate that the number of mobile phones throughout the world exceeded 1.5 billion in 2004 (CellularOnline, n.d.). Moreover, it is estimated that within a few years, about 70% of cell phones in the developed countries will have Internet access. Thus, mobile phones have become potential devices for serving as patient terminals in tele-monitoring applications. Mobile phones as well as communication technologies have undergone incredible changes and advancements during the last years. The amazing employment of mobile phones started in the 1980s when first-generation (1G) cellular systems were introduced. This technology was based on analog circuit-switched technology. Low data rates prevented this technology from being used for data transfer. Up to now, most of the 1G networks have been replaced by second-generation (2G) wireless networks, which are based on digital circuit-switched technologies. Several standards have been developed in different parts of the world: the Global System for Mobile Communication (GSM) technology in Europe, code division multiple access (CDMA) technology in the USA, and personal digital communication (PDC) in Japan. Second-generation wireless networks are digital and expand the range of applications to more advanced voice services and data capabilities such as fax and SMS at a data rate up to 9.6 kbps, which still makes it mostly impractical for extensive Web browsing and multimedia applications.

Through the years, several advanced techniques based on 2G networks were introduced such as general packet radio service (GPRS) and enhanced data rates for global evolution (EDGE). These technologies, also known as 2.5G networks, make it possible to use several time slots simultaneously when sending or receiving data, resulting in a significantly increased data rate (171 kbps for GPRS, 384 kbps for EDGE). The data packages are sent over the network using an IP (Internet protocol) backbone so that mobile users can assess services on the Internet.

The Universal Mobile Telecommunications System (UMTS) presents the third generation (3G) of wireless communication technology. The broadband, packet-switched transmission concept supports the transmission of text, digitized voice, video, and multimedia at data rates up to 2 Mbps. At the moment, users in real networks can expect performances up to 384 kbps for downloading and at least 64 kbps for uploading data. Third-generation systems are expected to have the following features: fixed- and variable-rate bit traffic, bandwidth on demand, asymmetric data rates in the forward and reverse links, multimedia mail storage and forwarding, the capability to determine the geographic position of mobile units and report it to both the network and the mobile terminal, and international interoperability and roaming (Tachakra, Wang, Istepanian, & Song, 2003).

In the course of rapid advances in communication technology and increased data rates, new multimedia services appeared that are the basis for further developments of mobile-phone-based applications for healthcare.

Short-Message Service

SMS allows sending and receiving text messages of up to 160 characters in length to and from mobile phones as defined within the GSM digital mobile-phone standard (Buckingham, 2000a). The text can comprise words or numbers or an alphanumeric combination. SMS is a store and forward service. This means that SMS messages are not sent directly from sender to recipient. Each mobile telephone network that supports SMS has at least one messaging centre to handle and manage the short messages.

To receive the text message on the server side, a mobile phone is connected to the server via standard interfaces such as RS232. Using the Hayes command set (Attention (AT) commands), the arrived messages can be accessed and stored in the database for further processing.

There are several reports on the use of SMS in medical application areas. SMS has been basically used for patient and appointment reminders. Outpatient clinics that are using SMS-based appointment reminder systems are seeing a reduction in missed appointments or "did not attends" (DNAs). It has been demonstrated that outpatient clinics that deployed SMS patient reminder systems saw DNA rates fall by as much as 30% even though less than 20% of patients chose to use the service (Research and Markets, 2005).

Besides the usage of SMS as a reminder service, SMS has also been applied in several medical application areas to transmit data from patients' homes to a monitoring centre. In 2004, Ferrer-Roca, Cárdenas, Diaz-Cardama, and Pulido presented the use of SMS for diabetes management. Twenty-three diabetic patients (18 to 75 years old) were asked to transmit data such as blood-glucose levels and body weight to a central server. The server automatically answered via SMS with a prerecorded acknowledgement, specific help, or warning messages when data were out of range. During an 8-month study period, an average of 33 SMS messages was sent per patient and month. Unfortunately, no medical benefit was reported. However, they conclude that SMS may provide a simple, fast, efficient, and low-cost adjunct to the management of diabetes at a distance.

On the other hand, medical benefit in asthma monitoring could be demonstrated by Ostojic, Cvoriscec, Ostojic, Reznikoff, Stipic-Markovic, and Tudjman (2005). Sixteen patients (24.6 +/- 6.5 years old) were asked to transfer peak expiratory flow (PEF) measurements at least three times a day during a 16-week study period. Patients randomized to the tele-monitoring group received a weekly adjustment by an asthma specialist based on the values sent to the monitoring centre. They reported that asthma overall was better controlled in the intervention group, according to their findings of reduced PEF variability.

Figure 3. WAP programming architecture. The application is stored on the content server and the requested pages are generated dynamically

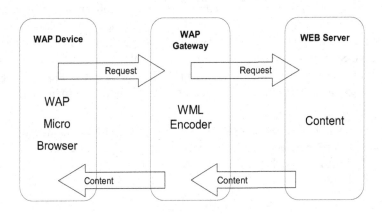

Wireless Application Protocol

WAP is an open global standard for communication between mobile phones or other mobile devices and the Internet (Buckingham, 2000b). WAP-based technology enables the design of advanced, interactive, and online mobile services, such as mobile banking, Internet-based news services, or even tele-monitoring applications.

The WAP standard is based on Internet standards like the hypertext markup language (HTML), extensible markup language (XML), and transmission control protocol/ Internet protocol (TCP/IP). Basically, it consists of the wireless markup language (WML) specification, which is a markup language derived from HTML. However, WML is strongly based on XML, so it is much stricter than HTML. WML is used to create Web pages including text, images, user input, and navigation mechanisms that can be displayed on a WAP micro browser.

The WAP architecture consists of a WAP device, a WAP gateway, and a content server (Figure 3). The handheld WAP device communicates with the content server, which stores information and responds to user requests. The gateway in between translates and passes information between the device and the server. To access an application stored at the content server, a connection to the WAP gateway is initialized. Thereafter, the WAP request is converted to HTTP (hypertext transfer protocol) and forwarded to the content server. Upon handling the request, the requested content is returned to the gateway, transformed into WAP, and sent back to the device to be displayed via the micro browser. The new version of WAP, WAP 2.0, is a reengineering of WAP using a cut-down version of the extensible hypertext markup language (XHTML) with end-to-end HTTP connection. This means that using WAP 2.0, the WAP gateway becomes dispensable.

A couple of authors (Hung & Zhang, 2003; Salvador et al., 2005) have already demonstrated tele-monitoring applications using WAP technology after WAP browsers became a standard feature of mobile phones. It is mentionable that besides demonstrating principal functionality in several pilot trials, only few clinical trials using WAP for patient monitoring have been mentioned in the literature yet.

For example, Italian researchers have shown that aftercare and patient communication might be improved if patients are asked to fill in daily questionnaires using WAP on their mobile phones (Bielli, Carminati, La Capra, Lina, Brunelli, & Tamburini, 2004). They developed the Wireless Health Outcomes Monitoring System (WHOMS), which allows structured questionnaires to be sent to the patient by the medical management team. Each day, an SMS message was sent informing patients of the survey and giving a link to a WAP site accessible through a standard GPRS connection. Users were asked to rate symptoms such as pain, lack of energy, and difficulty sleeping. The collected data were viewable by the doctors in a graphical format that highlighted the patients' states of health.

Although 42% of the patients failed to fill in the questionnaires mainly because of neophobia and unfamiliarity with the technology, the researchers concluded that health-outcome monitoring using mobile phones can be the method of choice for future developments in quality-of-life assessments. They address the need to develop more user-friendly communication terminals supported by upcoming technologies for mobile phones to increase the adoption rate.

In the course of our research, a lot of effort has been done to make WAP technology applicable for the broad usage in home monitoring applications and the management of chronic diseases.

The aim of the Cardio-Memory study (Scherr, Zweiker, Kollman, Kastner, Schreier, & Fruhwald, in press) was to evaluate whether WAP technology would be an acceptable, feasible, and reliable option to provide tele-monitoring for patients with chronic heart failure or hypertension. In the course of a clinical pilot trial, 20 patients (mean age 50 +/- 14 years) were enrolled. Each participant was equipped with a mobile phone, an automatic blood-pressure device, and a digital weight scale. Patients were asked to measure their blood pressure, heart rate, and weight every day. After accessing the system with the micro browser, the menu promptly routed the patients through subsequent entry templates generated by WML syntax (Figure 4). Eventually, all data were sent to the central database at the remote monitoring centre for further processing.

Authorized physicians could access the data via a secure Web site at any time. The Web site provided the data of each patient in numerical format and graphical trend charts, including patient-specific upper and lower parameter limits. Furthermore, it allowed physicians to set automatic reminders: These computer-generated SMS

Figure 4. Graphical user interface generated by WML guides the patient through the data-acquisition process

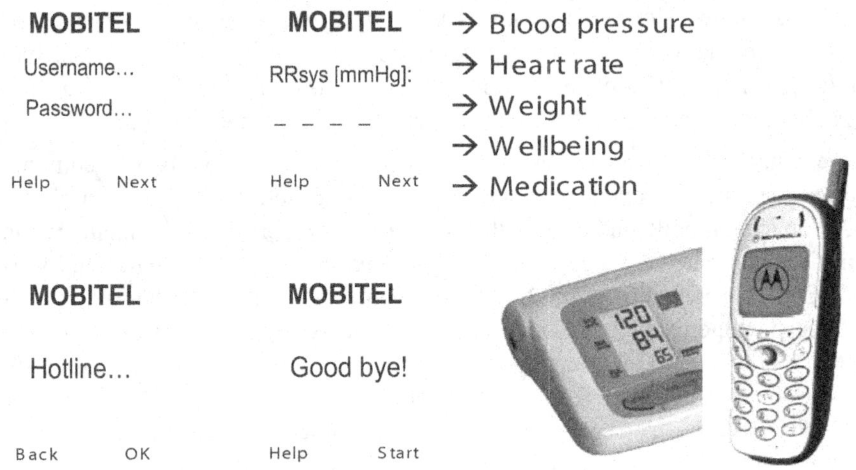

messages reminded patients to take their medications, weigh themselves, measure blood pressure and heart rate, and to transfer the respective data to the secure server at the monitoring centre. Furthermore, the physician could individually configure an automatic warning system for each patient. If the patient's values exceeded individually predefined vital parameter limits, physicians were notified immediately by a computer-generated SMS or e-mail warning. If an intervention (e.g., adjustment of medication dose) was indicated (for example, because a patient exceeded a vital parameter limit), the physician was able to contact the patient directly via mobile phone to confirm the parameters and ask the patient to make an adjustment in medication.

During the 3-month monitoring period, there were 2,040 data-transfer sessions (mean 102 +/- 43 per patient). Only 1 out of 20 patients dropped out after 25 days due to severely impaired vision. The Cardio-Memory system was evaluated through a questionnaire at the end of the study. Eighteen out of 20 patients rated the software as easy to operate, which allows us to conclude that overall patients' acceptance with the system was high. The entire process (measurements and data transfer) took approximately 3 minutes to be completed. Moreover, patients felt that computer-generated reminders about missing data increased their compliance with the treatment regime.

An example of a successful tele-medical intervention of a patient suffering from chronic heart failure is shown in Figure 5. Cardiac decompensation caused a weight

Figure 5. Trend chart of a patient monitored with Cardio-Memory

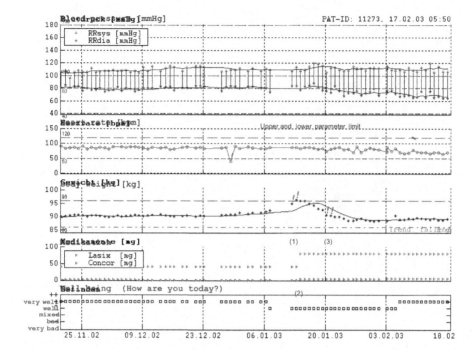

gain of more than 2 kg within 3 subsequent measurements and led to a patient alert (1). The physician contacted the patient and increased the diuretic dose (2). Subsequently, the patient lost weight and remained in stable conditions until the end of the observation period (3).

Following the promising results of the Cardio-Memory trail, an advanced study was set up in October 2003 in order to evaluate WAP technology in the management of chronic heart failure (Scherr et al., 2005). In a randomised, prospective, multicenter study, 240 patients who had been admitted to the hospital because of heart failures will be randomised to either pharmacological treatment (control group) or to pharmacological treatment plus tele-medical care (telegroup). Telegroup patients are provided with a mobile telephone, a digital weight scale, and a fully automated blood-pressure device. During the follow-up period of 6 months, patients are asked to send their self-measurements using a WAP application on the mobile phone on a daily basis. Up to now, 65 patients (45 male, 20 female; 64 ± 11 years old) from six centers have been randomised, and 44 patients have completed the study so far. Three patients from the telegroup dropped out due to being not able to handle the monitoring system equipment, in particular, the mobile phone.

Figure 6. A Java-based software application running on mobile phones supports the user in home-based data acquisition of diabetes-related key measurements

Flow chart

1 ... User authentication
2 ... Data entry (Blood glucose, bread units, insulin dose, ...)
3 ... Hypoglycaemia classification if occurred [1 to 4]
4 ... Physical activity if performed

Intermediate results indicate that WAP-based tele-medical surveillance of patients with a recent episode of acute heart failure significantly contributes to an improvement of functional status and may be a promising tool to improve heart-failure therapy and reduce emergency situations and hospitalizations.

Java-Based Software Application Running on Mobile Phones

Java 2 Platform, Micro Edition (J2ME), the small footprint version of the Java technology, is optimized to run on memory-constrained devices like mobile phones, PDAs, and TV-set-top boxes. Each device category receives its own profile that includes a set of category-specific application program interfaces (APIs) and a configuration that consists of a minimum set of APIs and a Java virtual machine. In particular, mobile phones support the mobile information device profile (MIDP), which includes APIs covering the user interface, networking, persistent storage (a record-oriented database), security, and messaging (Sun Microsystems, 2002).

The usage of a J2ME-based software application running on mobile phones has been evaluated in a pilot trial concerning the management of diabetes patients (Riedl, Kastner, Kollmann, Schreier, & Ludvik, 2005). The software has been designed to support the user in entering diabetes-related data like blood-glucose level, insulin dose, bread units, well-being, and activities with remote synchronization to the database at the central monitoring centre. The graphical user interface is shown in Figure 6.

After having logged in, the data is entered via the well-designed graphical user interface and stored directly in a local database on the mobile phone. From time to time, the user can initiate a synchronization process via HTTP request and GPRS data transfer. After the uploading of the data to the remote monitoring centre, the user is able to access several graphical representations, statistics, and trends on his or her data through a secure Web interface.

The software has been evaluated in a clinical pilot trial. Ten diabetes mellitus type 1 patients (36.1 ± 11.4 years) were asked to collect diabetes-related data (blood glucose, bread units, insulin dose, well-being, activity, etc.) at least 3 times and up to 10 times a day by using J2ME software application running on a NOKIA 7650 mobile phone. In case of less than 3 transmitted data records a day, an automatic reminder SMS message was sent to the patient. The total number of received messages was 10,053 (1,257 +/- 351 per patient). Average blood-glucose level during the initial and final 14 days was 138.5 +/- 63.7 vs. 137.1 +/- 69.2 mg/dl, and HbA1c values were 7.7 +/- 0.4% vs. 7.6 +/- 0.4%, respectively. Although no tele-medical intervention by the physician was yet established, corresponding parameters decreased due to an advanced adherence to therapy. Additionally, ubiquitous availability and the appealing user interface were well perceived by the patients resulting in a high success rate for data transmission and patient acceptance.

Semiautomatic Methods

Running a J2ME-based software application on mobile phones supports also the possibility to access integrated mobile-phone features like Bluetooth. Bluetooth wireless technology is a short-range radio technology that makes it possible to transmit data over short distances between mobile phones, computers, and other devices. Supporting standard measurement devices like blood-pressure meters or glucose meters with a Bluetooth communication module gives them the possibility to transfer data to the mobile phone in an automated way. The software running on the mobile phone handles data exchange and triggers further data transmission to a central monitoring centre (Husemann & Nidd, 2005).

Multimedia Messaging Service

MMS, an extension to the SMS, defines a way to send and receive wireless messages that may include images, audio, and video clips in addition to text. When

Figure 7. "MoniCam" principle: A special software application helps the user to take a photo of the display and send it to a central monitoring centre via MMS, where the values are extracted and stored in the database automatically

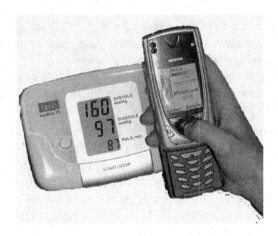

the technology will be fully developed, the transmission of streaming videos will be supported as well. A common current application of MMS messaging is picture messaging - the use of camera phones to take photos for immediate delivery to a mobile recipient.

MMS has been used in our research as a unique approach to the human-computer interface challenge based on digital-camera-enabled mobile phones. Values shown on the display of the measurement device were taken by a camera-enabled mobile phone. Special Symbian-based (Symbian, 2005) software, running on the mobile phone, helps the user to handle this data-acquisition process. All necessary configurations are made by the program automatically. If an online connection to the data carrier service (GPRS, UMTS) at a given location and time is not available, the mobile phone repeatedly tries to send the data until the message has been sent successfully. Data acquisition and transmission can be achieved with only two keystrokes.

At the monitoring centre, the MMS arrives as a photo attached to an e-mail. The subject of this e-mail contains the IMEI (international mobile station equipment identity) of the mobile phone. Thus, the photo can unambiguously be assigned to a patient. Other important parameters like capturing date and time are also stored within the e-mail.

Special software, running on the Web server, fetches the e-mail available on the mail server, extracts the photo, and moves it to the local file system. Subsequently, the incoming photo is registered to the database and assigned to the corresponding patient.

Thereafter, a special character-extraction algorithm starts to process the photo in order to extract the numerical values. In case of successful character extraction, the values are stored into the database and the report-processing unit is started. Graphs and trends are generated and limit checks are performed. Finally, all information is made accessible via the Web interface to authorized users. In case of unsuccessful character extraction or failed plausibility check, the user may receive a message via SMS to repeat the measurement.

This new method has been evaluated in a feasibility study on five users with four different blood-pressure devices and two different camera-enabled mobile phones (Schreier, Kollmann, Kramer, Messmer, Hochgatterer, & Kastner, 2004). The results indicate that the rate of correct value extraction varied considerably with respect to the type of the measurement device but was comparable for the two different types of mobile phone. For two types of blood pressure meters the method was capable of determining the correct values in well above 90% of the cases. The MMS-based transmissions of the taken photos succeeded in all cases.

Requirements on Mobile-Phone-Based Telemonitoring Application

Because of the ubiquitous availability of mobile phones, these devices may be considered to be the patient terminal of choice to provide an interface to home monitoring systems. It has to be taken into account, however, that elder patients are often very unskilled and not familiar with the handling of mobile phones. To avoid rejection of this concept as well as to motivate the patients in using such devices, consequentially, an easy-to-learn and easy-to-use system for data acquisition is essential. Based on our experience, the ideal human-computer interface would have the following properties.

High Usability

Usability tells us how well the users can use the system productively, efficiently, and pleasantly to reach the goals in a certain environment. Using mobile phones as home monitoring terminals is limited in terms of input interaction (small buttons to enter data as well as to navigate) and output interaction (small display size and poor resolution). To overcome these limitations, an appealing user interface, well-structured graphical design (metaphoric), and intuitive navigation are essential. The number of buttons to be used as the necessary number of button clicks for navigation and entering data should be reduced to a minimum. Moreover, the data-acquisition process should be easy to learn, intuitive, and well structured to decrease barriers against technical devices and improve users' satisfaction in using the system.

Low Cost

The main driving force for introducing new innovations into healthcare is to reduce costs. Hence, patient terminals for home monitoring applications should ideally be based on off-the-shelf technology available without significant set-up expenses or extra costs.

Error Resistant

User actions as well as data entries should be checked for plausibility on the earliest possible stage. For example, a data-entry plausibility check on the side of the patient will exclude errors at origin. A major aspect in handling errors is to provide the user with well-defined error messages that are also easy to interpret and to understand. Well-structured error handling can guide the user through difficulties without confusing him or her.

Off-Line Data Acquisition

It is a well-known fact that wireless networks especially sometimes lack availability (e.g., in buildings). To provide the user with the possibility to enter data at any time and any location, intermediate local data storage is desirable. When the network is available again, stored data can be synchronized with the central database.

High Flexibility and Adaptability

 Patient terminals should provide flexibility in terms of adding or removing parameters to the user interface, corresponding to changing conditions in the treatment of chronic diseases. For meeting the needs of an adjusted, personalized patient terminal, the system should support the possibility to interrogate different parameters depending on the time of day or the patient's history. The user interface for daily use should be configurable to different patients and conditions such as displaying the appropriate drug name to enter the correct dosage. Ideally, user interactions or setup procedures are no longer necessary.

Table 1. Overview of the properties and technologies (+ means the method complies with the requirement, +/- means it complies only partly, - means does not comply)

Requirement	Technology				
	SMS	WAP	J2ME	MMS	Automatic
Overall usability	-	+/-	+	+	+/-
Low cost	+/-	+/-	+/-	-	-
Error resistant	-	+/-	+	+	+
Off-line data acquisition	+	-	+	+	+/-
High level of flexibility/ adaptability	-	+	+/-	-	-
High security level	-	+	+	+/-	+
Bidirectional communication	+	+/-	+/-	+	+/-
Device independency	+	-	+/-	+	-

High Security Level

Handling data about the health status of patients, in general, requires a high standard of security. In most countries, end-to-end encryption is compulsory to meet security directives and laws. As a consequence, most standard communication technologies and protocols already provide some sort of data protection to guarantee that data cannot be accessed, stored, or manipulated while they are transmitted via Internet or wireless networks.

Bidirectional Communication

The ideal patient terminal combines the possibility to provide the user with an interface not only for entering data but also for receiving feedback, reminders, or medical advice on the same device.

Device Independency

Currently, numerous models of mobile phones from various manufacturers are available on the market to be used as patient terminals. Therefore, a software application that provides a graphical user interface highly demands independency of device specifications like operating system, display resolution, mobile Internet browser, menu navigation, and so forth to ensure high usability for a wide range of mobile phones.

Table 1 gives an overview of the requirements and the degree of compliance of the various technologies as discussed in this chapter and proposes a three-level classification based on our experience and the results of trials of other authors.

SMS

Using SMS technology for data acquisition is not very suitable due to the lack of usability and the fact that a graphical user interface cannot be provided. Basic skills in handling mobile phones as well as knowledge in using the keypad are essential to enter values in a predefined template. SMS for data transmission can be considered when only a single value has to be transmitted (e.g., PEF value in case of monitoring asthma patients; Ostojic et al., 2005). On the other hand, SMS is very suitable for sending messages, automatically generated reminders, or medical advices from the healthcare provider to the patient as an additional way of communication.

WAP

Several studies have been performed utilizing a WAP browser to provide the user with a user interface to enter data. Although a graphical user interface is provided, basic skills in handling a mobile phone and using the numeric keypad are important. Moreover, WAP lacks the possibility of client-side data-entry plausibility checks. The data have to be sent to the server before the values can be processed or checked for plausibility. In case of an error, the user is prompted to the page where the error occurred and asked to reenter the data.

On the other hand, WAP technology provides high flexibility and adaptability because WML scripts are server based and can be generated dynamically depending on user settings and requirements. This fact makes this method seem quite suitable for providing dynamical questionnaires such as quality-of-life assessments or to track medication intake, which is quite different from patient to patient. WAP technology may also be useful were a stepwise data input is required.

A disadvantage of WAP-based systems is that an online connection has to be established during the whole data-entering process. We experienced that this fact can lead to problems. For example, when the data connection is lost, users are confused by the corresponding error message. Thus, incomplete record sets are common and some patients were not able or willing to use the method for daily data acquisition.

Java

Java-based software is less affected by temporary lack of network availability because the data can be stored locally on the mobile phone and the data-transmission process may be postponed until network connectivity is available. Additionally, implementation of checking the plausibility of the entered data is also feasible, resulting in a lower error rate.

Moreover, Java technology allows one to improve usability through the design of user-friendly GUIs by using metaphoric elements. However, updating software applications running on the mobile phones is difficult and requires some additional user experience, although over-the-air application downloading (OAD) simplifies the way applications are delivered to customers.

In addition to standard TCP and HTTP, the Java environment supports the secure hypertext transfer protocol (HTTPS) using standard secure socket layer (SSL) to enable encrypted, secure connections. J2ME is particularly suitable for monitoring applications where frequent data acquisition is necessary (e.g., diabetes) and provides a fast and efficient method to enter, store, and transmit data.

Irrespective of which technology will be used, the most challenging part in developing mobile-phone-based applications for home monitoring is to guarantee interoperability. For example, WAP technology is strongly based on standards but every mobile-phone manufacturer speaks a slightly different WAP language. According to our experience, J2ME applications also change their looks depending on the type of the mobile phone.

MMS

Using a camera-enabled phone for data acquisition by taking a photo of the display provides a really easy and intuitive method for health-data acquisition at home. On the other hand, there is a huge effort on the server side to provide an image-processing algorithm to extract and interpret values correctly. This method lacks also in terms of flexibility and adaptability. Furthermore, costs for the MMS picture transfer are an obstacle for daily usage.

Automatic

Some measurement devices are already equipped with short-range wireless communication technologies like Bluetooth or infrared to transmit data to the mobile phone, which operates as a hub and a gateway to relay those data to a central monitoring centre. Because there is no mass market for such devices yet, they are usually considerably more expensive. Another weak point of this concept, according to our experience, is that it is sometimes not straightforward to establish communication between the mobile phone and the measurement device. Depending on the model, it may be necessary to navigate deep into the menu for setting up the connection. This makes such methods often unusable for technically unskilled people. However, once the system is set up correctly, data acquisition is fairly automated and error resistant. Using the mobile phone as a hub provides a bidirectional communication between patients and healthcare providers; hence, the mobile phone can receive SMS. Therefore, in the long run, automated systems will be the method of choice for elderly and unskilled or handicapped patients to transmit their self-measurements to the monitoring centre.

Developing Mobile Applications for Telemonitoring Applications

Because of small display sizes, limited resolution, and restricted possibilities for user interaction, navigation design and the implementation of software applications for mobile devices is quite different from software developments on other platforms like stationary desktop PCs.

Mobile software applications can be divided into two groups:

1. Highly goal driven.
2. Entertainment Focused.

Highly goal-driven services aim at providing fast replies to specific problems, whereas entertainment-focused services enable the user to pass the time, for example, by offering gossip, games, or sports results (Ramsay & Nielsen, 2001). Mobile applications for home monitoring are definitely goal-driven, that is, to provide the users with a method to enter data in the most suitable way. Therefore, a user-centered design process is necessary. This process involves a number of important phases.

Analysis

The initial phase of software development is the most critical one. In this stage, the requirements of the patient terminal should be clearly defined by determining the user-group characteristics and the monitoring scenario. Typical attributes of the user group are age, expertise, experience level, and physical limitations. The definition of the monitoring scenario comprises the parameter to be monitored, the way feedback is given, and the demand in terms of flexibility.

Thereafter, the use case of the specific monitoring scenario should be clearly described in order to define objectives and features that require close cooperation between technician and physician. Once the general use case has been defined, the technology has to be selected that best supports the requirements of the respective application. Table 1 can serve as a guide to get an overview of properties of available technologies. Features as well as limitations of the selected technology should be carefully balanced during the design process.

Design

According to our experience, the typical software application for home monitoring applications comprises the following four stages:

1. **Identification:** The user is asked to log onto the system with a unique user name and password combination to facilitate authentication and access control.

2. **Data acquisition:** Menus and input templates guide the user through the data-acquisition process.

3. **Transmission:** Entered data are either stored locally or are transmitted to a central database in a monitoring centre.

4. **Feedback:** To indicate that data storage or transmission has been executed successfully or the presence of special situations, feedback should be given to the user immediately after data transfer in a representative form.

The development of a consistent, easy-to-use application does not require coding at this stage. On the contrary, ideally, a conceptual framework is established, representing the application and its workflow on a metaphoric level (e.g., a storyboard).

Evaluation

The concept designed in Step 2 should be evaluated by both types of users: patients and physicians. Usability considerations should be made; for example, unnecessary button clicks or confusing workflows should be avoided. There is also the need to identify sources of errors. If significant changes are necessary, an iteration starting with the design step is required.

Implementation

There are several software-development kits (SDKs) available to develop software for mobile devices. However, irrespective of which SDK or programming language is adopted, developing software applications for mobile phones requires substantial experience in programming efficient and reliable software.

Testing

The developed software can be tested and debugged on emulators that are provided by the mobile-phone manufacturers. However, according to our experience, emulators are often error prone and sometimes special features are not supported. Hence, it is essential to test developed software on target devices themselves.

Future Trends

Current trends in chronic-disease management enforce shifting tasks from the clinic to patients' homes. This means that self-management and collaboration between patients and caregivers will become more and more important and may benefit from upcoming technologies.

Technical Advancements for Mobile-Phone-Based Patient Terminals

Phones enabled with radio-frequency identification (RFID) and near-field communication (NFC) may soon be used as patient terminals to provide an intuitive and easy-to-use way for health-data acquisition at home.

RFID tags are able to uniquely identify an object, animal, or person, or to store data. They have been introduced in the industry as an alternative to the bar code. Passive RFID tags are powered by the magnetic field generated by the reader. The tag's antenna picks up the magnetic energy, and the tag communicates with the reader in order to retrieve or transmit data.

In a current research project, we developed a scenario where RFID tags are used in a home monitoring environment. Objects to be tracked or identified are equipped with RFID tags. For example, tagging medication boxes with RFID tags provides an easy and intuitive method for the patient to indicate which medication has been taken simply by touching the box with an RFID-reader-enabled phone. Special software running on the mobile phone fetches the information from the tag, adds a time stamp, and initializes a transmission to the monitoring centre automatically. Hence, no cumbersome user interaction and configuration is needed.

Electronic Data Capture in Clinical Trials

The combination of mobile and Web-based technologies will improve clinical-trial efficiencies through increased data accuracy, higher data yield per patient, and real-time access to trial data (Stokes & Paty, 2002) by using mobile phones for timely and patient-centered data acquisition. Patient diaries as well as consequent monitoring of health parameters of interest can add meaningful information about the safety and efficacy of a treatment and can save time and money. Additionally, quality-of-life data will also play a central role in future clinical trials, which can be accessed easily by mobile-phone-based software solutions.

Lessons Learned

Self-management and cooperation between patients and healthcare providers are the basic aspects and strategies of efficient chronic-disease management. An important element in efficient self-management is the monitoring of health-related data reliably. Because of the ubiquitous availability of mobile phones, these devices can be used as patient terminals so as to provide the patient with a method to enter data easily as well as to receive feedback or medical advice from remote healthcare professionals.

The most critical part in this respect is the user interface. Software specifically developed to the needs of the respective patient group is required to guide them through the data-acquisition process. Mobile data services and transmission protocols like SMS, MMS, WAP, and HTTP can be used to exchange data and information between patients and their caregivers. These methods have already been evaluated in several clinical trials and feasibility studies, and medical benefit could be demonstrated as well. However, using mobile phones as patient terminals is limited due to small display, poor resolution, and small buttons for user interaction.

Up to now, there is no method that fulfils all criteria of an ideal patient terminal in terms of high usability, adaptability, flexibility, and low cost. Every method for entering data implies specific advantages and disadvantages. Hence, when designing a mobile-phone-based home monitoring system, the patient terminal that best fits into a particular monitoring application has to be chosen on an individual basis, depending on the requirements, the user group, and the medical demand.

References

Barold, S. S. (2003). Willem Einthoven and the birth of clinical electrocardiography a hundred years ago. *Cardiac Electrophysiology Review, 7*, 99-104.

Bielli, E., Carminati, F., La Capra, S., Lina, M., Brunelli, C., & Tamburini, M. (2004). A wireless health outcomes monitoring system (WHOMS): Development and field testing with cancer patients using mobile phones. *BioMed Central Medical Informatics and Decision Making, 4*, 7.

Brown, N. (1995). *A brief history of telemedicine.* Retrieved September 25, 2005, from http://tie.telemed.org/articles/article.asp?path=articles&article=tmhistory_nb_tie95.xml

Buckingham, S. (2000a). *What is SMS?* Retrieved September 25, 2005, from http://www.gsmworld.com/technology/sms/intro.shtml

Buckingham, S. (2000b). *What is WAP?* Retrieved September 25, 2005, from http://www.gsmworld.com/technology/wap/intro.shtml

Celler, B. G., Lovell, N. H., & Basilakis, J. (2003). Using information technology to improve the management of chronic disease. *Medical Journal of Australia, 179*, 242-246.

CellularOnline. (n.d.). *Stats snapshot.* Retrieved September 25, 2005, from http://www.cellular.co.za/stats/stats-main.htm

Cleland, J. G. F., Louis, A. A., Rigby, A. S., Janssens, U., Balk, A. H. M. M., & TEN-HMS. (2005). Noninvasive home telemonitoring for patients with heart failure at high risk of recurrent admission and death: The Trans-European Network-Home-Care Management System (TEN-HMS) study. *Journal of the American College of Cardiology, 45*, 1654-1664.

Epping-Jordan, J. E., Pruitt, S. D., Bengoa, R., & Wagner, E. H. (2004). Improving the quality of healthcare for chronic conditions. *Quality and Safety in Healthcare, 13*, 299-305.

Ferrer-Roca, O., Cárdenas, A., Diaz-Cardama, A., & Pulido, P. (2004). Mobile phone text messaging in the management of diabetes. *Journal of Telemedicine Telecare, 10*, 282-285.

Field, M. J. (Ed.). (1996). *Telemedicine: A guide to assessing telecommunications for healthcare.* Washington, DC: National Academy Press.

GESICA Investigators. (2005). Randomised trial of telephone intervention in chronic heart failure: DIAL trial. *British Medical Journal, 331*, 425-427.

Glasziou, P., Irwig, L., & Mant, D. (2005). Monitoring in chronic disease: A rational approach. *British Medical Journal, 330*, 644-648.

Hung, K., & Zhang, Y. T. (2003). Implementation of a WAP-based telemedicine system for patient monitoring. *IEEE Transactions on Information Technology in Biomedicine, 7*, 101-107.

Husemann, D., & Nidd, M. (2005). Pervasive patient monitoring: Take two at bedtime. *ERCIM News, 60*, 70-71.

Meystre, S. (2005). The current state of telemonitoring: A comment on the literature. *Telemedicine Journal and E-Health: The official journal of the American Telemedicine Association, 11*, 63-69.

National Library of Medicine. (2005). *MeSH descriptor data* (C23.550.291.500). Rockville Pike: Bethesda. National Library of medicine.

Ostojic, V., Cvoriscec, B., Ostojic, S. B., Reznikoff, D., Stipic-Markovic, A., & Tudjman, Z. (2005). Improving asthma control through telemedicine: A study of short-message service. *Telemedicine Journal and E-Health: The official journal of the American Telemedicine Association, 11*, 28-35.

Ramsay, M., & Nielsen, J. (2001). *WAP usability.* Fremont, CA: Nielsen Norman Group.

Research and Markets. (2005). *Mobile and wireless services for outpatients.* Retrieved September 25, 2005, from http://www.researchandmarkets.com/reports/297065/297065.htm

Riedl, M. , Kastner, P., Kollmann, A., Schreier, G., & Ludvik, B. (2005). *DiabMemory: A smart phone based data service for intensified insulin therapy in patients with type 1 diabetes mellitus. A pilot study.* Proceedings of American Diabetes Association's 65th Annual Scientific Sessions, San Diego, CA.

Salvador, C. H., Carrasco, M. P., Gonzalez de Mingo, M. A., Muñoz Carrero, A., Márquez Montes, J., Martin, L.S., et al. (2005). Airmed-cardio: A GSM and Internet services-based system for out-of-hospital follow-up of cardiac patients. *IEEE Transactions on Information Technology in Biomedicine, 9*, 73-85.

Scherr, D., Fruhwald, F. M., Zweiker, R., Kastner, P., Schreier, G., & Klein, W. (in press). Mobile phone based surveillance of cardiac patients at home. *Journal of Telemedicine and Telecare.*

Scherr, D., Kollmann, A., Hallas, A., III, Krappinger, H., Auer, J., Kastner, P., et al. (2005). Telemonitoring for heart failure patients following acute decompensation: First results on influence of the system on functional status and heart failure therapy. *European Heart Journal, 26*(Suppl.20).

Schreier, G., Kollmann, A., Kramer, M., Messmer, J., Hochgatterer, A., & Kastner, P. (2004). Mobile phone based user interface concept for health data acquisition at home. In J. Klaus (Ed.), *Computers helping people with special needs* (pp. 29-36). Springer.

Stokes, T., & Paty, J. (2002). Electronic diaries, Part 1: What is a subject diary, and how do regulations apply? *Applied Clinical Trials.* Retrieved September 25, 2005, from http://www.actmagazine.com/appliedclinicaltrials/article/article-Detail.jsp?id=83521

Sun Microsystems. (2002). *Java™ 2 Platform, micro edition.* Retrieved September 25, 2005, from http://java.sun.com/j2me/docs/j2me-ds.pdf

Symbian. (2005). *Symbian OS: The mobile operating system.* Retrieved September 25, 2005, from http://www.symbian.com/

Tachakra, S., Wang, X. H., Istepanian, R. S. H., & Song, Y. H. (2003). Mobile e-health: The unwired evolution of telemedicine. *Telemedicine Journal and E-Health: the official Journal of the American Telemedicine Association, 9,* 247-257.

Wagner, E. H. (1998). Chronic disease management: What will it take to improve care for chronic illness? *Effective Clinical Practice: ECP, 1,* 2-4.

World Health Organisation. (2005). *Facts related to chronic diseases.* Retrieved September 25, 2005, from http://www.who.int/dietphysicalactivity/publications/facts/chronic/en/

Zundel, K. M. (1996). Telemedicine: History, applications, and impact on librarianship. *Bulletin of the Medicine Library Association, 84*(1), 71-79.

Appendix: Austrian Research Centers GmbH-ARC

Austrian Reseach Center GmbH-ARC (ARC-sr) is Austria's largest centre for applied research and development, employing over 500 highly qualified specialists. The corporation's portfolio is strongly oriented toward national and European-wide research projects and development programs. ARC-sr comprises nine business divisions. The Division of Biomedical Engineering comprises research teams in the fields of e-health systems and smart biomedical systems, and cooperates with all major Austrian medical and technical universities in Vienna, Graz, and Innsbruck as well as a number of additional medical institutions and industrial partners within Austria and abroad.

The E-Health Systems Research Team focuses on e-health-related applications and research and development projects according to its mission: "to provide new connections between patients, physicians and other healthcare partners." Accordingly, we developed several home monitoring solutions utilizing modern information and communication technologies and a 24/7 monitoring service centre to establish these new connections between physicians and patients. The integration of different devices and monitoring indications as well as standardization are major objectives. The ultimate goal of our tele-, home, and health-monitoring concept is to empower physicians and patients with advanced data-acquisition methods for prevention, diagnosis, and therapy management.

Chapter XI

Considerations for Deploying Web and Mobile Technologies to Support the Building of Patient Self-Efficacy and Self-Management of Chronic Illness

Elizabeth Cummings, University of Tasmania, Australia

Paul Turner, University of Tasmania, Australia

Abstract

This chapter examines issues relating to the introduction of information and communication technologies that have emerged as part of planning for the Pathways Home for Respiratory Illness project. The project aims to assist patients with chronic respiratory conditions (chronic obstructive pulmonary disease and cystic fibrosis) to achieve increased levels of self-management and self-efficacy through interactions with case mentors and the deployment of ICTs. The chapter highlights that in deploying ICTs, it is important to ensure that solutions implemented are based on a detailed understanding of users, their needs and complex interactions with health professionals, the health system, and their wider environment. Achieving benefits from the introduction of ICTs as part of processes aimed at building sustainable self-efficacy and self-management is very difficult, not least because of a desire to avoid simply replacing patient dependency on health professionals with dependency on technology. More specifically, it also requires sensitivity toward assumptions made about the role, impact, and importance of information per se given that it is often only one factor among many that influence health attitudes, perceptions, actions, and outcomes. More broadly, the chapter indicates that as ICT-supported patient-focused interventions become more common, there is a need to consider how assessments of benefit in terms of a cohort of patients inform us about an individual patient's experience and what this implies for terms like individualized care or patient empowerment (Muir Gray, 2004). At this level, there are implications for clinical practice and one-size-fits-all care-delivery practices. This collaborative project involves a multidisciplinary team of researchers from the University of Tasmania's School of Medicine, School of Nursing and Midwifery, and School of Information Systems. The project is supported by the Tasmanian Department of Health and Human Services and funded by the Commonwealth Department of Health and Ageing, and is due for completion in June 2008.

Introduction

The crisis in healthcare across the developed world is, ironically, partly due to the success of medical innovations in fighting disease and increasing life expectancy. Aging populations are dramatically changing the nature and demand for medical procedures, medications, and healthcare services such that the need for high-quality, cost-effective approaches to the growth in chronic and/or complex medical conditions has been widely recognized. As part of the response to this need, a number of approaches that empower patients to participate directly in their own care are increasingly being explored as a means of improving disease treatment, management, and education. Underpinning these approaches are assumptions that patients are willing and able to take on these new responsibilities and that when they do, the result will be positive in terms of quality of care and health outcomes. While most evaluations to date report some benefit, the variety of methodologies and assessment procedures used make comparisons of efficacy difficult and highlight the complexity and uncertainty associated with supporting the self-management of chronic illness (Warsi, Wang, LaValley, Avorn, & Solomon, 2004).

E-health initiatives have also been identified as a critical component in the development of responses to the health crisis. By improving opportunities for information access, delivery, and update, ICTs have strong support. However, there is increasing awareness that the design, development, and deployment of ICTs also raise numerous socio-technical, clinical, and legal challenges that influence the realization of benefits. Many approaches to the deployment of ICTs in the health domain continue to make problematic assumptions about how ICTs will actually benefit patients, health professionals, and the healthcare system as a whole. More broadly, as meta-analysis of research recording positive benefits from the introduction of ICTs into health shows, many measures of success have little to do with improvements in patient care or outcomes (Wyatt, 2004).

Combined, these discussions highlight that the development of technology to support self-efficacy and self-management of chronic illness is highly complex. However, from a practical perspective, it is clear that understanding the users is important not just for designing approaches to build self-efficacy and self-management, but also for considerations of how ICTs should be deployed: "finding out prior to design what the unique requirements are, and designing to support them, is much more cost-effective in the long run than finding out after launch that your design does not meet requirements" (Mayhew, 2001).

This chapter adopts a patient-centered approach in its examination of issues around the deployment of Web and mobile technologies to support the building of patient self-efficacy and self-management of chronic illness. From a patient-empowerment perspective, this work draws on a range of approaches advocating how to build self-efficacy and self-management based on existing models of chronic-disease management. From a technology perspective, this work is informed by theoretical insights drawn from a range of approaches that indicate that successful design and deployment of ICTs rely on understanding users' needs and ensuring technology is both easy to use and useful (Singh, Turner, Burke, & Castro, 2003). Alongside generating practical insights for those engaged in deploying ICTs with patients, this chapter also aims to point toward the need for broader discussions on the implications of patient-focused interventions for current clinical care-delivery practices.

Background

Patients' attitudes and willingness to participate more actively in their own care appear to be changing. A survey on the United Kingdom's National Health Service (NHS) conducted by the consumers' association in October 2003 and involving 2,000 respondents (drawn from the general public, and nationally representative and demographically weighted) revealed that patients increasingly want more control over their treatment and care (Granger, 2003). From the 2,000 participants, 33% interacted regularly with the NHS and 76% did so at least once in the previous year, with 40% having experienced a relevant event (problems with appointments, problems with their records, repetition of data, missing letters). In identifying the most significant benefits wanted by patients, the survey found the following:

- 63% wanted access to their records to see recent test results and 60% to see medical history.
- 68% wanted access from home.
- 60% wanted their GP (general practitioner) to be able to book instant hospital appointments.
- 53% wanted repeat prescriptions without having to go to the surgery.

This survey supports the perspective that there is a growing interest amongst patients to become more empowered and engaged in decision making around their own care. More specifically, there is also a growing body of literature on chronic-disease management that advocates a horizontal, integrated approach across hospital and community settings, with the inclusion of coordinated care and partnerships with patients. In this regard, supported early discharge and home-based care of patients with exacerbations of chronic obstructive pulmonary disorder (COPD) have shown promising results, with positive responses from clients, caregivers, and healthcare providers, as well as reductions in cost.

Within the technology domain, numerous approaches have also recognized that involving users is an important aspect of design and deployment. Indeed, there is now a large volume of research into the adoption and use of technologies that has revealed that to increase the probability that a consumer technology will be successful, it is important that it meets the following criteria.

- To be easy to use.
- To provide relative value in terms of cost, convenience, a mix of channels, or better ways of conducting the activity.
- To have acceptable social and cultural meanings.
- To support the generation of trust (Singh et al., 2003).

Contribution of this Research

In this context, the Pathways Home for Respiratory Illness project aims to assist patients with chronic respiratory conditions (COPD and cystic fibrosis [CF]) to achieve increased levels of self-management, self-efficacy, and empowerment in relation to their conditions through interactions with case mentors (community health nurses, CHNs).

The project incorporates aspects of a number of different models of chronic-disease management, including the Stanford model, Flinders cue and response model, and the Whitehorse Division of General Practice Good Life Club. This type of approach has been shown to work well to capture those patients who may not wish to, or physically cannot, attend group sessions, and contrasts with the conventional models of medical intervention.

This patient-focused approach is premised on the view that where possible, patients should play a central role in decisions about their own health. At the broadest level, this approach is underpinned by the perspective that providing evidence-based knowledge to patients will enhance their ability to participate in decisions about their own care and contribute to the development of an increasingly effective patient-centered healthcare system (Hill, 1998). Within the area of chronic illness, two important elements of the patient-centered approach are the concepts of self-efficacy and self-management.

In this context, following Bandura (1994), self-efficacy can be defined as follows:

people's beliefs about their capabilities to produce designated levels of performance that exercise influence over events that affect their lives. Self-efficacy beliefs determine how people feel, think, motivate themselves and behave. Such beliefs produce these diverse effects through four major processes. They include cognitive, motivational, affective and selection processes. (p. 71)

Aligned to self-efficacy is the concept of self-management, which involves individual chronically ill patients working in partnership with their caregivers and health professionals to manage their illnesses. Adapting the Flinders Human Behavior & Health Research Unit (n.d.) approach, the aim in this project is to ensure that patients are able to self-manage to the extent that they can achieve the following:

- Know their condition and various treatment options.

- Negotiate a plan of care, that is, structure a care plan, and review and monitor the plan.

- Engage in activities that protect and promote their health.

- Monitor and manage the symptoms and signs of the condition

- Manage the impact of the condition on physical functioning, emotions, and interpersonal relationships.

From a technology perspective, developing and deploying information systems to support self-efficacy and self-management amongst chronically ill patients presents numerous challenges. Most significantly, this cohort of patients exhibit diverse levels of physical and psychological capacities as a result of their illness, as well as a wide range of abilities, experiences, support mechanisms, and interests in relation to the following:

- Participating in the project.
- Building self-efficacy and self-management competencies.
- Adopting and utilizing ICTs.

Critically, the project team was keen to avoid the possibility that any ICT systems introduced should end up simply replacing patient dependency on health professionals with a dependency on the technology, such that patients end up undertaking the monitoring of their symptoms without actually developing the self-efficacy and self-management skills necessary to respond to changes in their illness. As a result, the project team made considerable efforts to identify and accommodate the range of patient characteristics from amongst potential users of the system in the design, deployment, training, and use of the ICTs developed. In essence, this involved providing a variety of different accessibility tools and the provision of an extensive range of data-entry methods to enable accessibility for all members of the disease cohort. This is demonstrated by the wide variety of Web and mobile devices, applications, and interfaces made available to different patients as part of the project. Noticeably, major distinctions were apparent between the two main types of patient groups (COPD and CF).

From this experience, it is argued that for an ICT system to be truly patient centered, its actual purpose, the characteristics of the end users, and their contexts must be considered along with ensuring that the system is easy to use, fulfills a perceived need, and presents a clear value proposition for adoption and utilization. In this regard, information-systems researchers within the team have spent considerable amount of time and effort in understanding, interacting with, and training all participants in the adoption and use of the technology.

More specifically, this resulted in the following aspects of the tailored solution developed:

- At the level of systems architecture, the project relied on four application modules: the research-management system, the participant portal, the mentor portal, and the workflow system. The research-management system, participant portal, and mentor portal are all Web-based applications built using Oracle HTMLDB 1.5 (Application Express). The applications operate on a Mac OS X Server (V10.4) running the Oracle 10G Database environment and HTTP (hypertext transfer protocol) server. The Web-based applications were secured using the Oracle HTMLDB built-in authentication scheme, with user sessions accessed via HTTPS and encrypted using a 128-bit SSL (secure socket layer) certificate. Workflow was implemented using XForms and Microsoft's Infopath 1.5 Forms.

- At the level of hardware and software development, project participants were provided with the following equipment. Mentors received laptop computers running a full suite of Microsoft applications including InfoPath 1.5 forms. COPD patients received desktop computers with accessibility options tailored to users' needs including the use of trackballs. CF patients will receive handheld wireless-enabled pocket PCs (personal computers).

- At the level of education and training on self-efficacy and self-efficacy mentoring, all team members and mentors participated in an intensive 2-day workshop on training in self-efficacy, mentoring, and capacity building in self-efficacy.

- At the level of education and training on information systems, mentors and patients were given an initial assessment to identify skills and capabilities. This was followed by the tailored training of mentors (community health nurses) and patients on an individual basis to stimulate adoption and usage.

- From the perspective of the research team, the database was designed to collect all information from the forms generated and used by the mentors: all the information entered by patients in their daily diaries and ongoing action plans as well as data and Web logs for use to generate usage-patterns statistics. A document-management system and archive of qualitative field notes was also maintained.

Key Issues and Considerations: Development of the Approach

Due to the closely intertwined nature of the technology development and building of self-efficacy and self-management, a more detailed explanation of the project as a whole is provided prior to presenting the key technological considerations.

During the initial phases of the project, the methodology was developed. The first phase of the project consisted of the formation of a multidisciplinary team, an evidence-based review of the literature, and an iterative process of discussion among the team, culminating in the identification of a preferred methodology for the development of self-efficacy. This process culminated in training and project-development workshops involving community health nurses, hospital-based respiratory nurses, physiotherapists, respiratory medical specialists, and information-systems researchers.

As this is a collaborative, multidisciplinary piece of research, it has been developed to facilitate multiple research objectives. To this end, the research is a balanced allocation-controlled study with independent, objective, and concealed recruitment into study groups. Recruitment is to be undertaken for a period of 12 months, with each participant being actively engaged in the project for 12 months post recruitment.

Participants in the project are recruited while hospitalized with an acute exacerbation of COPD; those in the CF group will be recruited on a voluntary basis. Participants are allocated to the intervention or control (usual-care) group according to domicile. The enrollment process involves establishing baseline indicators that can then be used to compare with the quarterly evaluations to be undertaken on both intervention and control groups throughout their participation period.

Following discharge, intervention-group participants are linked with a CHN mentor who acts in partnership with them to facilitate their self-management during the 1 year that they are involved in the study. Mentors have been prepared for their coaching role by being trained in the trans-theoretical model of change (TTMC) and motivational interviewing (MI), as well as in the utilization of the IT supports. The TTMC aims to develop effective interventions to promote health-behavior change, while MI is a directive, patient-centered counseling style that assists participants in changing their behavior while respecting their choices about the change. Mentors visit participants in their home on two occasions early after hospital discharge to establish contact, initiate the rapport-building process, and to perform initial assessment of the situation and initial orientation of the patient to self-monitoring. They then maintain contact at regular (weekly-monthly) intervals via the telephone. The mentor encourages the client to recognize adverse health behaviors and to formulate new healthy behaviors, structuring these discussions into a written action plan.

The mentoring process has been augmented by the development of information systems to enable participants to closely monitor their diseases on a daily basis using their newly developing skills of self-management to respond appropriately. This approach is particularly suited to these participants because of their breathlessness, lack of mobility, and geographical dispersion, which limits their attendance at centralized or group activities. Due to the variation in IT capability and experience, and to provide an opportunity for them to become comfortable with daily self-monitoring and reporting processes, participants initially use paper-based monitoring systems (daily diary) and progress at their own pace to telephone and/or Web-based systems.

The project seeks to support self-monitoring and recording of symptoms (preferably in an electronic format), which can then be viewed in a graphical longitudinal form by the patient and automatically transferred to a repository for viewing by clinicians. This system has been developed to assist participants, and to a lesser degree health professionals, in the early identification, comprehension, and initiation of early action in relation to alterations in their condition.

IT Considerations

The IS researchers in the project team have been involved with the project from the time that funding was confirmed. Thus, they have been integral to the complete project-development cycle. This early and close involvement has resulted in the intermingling of ideas throughout the development cycle and a deep understanding of all aspects of the project. Throughout this process, the IS researchers have been careful to encourage and support the team to identify their specific requirements and those of the participants rather than limiting the requirements to those perceived to be technically simple. The only limitations resulted from budgetary constraints and the desire to develop the project for expansion and sustainability. This resulted in the need to conform to local health-department systems and intended state and national initiatives.

Through the expertise of the project team and initial brief interviews with COPD sufferers, an understanding of the potential range of technological experience and expertise of this group was developed. The picture that emerged was of a group of elderly people with minimal exposure to or trust of technology. The majority had telephones with very few having mobile phones or computers. As the second participant cohort, CF sufferers are much younger and generally more technologically savvy. This information coupled with the original project objectives and information gained through the initial stages of development led to the identification of the need to provide a range of technologies for use by the participants so they can move through different technologies as their skills and technology use evolves.

The technologies identified for development included everything from paper-based data entry, telephones using a call centre interactive voice response (IVR), to full Web-based data entry on desktop PCs and via mobile devices (personal digital assistants, PDAs). While it is anticipated that the CF cohort will be the main end users of the mobile technologies, every effort is made to make all technology options available to all participants. As these groups frequently have limited financial resources, all hardware requirements are provided by the project.

The Participants

Patients participating in the project are interviewed by one of the IS researchers as part of the enrollment process. Initially it was anticipated that this would be a data-gathering exercise, but with increasing experience, it was discovered that more benefit was gained by using this as a relationship-building exercise. With participants being more comfortable with the IS researcher, it has made them more comfortable with the idea of these people entering their homes at a later date to install technologies and conduct training.

Individualized training will be provided for participants in their homes to support the adoption of new technologies. This is in part because the COPD participants are primarily sick, elderly, and not particularly mobile, but also because people learn better in a familiar, non-threatening environment. For the CF population, individualized training is also essential as this group should not be gathered together for cross-infection purposes. The concept of individual training in these instances is supported by the principles of self-efficacy in education (Bandura, 1994).

The CHN Mentors

Engagement with the CHN mentors by the IS researchers began early in the development stage. It was quickly identified that this group was quite diverse in age and computer experience. The IT infrastructure available to this group was found to be minimal, and also different groups of CHNs displayed considerable variation in terms of high-level management support. These features all had significant implications for the development of the system.

New laptop computers were provided for those sites where the managers identified the need. This created some discussion at one site where the manager had decided that the CHNs did not currently use computers and would therefore not wish to use a laptop. This centre subsequently ordered and received a laptop once the manager concerned moved on to another position.

The identification of IT experience and self-efficacy amongst the CHNs was undertaken through a preliminary survey, with the intention that this will be repeated a number of times over the duration of the project to assess any increases amongst CHNs in self-efficacy. This survey indicated that many of the CHNs actively avoided using computers and only a very small number had previously had good experiences with technology. This demonstrated the potential for poor adoption of technology within this group and therefore additional knock-on effects in relation to patients' adoption and use of technology solutions available. Again, it became evident that a significant investment of time was required in gaining the confidence of this group prior to the discussion of any technology. The IS researchers attended many information sessions to gain exposure to the CHNs. From experience, minimal discussion of technology was undertaken at these meetings as the focus was upon familiarization and relationship building. This has resulted in a good rapport being established with the IS researchers and the CHNs—an essential element in the project.

Training of the CHNs for use of online forms has been undertaken in an incremental manner. This has involved a number of walk-throughs demonstrating the forms and requesting feedback. This information has then been fed back into the iterative design of the forms. This process has resulted in forms being designed with significant input from the CHNs, and they have incrementally been exposed to the technologies being implemented. The CHNs have direct access to the IS researchers and can at any time contact them for additional small-group or individual support or training.

The Project Team

Despite, or as a result of, the close working relationship between the IS researchers, clinicians, and nursing researchers, the identification of requirements was not a simple task. The multidisciplinary nature of the group and the project has required a mixture of research paradigms. Traditional power structures have been confronted and a balance of personalities has also offered challenges that are perhaps broader than those confronted in other types of development where a single discipline is involved.

During the development of the project, a number of skills requirements for the project team were identified. Training in a number of software applications was undertaken, particularly for the project officer. Training has taken into account the maintenance requirements for the system. This has seen the project officer and others actively involved in the development of online forms as well as learning how to maintain them. Further understanding of data-management principles may be required as the project is bedded down.

The System

The information-system model being developed is one that is compatible with the Department of Health and Human Services (DHHS) IT infrastructure. Participants' data will be available to be viewed by individual participants and their primary and secondary mentors only. However, should a participant desire their GP or other clinicians access their data, this can occur with the participant present. It is intended that the possibility of healthcare providers being given access will be investigated as the trial progresses. Development of the system is being undertaken in consultation with DHHS staff and Health*Connect* to ensure compatibility and scalability of the model.

Evaluation

Evaluation tools, as part of the research project, will monitor a range of elements relating to respiratory physical symptoms (sputum, breathlessness, cough, physical activity, medication use, spirometry, weight) as well as a sense of well-being (SF-36v2), the level of self-efficacy (Stanford self-efficacy for managing chronic disease six-item scale), and depression and anxiety scales (Hospital Anxiety and Depression

Scale). Participants in both the intervention and control groups will be visited by the research officer at baseline (in the hospital) and then every 3 months over the 1 year of the participant's involvement in the study. This will allow comparisons to be made between participants in the intervention group and those who have received usual care for quality of life, symptom levels, health-resource usage (e.g., doctor visits, drug use, readmission rates), and level of self-efficacy. Evaluation of the IT and CHN aspects of the project will also occur through focus groups with CHNs and qualitative feedback from participants and their caregivers.

It is hypothesized that this model will improve and optimize the quality of life for people with chronic respiratory conditions and will slow the progression of their disease by initiating early treatment of exacerbations, thus avoiding unplanned hospital admissions and presentations to emergency departments. It is anticipated that the model will be transferable to other chronic respiratory conditions. The project will build capacity to support clients with a range of chronic diseases by forging links between acute- and chronic-care providers (utilizing IT networks), enhancing skill levels of CHNs in effective management of chronic illness, facilitating self-management among participants, and building a system of integrated care based on partnership between hospital and community services.

Next Steps and Future Directions

The health system has evolved to be responsive to acute conditions. There is a rising prevalence of chronic and complex conditions now, and the current health system cannot cater to this. Chronic conditions require treatment at different levels including psychological and lifestyle changes as well as the physical and treatment regimens themselves. Patients need to adapt to being self-caregivers in partnership with health professionals so as to improve the management of their illness and the ability of the health system to respond to the growing challenges.

The role of medical care is changing and it is now important to ensure that patients with chronic illnesses "have the confidence and skills to manage their condition; the most appropriate treatments to assure optimal disease control and prevention of complications; a mutually understood care plan; and careful, continuous follow-up" (Wagner, Austin, Davis, Hindmarsh, Schaefer, & Bonomi, 2001, p. 66).

With the development of a more sustainable model of care for chronic illness, the principles of self-management become increasingly important. Adopting these principles requires supporting the chronically ill and their caregivers to make the move from the traditional dependence on clinicians for decision making to a team approach that is based upon a patient-centric model. Self-management requires an understanding on the part of the patient of the need for self-care, and the importance of the maintenance of their condition to aid in prevention of exacerbations and the subsequent requirement for acute care. To support this paradigm shift, chronic-illness sufferers need to be provided with appropriate information, and symptom-recognition and problem-solving skills to empower them to develop the self-efficacy to make appropriate decisions regarding their care needs (Bodenheimer, Lorig, Holman, & Grumbach, 2002).

The provision of appropriate information can assist patients in how to determine the appropriate balance between self-reliance and when to seek professional help. Aligned to this is the recognition of the shortage of health professionals and the need to find ways to address the growing demand for services. Critically, this paradigm shift cannot and should not attempt to replace clinician consultations, but it can make them more productive.

The rise in the use of ICTs in health coincides with the desire of consumers for more information and to assume more responsibility for their own health:

Patients are the experts in their experience of a condition and coping with it. Capturing this experience and using it to benefit others as well as improve the quality of care is vital to improving the whole healthcare process. The development of e-health utilizing the Internet could be pivotal in this regard. (Detmer, Singleton, MacLeod, Wait, Taylor, & Ridgwell, 2003, p. 13)

Eysenbach (2000) demonstrates this trend toward increasing information requirements and increasing use of ICTs.

However, care must be taken to ensure equity of access to healthcare and health information. Eysenbach (2000) identified the "inverse information law." This is where access to appropriate information is particularly difficult for those who need it most. More specifically, it also requires sensitivity toward assumptions made about the role, impact, and importance of information per se given that it is often only one factor among many that influence health attitudes, perceptions, actions, and outcomes.

The use of mobile and Web technologies have the potential to provide significant support for information exchange and healthcare reform. It is essential that when considering the introduction of ICTs, the characteristics of the whole patient cohort are considered to ensure that an appropriate choice of technologies is deployed. This must include an understanding of the cohort's range of physical, economic, emotional, and experiential abilities or limitations. In considering the employment of technologies, it is also essential that the basic well-utilized technologies such as fixed telephones are not forgotten in the enthusiasm for frontier, next-generation solutions. In many situations, the introduction of a call-centre service may have greater potential for uptake, and so provide greater benefits than a state-of-the-art system.

Through the process of implementing Web and mobile solutions to support the development of self-efficacy and self-management skills, it has become evident that the actual role of the ICTs and the evaluation of them are extremely difficult. When the process of introducing ICTs needs to be transparent and their use is not the primary focus of the project, how does one identify the benefits that can be attributed directly to the ICTs and how should they be evaluated? Where does the role of the ICT start and stop, and where does dependency shift from the clinician to the development of genuine self-efficacy and self-management, if at all?

Clearly these questions are difficult to provide definitive answers to at this stage of the project, but a number of conventional approaches have been considered. These include the main approaches used in the measurement of technology adoption and acceptance that variously focus on end users' perceptions, satisfaction, and usage patterns, and the ease of use and/or usefulness of the technology studied. While these approaches vary in scope from broad-based theories such as the diffusion of innovations (Rogers, 1995) to attitudinal approaches including the theory of planned behavior (Ajzen, 1985) and the technology acceptance model (Davis 1989), it is evident that they have relevance. This is particularly the case where more recent approaches have seemed to combine a number of theories (Karahanna, Straub, & Chervany, 1999; Van Akkeren & Cavaye, 1999). Significantly, these newer approaches have highlighted differences between users' pre- and post adoption attitudes and beliefs, including the issue of discontinuance (Karahanna & Limayem, 2000; Karahanna et al.). In the context of this and other health projects, there is also a critical requirement for the evaluation of the impact of the systems upon patient health outcomes. More broadly, this chapter indicates that as ICT-supported patient-focused interventions become more common, there is a need to consider how assessments of benefit in terms of a cohort of patients inform us about an individual patient's experience and what this implies for terms like individualized care or patient empowerment (Muir Gray, 2004). At this level, there are implications for clinical practice and one-size-fits-all care-delivery practices.

Conclusion

This chapter has examined issues relating to the introduction of ICTs that have emerged as part of planning for the Pathways Home for Respiratory Illness project. The project aims to assist patients with chronic respiratory conditions (chronic obstructive pulmonary disease and cystic fibrosis) to achieve increased levels of self-management and self-efficacy through interactions with case mentors and the deployment of ICTs.

From the above discussion, the following is evident:

- ICT development and deployment considerations need to be based on a detailed understanding of users, their needs and complex interactions with health professionals, the health system, and their wider environment.

- Whilst this project is still in progress, mechanisms are in place to enable the team to implement and evaluate both the role and impact of building self-efficacy and the influence and contribution of ICTs.

- The work to date has already contributed to a realization of the need to reconceptualize the role of information and ICTs, and the role of patients and their interactions with health professionals as part of process of developing new paradigms for patient-centered healthcare.

- There evidently is a role for Web and mobile technologies in the changing healthcare environment, but these must be flexible in their design and implementation.

- However, this important role in supporting self-efficacy and self-management must be done in a manner that avoids building dependence on the technology as opposed to genuine patient empowerment. For the research team, this continues to be the main challenge.

The authors look forward to presenting further results and insights from this project as it continues until its completion in 2008.

References

Ajzen, I. (1985). From intentions to actions: A theory of planned behavior. In J. Kuhl & J. Beckmann (Eds.), *Action-control: From cognition to behavior* (pp. 11-39). Heidelberg, Germany: Springer.

Bandura, A. (1994). Self-efficacy. In V. S. Ramachaudran (Ed.), *Encyclopedia of human behavior* (Vol. 4, pp. 71-81). New York: Academic Press.

Bodenheimer, T., Lorig, K., Holman, H., & Grumbach, K. (2002). Patient self-management of chronic disease in primary care. *Journal of American Medical Association, 288*(19), 2469-2475.

Davis, F. D. (1989). Perceived usefulness, perceived ease of use, and user acceptance of information technology. *MIS Quarterly, 13*(3), 319-340.

Detmer, D. E., Singleton, P. D., MacLeod, A., Wait, S., Taylor, M., & Ridgwell, J. (2003). *The informed patient: Study report.* University of Cambridge, Judge Institute of Management. Cambridge: UK.

Eysenbach, G. (2000). Recent advances: Consumer health informatics. *British Medical Journal, 320*(7251), 1713-1716.

Flinders Human Behaviour & Health Research Unit. (n.d.). *What is self-management?* Retrieved September 19, 2005, from http://som.flinders.edu.au/FUSA/CCTU/Home.html

Granger, R. (2003). *National programme for IT in the NHS in England.* Retrieved July 21, 2004, from http://www.connectingforhealth.nhs.uk/publications/

Hill, S. (1998). *Using the evidence: Empowering consumers of healthcare.* Retrieved September 19, 2005, from http://www.menziesfoundation.org.au/conferences/ebm/ebm09hill.htm

Karahanna, E., & Limayem, M. (2000). Email and v-mail usage: Generalizing across technologies. *Journal of Organisational Computing and Electronic Commerce, 10*(1), 27-143.

Karahanna, E., Straub, D., & Chervany, N. L. (1999). Information technology adoption across time: A cross-sectional comparison of pre-adoption and post-adoption beliefs. *MIS Quarterly, 23*(2), 183-213.

Mayhew, D. J. (2001). *Investing in requirements analysis.* Retrieved July 1, 2005, from http://www.taskz.com/ucd_invest_req_analysis_indepth.php

Muir Gray, J. A. (2004). Self-management in chronic illness. *The Lancet, 364*, 1467-1468.

Rogers, E. M. (1995). *Diffusion of innovations* (4th ed.). New York: The Free Press.

Singh, S., Turner, P., Burke, J., & Castro, M. (2003). The discovery phase of user-centred design: Putting users first in the design of smart Internet technologies. *Fourteenth Australasian Conference on Information Systems.* Retrieved December 12, 2003, from http://www.we-bcentre.com/acis2003/

Van Akkeren, J., & Cavaye, A. L. (1999, December). Factors affecting entry-level Internet technology adoption by small businesses in Australia: An empirical study. In *Proceedings of the 10th Australasian Conference on Information Systems (ACIS)*, Wellington, New Zealand.

Wagner, E. H., Austin, B. T., Davis, C., Hindmarsh, M., Schaefer, J., & Bonomi, A. (2001). Improving chronic illness care: Translating evidence into action. *Health Affairs, 20*(6), 64-78.

Warsi, A., Wang, P. S., LaValley, M. P., Avorn, J., & Solomon, D. (2004). Self-management education programs in chronic disease: A systematic review and methodological critique of the literature. *Archives of Internal Medicine, 164*(15), 1641-1649.

Wyatt, J. (2004, July). Evaluating health informatics research [Keynote address]. In *Proceedings of the Health Informatics Conference*, Brisbane, Australia.

Chapter XII

PDAs as Mobile-Based Health Information Deployment Platforms for Ambulatory Care:
Clinician-Centric End-User Considerations

Jason Sargent, University of Wollongong, Australia

Carole Alcock, University of South Australia, Australia

Lois Burgess, University of Wollongong, Australia

Joan Cooper, Flinders University, Australia

Damian Ryan, South Eastern Sydney and Illawarra Area Health Service (SESIAHS), Australia

Abstract

This chapter discusses the broad theme of clinician-centric end-user acceptance toward the adoption of personal digital assistants (PDAs) as mobile-based health information deployment platforms within ambulatory care service settings. Personal digital assistants, ambulatory care, and point of care are defined and the interrelatedness of each discussed. Issues, controversies, and problems such as mapping existing workflows, security, and change management are identified, and solutions are suggested for the process of transforming predominantly paper-based ambulatory care systems into electronic point-of-care (ePOC) systems. A current research and development project, the ePOC PDA project, is used as a case study to highlight discussion points. The purpose of this chapter is to illustrate end-user implications and considerations when introducing ePOC systems into ambulatory care service settings and highlight ways and means of improving future levels of acceptance and support of ePOC systems for clinician end users.

Introduction

A paradigm shift within community-based healthcare delivery is under way with regard to clinical-information access and diffusion. Personal digital assistants (PDAs) as mobile health information system deployment platforms are set to move beyond traditional wired networks within bricks–and-mortar hospital walls and increasingly find a place in ambulatory care service settings. Driving this shift is the dichotomy of increased demand for community-based healthcare services (from a growing, aging population) and government e-health technology implementation initiatives (such as the Australian federal government's Health Connect) to enable healthcare services to meet future needs. A convergence of features (computing, telecommunications, and multimedia) into a single device (PDA), increasingly ubiquitous wireless access throughout metropolitan and regional communities, and increased familiarity and acceptance of mobile devices in general as a result of the trend toward mobile computing users ("road warriors") point toward PDAs as ideal platforms for ambulatory care information systems.

PDAs for ambulatory care offer benefits such as the ability for the collection, delivery, and exchange of timely information (both text and images) at the point of care (Walsh, Alcock, Burgess, & Cooper, 2004), leading to a more efficient healthcare system (NSW Health, 2001). PDAs deployed in such contexts empower clinicians, improve decision making, and facilitate improved levels of patient care at the point of care. However, effective management of the development and integration of such mobile systems with regard to end-user acceptance is essential if proposed benefits from this paradigm shift are to be realized. A technically sound, elegant system solution does not in its own right constitute a successful system. Failing to handle correctly the people side of the system has turned technically sound systems into implementation failures (McNurlin & Sprague, 2005). Therefore an understanding and appreciation of the myriad implications of electronic point-of-care (ePOC) systems upon clinicians, the intended end users of such systems, is essential.

This chapter discusses the broad theme of clinician-centric end-user acceptance toward the adoption of PDAs as mobile-based health information deployment platforms within ambulatory care service settings. This is achieved through addressing the following objectives from a clinician-centric (end-user) perspective:

1. Define PDAs, ambulatory care, and point of care.
2. Understand the importance of mapping current paper-based systems, workflows, data requirements, and work practices of ambulatory care end users (clinicians) to any proposed mobile electronic point-of-care system implementation.
3. Identify planning, development, and implementation issues that may impact end-user acceptance and adoption of PDA-based mobile health information systems.
4. Address these issues and propose solutions through a discussion of technology-acceptance frameworks, training, data standards, security, and privacy policies for end users within ambulatory care service settings.
5. Identify other issues and discuss future trends such as emerging technologies and applications, and adapting agile programming techniques to managing user acceptance iteratively throughout the life cycle of mobile-based health information system implementation and integration projects.

Combined, these objectives assist in understanding the potentially supportive role PDAs may provide in the delivery of community health services, particularly ambulatory care services, and will help to conceptualize the full implications of mobile-based health information systems upon end users.

Background

Personal Digital Assistants

PDAs are small, mobile, handheld computer devices that may be programmed for specific functions, such as storing or retrieving information. More specifically, a PDA is a lightweight, usually pen-based computer (Answers.com, n.d.). The term personal digital assistant was coined in 1992 by John Sculley (then CEO [chief executive officer] of Apple), referring to the Apple Newton (Wikipedia, n.d.). Nowadays, the generic terms of popular PDA products are commonly used (searchCIO.com, n.d.), such as Palm or iPAQ to refer to a PDA device.

The lightweight and handheld characteristics make PDAs ideal as mobile-based health information system delivery platforms, while the pen-based characteristic illustrates the utility of the device by enabling data entry through the use of a stylus (pointing device) or handwriting recognition. Increasingly, technology convergence and the trend toward mobile computing have driven demand for these devices. As a result, PDA usage has moved beyond that of an electronic organizer (calendar, contact list, and simple note taking) to now include wireless data access, Web browsing, multimodal telecommunications, and image capture. Initially, the device was designed as an extension of the desktop PC (personal computer). However, the utility of PDAs has extended the device's usage. From a technical perspective, several network connectivity architectures are suitable for enabling the PDA to connect to, download, and upload information from a variety of sources (the process of synchronization). PDAs are well suited to short-range personal networks, medium-range wireless local area networks (WLANs), kiosks, and wireless Internet. The challenge now for community-based healthcare delivery services such as ambulatory care and hospital in the home (HITH) is the process of transforming predominantly paper-based disparate systems into an integrated mobile health information solution, deployable upon PDA devices.

Ambulatory Care and Hospital in the Home

HITH patients are those who, without the provision of the HITH service, would require inpatient care by the nature of their medical or social condition (Grayson, 1996). HITH has been considered as a variant of home-based care, outpatient or ambulatory care, or part of the acute hospital sector (Montalto, 2002). Montalto indicates several factors that have fostered hospital in the home. These factors include economics (an increased demand for accessible and high-quality services, technology associated with hospital care, and increased skill requirement for staff in hospitals), reduced clinical reliance on bed stays for observation, diagnosis or rehabilitation (supplanted by observation technology possible for application on an outpatient basis), and importantly, an emphasis on patient satisfaction.

The mission of Ambulatory Care Australia is "to support Australian healthcare providers as regional & international health industry leaders in the development, trialing and evaluation of ambulatory care initiatives involving the convergence of healthcare, technologies and communications" (Victorian Government Health Information, n.d.). According to Montalto (2002), the integration of hospital services is slowly developing into a valued resource: "There is recognition of the role of community medical and nursing practice in relocating some care previously delivered in a hospital, or in providing care which avoids hospitalization altogether." Montalto views "substitution of this kind as an extension of the notion of integration" (p. 9). Within healthcare settings, Schou (2001) notes "mobile computers only provide value if they do at least one of three things: increase the level of patient care; increase caregiver productivity; help eliminate mistakes." PDAs, or more correctly, the combination of device, application, and information diffusion, enable each of Schou's three value constructs to be achieved in ambulatory care environments. Physicians have incorporated PDAs into their practice settings for some time (McLeod, 2003). Versel (2002) goes further by stating, "PDAs appear to be the driving force behind the increased rate of adoption of Information and Communication Technologies (ICT) by healthcare workers" (p. 13). Schou concludes, "Function-rich and diverse applications make point-of-care data access a healthcare reality."

Transitioning from POC to ePOC

Generally, POC in ambulatory care services refers to the provision of patient care and support in a patient's home environment (which may include an aged care facility). ePOC can be further defined whereby the process of accessing, collecting, and modifying or updating patient data occurs by electronic means (commonly by secure wireless connection) external to a traditional bricks-and-mortar ambulatory care physical infrastructure, usually at the patient's residence.

As illustrated in Figure 1, the practicalities of electronic point of care become problematic for ambulatory care clinicians once they move beyond the wired information network infrastructure of their office, hospital, or outpatient facility. At the point of care (patient's residence), ambulatory care clinicians currently revert to using paper-based records, reviewing printed documents and transcribing patient data gathered at the point of care. These newly amended records must be manually entered back into centralized medical record databases upon the clinician's return to the ambulatory care central office.

A PDA-based mobile health information management system provides an opportunity to leverage mobile technologies for the support of ambulatory care clinicians and in the process, streamline clinicians' and administration workflows, and ultimately improve levels of patient care. It is proposed that electronic point of care, as can be provided through handheld devices such as PDAs, will enable more efficient and effective delivery of ambulatory care services, with consequent improvements in healthcare. The ePOC PDA project is an example of a current e-health research and development initiative that is testing such hypotheses.

Figure 1. Problematic POC electronic patient-data availability

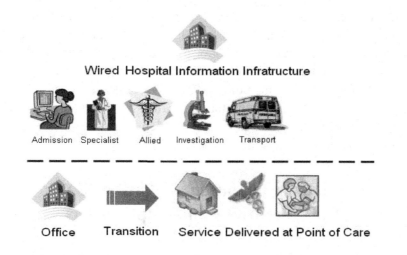

The ePOC PDA Project

The ePOC PDA project involves the design and development of a prototype PDA-based point-of-care information delivery system for The Ambulatory Care Team (TACT), Northern Illawarra. The ePOC project is funded by the Australian government's Australian Research Council (ARC) and is part of their industry linkage scheme, an agreed partnership between researchers and industry, government and community organizations, as well as the international community (Australian Research Council, n.d.). The project timeframe is 3 years (June 2004-June 2007). Presently, a limited-functionality prototype is nearing completion, which will enable the project team to begin data access, collection, and diffusion of clinical information at the point of care. A series of field trials involving a prototype ePOC system is scheduled for commencement in April 2006. Such field trials will continue in tandem with the development of health standard messaging and further transformations of paper-based workflows into ePOC systems.

The philosophy of the project is to utilize generic, reusable, and scalable components that have the capacity to exchange or send data between existing legacy-area health systems (such as hospital or community, patient, pathology, radiology, and medical reference database systems) via a suite of HL7- (Health Level 7) compliant middleware applications (HL7 is a standards-developing organization [SDO] whose domain is clinical and administrative data; HL7, n.d.). The delivered ePOC system will specifically assist in the management of client (patient) information, such as personal information, clinical information, electronic assessment, and healthcare plans in an ambulatory care (point-of-care) environment. In addition to the development of a product, one of the research outcomes of this project is to address the issue of levels of uncertainty for end users, including perceived usefulness and ease of use of proposed technology applications that are mandated. For this to be achieved, however, it is necessary to examine the problems and implications relating to this clinician-centric end-user acceptance approach.

Clinical Information at Point of Care: Considerations and Implications

A Multidimensional Approach to Electronic Point-of-Care System Development

A multidimensional approach needs to be taken toward the planning, development, and integration of mobile health information systems for ambulatory care services. System planning has become business planning; it is no longer just a technology issue (McNurlin & Sprague, 2005). Several issues must be addressed during system development for ambulatory-care-specific applications. Such issues include, but are not limited to, the following:

- Mapping data flows, requirements, and work practices of current systems.
- Limitations of PDA devices.
- The utility of data at point of care.
- Standards and protocols.
- Resistance to change.

Mapping Data Flows, Requirements, and Work Practices of Current Systems

Correctly mapping data flows, requirements, and work practices assists in the design of transforming paper-based systems to electronic systems. Ambulatory care systems are predominantly paper based with clinicians limited by what they can effectively carry during patient treatments at the point of care. A (paper) patient file consists of a number of pieces of documentation: referral, medications, occupational health and safety (OH&S), and so forth. From a human-centric (clinician end-user) perspective, considerable consideration must be given to the transformation from a paper-based to an electronic system to ensure alignment with often entrenched work practices. The level of correctness in mapping existing work flows to any new system will influence end-user acceptance.

During the initial and later stages of the ePOC project, considerable attention has been given to documenting all aspects of workflow and identifying where PDA technology might provide advantages. End-user acceptance should be a key consideration in determining initial wish lists of the proposed PDA application and more realistic expectations of the likely delivered system as a result of iterative consultation (including information sessions, focus groups, and project seminars). During iterative consultations, issues such as the physical limitations of the PDA device, the utility of data at the POC, and protocols and security can be addressed.

Limitations of PDA Devices

Human-Computer Interaction (HCI)

The beneficial characteristic of the size of PDAs in terms of the device's portability also represents a limitation in regard to screen size. The screen size is limited to the physical size of the device. According to Duncan Bremner (n.d.), engineering manager at Intel's silicon design facility in Glasgow, "Typically the screen size of the device will determine what sort of data it can be expected to deal with. Today the standard display sizes tend to be 2 inches for mobile phones, 4 inches for picture phones and PDAs/handheld devices." Many possible solutions exist as a means of overcoming screen-size limitations, and these are discussed in the future-trends section of this chapter. Briefly, by designing compact, menu-driven screens for data collection or the use of image maps for display, end users can be provided with appropriate levels of meaningful clinical information without crowding the limited size of each ePOC PDA screen with information. Determining the most suitable level of text and image combination has been a lengthy and iterative process of prototype screen designs, which, guided by the ePOC project's research philosophy of end-user involvement throughout the entire life cycle of the project, has been driven by clinician end users.

The issue of size was discussed during structured focus groups as part of the pre-implementation phase of the ePOC PDA project. Focus-group participants were generally unconcerned about screen size. Rather, the HCI between end user and device in regard to entering large amounts of text (such as clinical patient notes) was highlighted as problematic. Indications suggested the overall size of the device and portability, however, were viewed in a positive manner as these features of the PDA enabled clinicians to carry the device virtually hidden in a garment pocket or consultation brief case. The reasoning behind this positive perception of the PDA's size was in relation to potential personal security risks posed to clinicians who often are required to deliver ambulatory care services in less than secure, remote locations. Carrying a computer laptop or tablet PC of considerable size from the clinician's vehicle to the patient's home was viewed as "asking for trouble" in some circumstances. Given time during the development phase of any mobile PDA-based implementation, potential negative influences upon the level of acceptance of the device by clinician end users imposed by identified limitations of PDAs can be addressed.

The Utility of Data at the Point of Care

Providing access to data and information resources at the point of care has advantages in enabling timely responses from ambulatory care staff. Disparate databases can prove problematic if they are not integrated with central electronic records. It is essential that ambulatory care systems work within the constraints of existing electronic medical record (EMR) systems. Where specialized reference resources (e.g., Monthly Index of Medical Specialties (MIMS)) are provided via a PDA, some research has suggested that such provision may not be used unless there is clear value to the healthcare professional. One study (Smordal & Gregory, 2003) suggest that nonuse was a significant problem where some information resources were concerned, and that the PDAs might well be regarded less as personal digital assistants and more as "... gateways in a complicated web of interdependent technical and social networks" (pp. 327-328).

The concept of ownership of a clinician's documentation has also been identified as an issue when traditional paper-based systems are transformed into a component of an electronic system. Aydin and Forsythe (1997) discuss public vs. private clinical notes. Their premise is that handwritten notes, while often illegible, feel much less public; that is, they belong to the clinician. Dictated notes, being easier to read and more accessible, are likely to be considered public rather than private. This they believe is signaling a shift from private to public. A component of Aydin and Forsythe's study mentions the suggestion of one clinician to another regarding the act of dictation in front of a patient and the degree to which the clinician felt comfortable doing this activity. Introducing another piece of technology to the patient-clinician interface may result in the same levels of discomfort. This should not, however, be a defining issue in the full acceptance of PDAs in ambulatory care environments. The degree of discomfort, if any, between patient and clinician will diminish over time.

Protocols and Standards

Protocols

A process of following defined protocols of information (patient data) management should facilitate the smooth transition from paper-based to ePOC systems. Well-established work-practice protocols are also an important consideration in regard to achieving end-user support and acceptance of ePOC systems as the ability for a clinician to capture data, make use of the data, and deliver updated records back into established electronic systems post patient visit in a timely (or preferably improved) manner highlights the benefits of using an electronic system. Similarly, a consideration for enhancing end-user acceptance of PDA-based mobile health information management systems by ambulatory care clinicians is the utility of software applications that closely mirror paper-based systems currently in place. If software designed for the PDA mirrors familiar data flows in a standardized format and enables work-practice improvements and efficiencies, clinicians are more likely to demonstrate higher levels of end-user acceptance. The observation by Monahan (2002) that "…even more limiting to PDA perfusion is the software…applications for the medical market have grown up independently and often don't work together" need not affect end-user acceptance levels if the PDA-system software is designed in consultation with end users and tailored to the particular needs of ambulatory care clinicians.

Standards

Security of Electronic Patient Data, Including Away from Centralized Locations

The Mobile Healthcare Alliance in Washington, D.C. names three urgent priorities: data security, interoperability, and interference in regard to the usage of mobile devices for the access and diffusion of patient data (Monahan, 2002). Adopting agreed standards for encryption is one method for addressing security; however, the concept of security goes further when PDAs are utilized as mobile health information management system platforms. From an end-user perspective, security often infers the process of logging onto the PDA device or losing the PDA while traveling between the centralized hospital environment and a patient's place of residence.

Storage and transmission of patient data must conform to established regulatory and ethical standards in regard to security. The development of an ePOC system should follow such standards, therefore alleviating any misgivings by end users toward using a newly integrated technology application to perform clinical tasks. The process of digitizing patient data, being new to the TACT end user, assumes inherent risks that when addressed by protocols and standards, need not become an issue.

Health Level 7 (HL7)

The mission of HL7 (n.d.) is to:

provide standards for the exchange, management and integration of data that support clinical patient care and the management, delivery and evaluation of healthcare services. Specifically to create flexible, cost effective approaches, standards, guidelines, methodologies, and related services for interoperability between healthcare information systems.

While not an overt consideration for end users, the development of ePOC middleware will comply with HL7 standards. The decision to base middleware on these standards supports the notion of developing ePOC as a generic, scalable, and reusable application, well suited to adoption by ambulatory care or similar mobile healthcare service providers.

Resistance to Change

Resistance toward change to any new system is a nontrivial matter. In regard to electronic point-of-care systems, resistance toward change is not solely based around the fundamentals of change management, but also on the broader recognized barriers to the acceptance of the HITH ambulatory care model. Montalto (2002) indicates that these barriers include the following:

- The need to centralize technology and complex care in order to increase expertise and quality through teaching and research.
- Efficiency in managing people with multiple and complex health needs at one site.
- The role of the hospital as a safety net for deficiencies in community-based care.

Change management is the process of assisting people to make major changes in their working environment (McNurlin & Sprague, 2005). Implementing a change-management strategy will help in addressing and alleviating many organizational, service, and end-user concerns regarding the implementation of ePOC systems. Such a methodology is proactive rather than reactive.

This sampling of issues indicates the difficult task faced by mobile e-health system designers and highlights intricate considerations that may potentially impact the design and implementation (and ultimately the level of end-user acceptance) of PDAs as deployment platforms for mobile-based health information systems within ambulatory care. A well-defined process of developing solutions and mitigation strategies will assist in addressing and overcoming many of these limitations previously identified. Areas of particular interest to end users that present possible solutions to overcoming identified limitations, and in turn transform current ambulatory care systems, include technology-acceptance frameworks, training, and conducting systems development using an iterative, consultative approach.

Toward Achieving Clinician Acceptance of ePOC Systems

Three levels of perceptions in regard to proposed technology implementations in healthcare settings must be acknowledged to achieve best fit: perceptions at an organizational level (government health departments), at the service level (community health services), and at the end-user level. At the organizational level, the New South Wales Department of Health has demonstrated its appreciation of e-health applications by investing in CHIME (Community Health Information Management Enterprise; NSW Health, n.d.). The department has dedicated significant resources to the development of an EMR. E-health is demonstrably high on their agenda, and South Eastern Sydney and Illawarra Area Health Service (SESIAHS) has adopted this NSW Department of Health philosophy.

At the service level (SESIAHS), health technology is prominent in non-clinical applications already, with a prime example being the Picture Archiving Communication Systems (PACS) for radiology services, which is considered a state-of-the-art e-health system. SESIAHS has additionally invested in CHIME in community health with POC applications. In this capacity, SESIAHS has invested resources into ePOC, which illustrates commitment to e-health, particularly in the area of the health service level.

ePOC for Ambulatory Care Services

In the current public health environment of limited resources, one option to increase capacity with the same resources is to increase efficiency. Thus, having an ePOC system is one way of achieving this outcome. ePOC, deployed using PDAs, is intended to achieve the following:

- Be easy to use and integrate with work practices.

- Be used at the point of care and within a wired hospital or clinical outpatient facility.

- Push information to healthcare and deliver timely information for decision making.

- Utilize generic middleware, interfaces, and healthcare application software.

- Incorporate security, authentication, and privacy solutions, including policies.

- Address communication technology interface issues (Walsh, 2003).

A proposed ePOC example configuration is illustrated in Figure 2.

Figure 2. Example ePOC Configuration: PDA, communications, and database

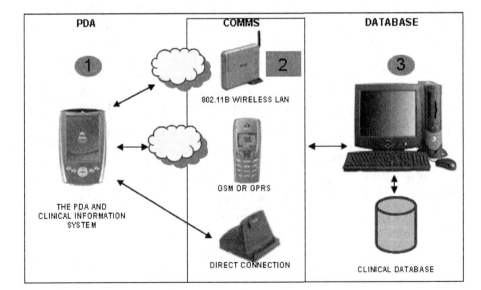

According to Moore (2001), "Practitioners need to access large volumes of information quickly, and they can't keep running back to centralized computers while they're taking care of their patients" (p. 110). This scenario holds true for community-based healthcare clinicians such as ambulatory care team members who visit patients away from centralized hospital settings. Moore's description of practitioners' information requirements adds substance to the need for the development of mobile- (PDA) based health information systems for community-based healthcare service providers. The need for a mobile information base is relevant throughout the healthcare sector. However, the focus of the ePOC PDA project is supporting the healthcare staffs' activities and less on the issues of quality of the underlying patient record data that we consider to be the responsibility of the Electronic Patient Record (EPR) (Raatikainen, Christensen, & Nakajima, 2002).

Conceptualizing the Benefits of Electronic Point of Care

There are many benefits for ambulatory care staff in having an electronic point-of-care health information management system. These benefits include, but are not restricted to, real-time communication of information between staff via PDA phone, text, and e-mail; and increased sense of personal security and OH&S. As previously discussed, the point-of-care environment has the potential to be an isolated environment at times, especially after hours. One often-underestimated advantage of ePOC is in OH&S benefits. Currently, nursing staff must carry up to eight sets of patient files to and from their vehicle to the patient's residence. Even with the assistance of trolleys, this is still a cumbersome exercise. Replacing these sets of patient files with a single PDA device has obvious benefits. Perhaps most importantly, from a service perspective, ePOC will enable clinicians to deliver contemporaneous information to patients, facilitate better access to requested ambulatory care services, provide more efficient POC illness management, and therefore deliver improved health outcomes. Additionally, the ability to document a patient's episode of care, at the point of care, will include the following:

- Patient care plans, assessments, progress notes, and variances.
- Ability to write and send admission or discharge letters.
- Direct electronic referrals to community health systems.
- Statistical data.

Other Benefits: Tangible and Intangible

Other benefits are expected from the implementation of ePOC within TACT at the organizational, service (staff), and patient levels. For the organization, better time management, anticipated reduction in workload leading to increased productivity, improved communication between members of the multidisciplinary team (TACT) and other service providers, improved statistical data collection, and the development of research partnerships with tertiary higher education and research institutions such as the University of Wollongong are seen to be areas of major benefit. The greatest benefit of an electronic point-of-care system, however, is the possibility of empowering TACT staff. Information is power, and ePOC will allow access to more health information, when required, at the point of care.

Empowering Clinicians at Point of Care

From an end-user (clinician) perspective, an electronic point-of-care system provides the means by which a multitude of information sources can be utilized to determine or verify the correct course of action for the current episode of care delivered to a patient. A PDA-based mobile system enables the clinician to perform a wide range of empowering (value-adding) functions at the point of care. For TACT staff, the major benefit from ePOC is anticipated to be access to clear, consistent information, including the following:

- Protocols.
- Reference material; one example may be the ability to browse online medical and drug information databases because "there is a general belief that the usefulness of drug information and compliance with formulary systems are enhanced if this information is available at the time of prescribing" (McCreadie, Stevenson, Sweet, & Kramer, 2002; p. 1340).
- Previous and current medical records.
- Pathology or x-ray results.
- Street maps and directions.
- Security, such as being able to send a distress signal to other staff or police.
- Track whereabouts back at the head office.

Clinician Empowerment: An Example from the Field

Wound care is a particular area of ambulatory care that has the potential to be improved through the use of a point-of-care system. Presently, it is commonly accepted practice for a clinician at the point of care to communicate by telephone with the medical director of TACT to discuss the condition of patients' wounds. The clinician at the point of care must descriptively explain the condition of the wound. A PDA-based point-of-care system takes advantage of enabling the clinician to search the Web for information on types of wounds and their treatment. Furthermore, by capturing an image of the wound, the clinician can transmit the image over the Internet back to the base (TACT office) where the medical director or other senior team member can view the wound and discuss the correct method of care with the clinician over the PDA telephone in real time. This scenario is an obvious improvement over the practice of having the patient returned to the hospital or ambulatory care outpatient facility. Training in the functionality of the PDA ensures TACT clinicians are proficient in capturing images, connecting to and browsing the Web, and making voice calls. The result is a transformation of work practice and efficiencies, and improved levels of patient care.

Traditional User-Acceptance Methodologies

A priori studies have explored socio-cognitive technology-acceptance models and frameworks as ways and means of increasing user acceptance of technology integration. Such models and frameworks include the technology-acceptance model (TAM; F. D. Davis, 1985), extended technology-acceptance model (TAM 2; Venkatesh & Davis, 2000), the theory of reasoned action (TRA; Azjen & Fishbein, 1980), and the theory of planned behavior (TPB; Azjen, 1985). A fundamental difference between variants of technology-acceptance models and the approach proposed in the ePOC project is the time frame in which review of the information system being studied occurs.

TAM and TAM 2 are tools that evaluate the perceived ease of use of technology application adoption (such as PDAs); however, the time when the process of managing an information system adoption occurs is made by reviewing prior actual adoptions, investigating variance of perceptions, and applying the results to subsequent implementations. Liang, Xue, and Byrd's (2003) study of the usage of TAM 2 to predict actual PDA usage among healthcare professionals is such an example of this traditional technology-acceptance approach, in which there is a review of past implementations and findings are applied to the new or next implementation. F. Davis and Venkatesh (2004) hypothesize that stable and representative measures of perceived usefulness only require that potential users be informed of what a system will be designed to do, that is, its intended functionality, and do not require hands-on interaction with a working system.

Mandated Technology Implementations

From the perspective of end users, the mode of technology adoption, voluntary vs. mandated, must be considered in planning system implementations, particularly when the proposed system is intended to replace paper-based systems. Such existing legacy systems invariably involve entrenched work practices. A voluntary-use environment is defined as "one in which users perceive the technology adoption or use to be a willful choice; a mandated environment is where users perceive use to be organizationally compulsory" (Brown et al., 2002, p. 284). In these environments, the new system must be used to complete job tasks that are tightly integrated with the tasks of multiple workers (Brown, Massey, Montoya-Weiss, & Burkman, 2002).

Perry, O'Hara, Sellen, Brown, and Harper (2001) indicate "the greater unpredictability—or heterogeneity (Krestoffersen & Ljungberg, 1999)—of the contextual constraints within which mobile work must take place means that mobile workers have less control over the configuration of their environment, and therefore the way they manage their work." The ePOC PDA project contests Perry et al.'s premise through the hypothesis that a nontrivial, consultative, user-centric approach toward design and development of the ePOC system will deliver control over configuration and therefore the way ambulatory care clinicians manage their work.

Adopting a Consultative Approach

A consultative, iterative approach to project management that addresses system implications upon end users is particularly well suited to technology-integration projects within small to medium-sized community-based healthcare services, such as ambulatory care services. When the mode of adoption and implementation is mandatory (the decision to adopt the proposed technology has been made by management or users other than intended end users), a consultative approach offers the potential to improve levels of end-user acceptance by providing a sense of diminished fears of ill-fitting, poorly designed systems, where anecdotal evidence purports end users being "lumped" with systems that fail to meet their particular needs. A consultative approach provides end users with a sense of system ownership, similar to Krall's (1998) concept of clinician "buy-in." Subsequently, results achieved in questionnaires and surveys conducted as part of the research methodology for the ePOC PDA project confirm increased levels of acceptance and support for a proposed, mandated ePOC system for ambulatory care (Sargent, Burgess, Cooper, & Ryan, 2005).

Training: Enhancing End-User Acceptance of ePOC Systems through Consultation

A proactive approach to enhancing end-user acceptance of ePOC systems is through training during the pre-implementation phase of system development. For TACT members, the level of computing skills and familiarity with mobile devices in general varies across the spectrum from average to quite limited or no experience. Acknowledging this spread, a tandem training and education program for PDA usage and the evolving ePOC system assists in developing the skills base of all TACT staff and enabling a better understanding of the processes involved in the proposed ePOC system. This facilitates higher levels of clinician acceptance of the final delivered product as input received by clinicians is iteratively incorporated into the ongoing development of the ePOC system.

It is hoped by taking this consultative approach, user-acceptance levels approaching 100% will be achieved for clinician end users of the mandated ePOC system. The size of TACT is the dependent factor that enables such a high level of acceptance to be predicted. The ePOC PDA project is envisaged to act as a test bed for this approach. Other outcomes and future research of the current project include the development of a framework for enhancing end-user acceptance, developing an instrument (scale) for measuring end-user acceptance, dismissing the preconceived notion of community healthcare clinicians' low level of user acceptance of new technology, and the refinement and validation of a proposed technology-acceptance framework in a future study (larger test bed, larger and more complex organization).

Future Trends

The concept of utilizing PDA-based mobile health information management systems for ambulatory care has begun. Implementation projects such as the ePOC PDA project are demonstrating the potential benefits of the alignment of mobile technology within community-based healthcare service settings. Advancements in technology, refinements in workflow analysis, and work practices that emanate from such projects will drive further changes in the area of mobile healthcare delivery. Future trends such as alternative data-entry methods, decision-support systems, Web services and service-oriented architectures, and GIS (Geopgraphical Information System) and GPS (Global Positioning System) applications are just some areas that may play a part in the evolutionary component design of ePOC systems in the coming years. Additionally, moving toward an adapted agile programming approach in regard to iterative consultation throughout the traditional development life cycle offers the potential to positively impact end-user acceptance of mobile-based health information management systems.

Alternative Data-Entry Methods

Generally, the stylus as a method for data input and touch-screen navigation through assorted PDA applications is adequate. End users familiar with navigating online forms and Web-based applications are well accustomed to check boxes, drop-down menus, and radio buttons. However, Monahan (2002) sees the stylus as an example of "clumsy options for data entry." This manifests as a potential drawback in the wider adoption of PDAs when extensive numbers of individual characters or blocks of data are required as input. This issue of data entry can addressed, and subsequently the utility of PDAs as mobile health information system deployment platforms further enhanced, by an assortment of peripheral devices and software applications such as keyboards and voice recognition.

Figure 3. Foldout wireless keyboard peripheral for PDA

Keyboards

Manufacturers including UK-based O_2 have recently released PDAs with inbuilt keypads (O_2 XDAIIs). However, the ease of use is still limited when compared with PC and notebook computer keypad sizes, which end users have become accustomed to using. Foldout keyboards (also referred to as expandable keyboards) provide a solution to data entry for PDAs. These keyboards, as depicted in Figure 3, offer a compromise in the size and ease of use between full-sized keyboards and the limited (mini) keypads of the latest PDAs.

An additional data-input option is the use of a virtual keyboard as illustrated in Figure 4. The virtual keyboard requires only a flat surface on which to beam the keypad onto. Once configured with compatible PDAs, keystrokes are registered as the user types onto the virtual keys on the flat surface. The advantages of the virtual keyboard are in its application in sterile healthcare settings (there is no physical keyboard to touch), its price, and the ability to provide another option for data input. However, keystroke error rates and the suitability of the device under varying ambient lighting conditions tend to limit its utility. In such cases, investigating other options for efficient data input, such as voice-recognition software, is prudent.

Figure 4. Virtual keyboard

Voice Recognition

The process of writing (or dictating) clinical notes takes considerable time for ambulatory care clinicians at the point of care. Voice-recognition software applications that run on PDAs offer a viable alternative to scribing patient notes. Entering considerable blocks of text is impractical using a stylus as it is tiresome and error prone. VoiceRecognition.com (2005) sees "an exponential trend toward better speed and accuracy each and every time a new upgrade is released." ScanSoft (Dragonsys, 2005b), a leading developer of voice-recognition software, including Dragon Naturally Speaking 8 Medical, indicates accuracy levels of 99% (comparable to many manual transcriptions) for Version 8.0 of its medical-based voice-recognition software. Additionally, healthcare clinicians benefit from specialty vocabularies such as radiology, general medical, ob-gyn, pathology, orthopedic, emergency, mental health, and oncology (Dragonsys, 2005a).

Taking an Agile Approach to Project Management

Apart from technology applications, another future trend may be in the area of adapting agile programming techniques to system development of ePOC systems for organizations of similar size to TACT. The agile aspect refers to the intentional intervention by system researchers, designers, and developers at points throughout the traditional system-development life cycle with the purpose of addressing end-user concerns proactively rather than reactively at the end of a project. Sargent et al. (2005) suggest such an approach leads to far higher levels of end-user acceptance and ongoing support for systems such as ePOC as end users partly drive the development of the proposed system. This leads to achieving a best fit of the proposed system with end users' expectations. According to Goulielmos (2003), IS developments are still regarded as a technical effort and not as a socio-technical process that unfolds within an organization. An agile approach addresses this issue as system-integration projects unfold and go further by focusing upon implications and considerations from a clinician-centric end-user perspective.

Other Future Trends

Several other future trends may emerge that could impact ePOC systems. Although the following applications are well established in their own right, the adaptations for use as part of an ePOC system are possible:

- Decision-support systems.
- Web services and service-oriented architectures.
- GIS and GPS applications in developing a patient scheduling and management subsystem that prioritizes patients based on the criticality of care. Such a system could determine the most efficient caseload, management, and travel route. Utilizing GPS and deploying location-based services, such a system would benefit mobile ambulatory care services through updating and monitoring of patient visits by clinicians in a real-time environment.

Conclusion

The design and integration of PDA-based point-of-care health information management systems for ambulatory care is in its infancy. The demonstrated needs for such systems has been established through the exploration of transforming paper-based systems, improving workload efficiencies for clinicians, and developing solutions to address necessary increasing resource deployment under tightening budgetary constraints experienced by the healthcare sector in Australia. Through careful design and development of ePOC systems by taking an iterative, consultative approach to addressing clinician-centric end-user concerns, levels of acceptance and ongoing support by clinicians are improved, ensuring a smooth transition from current paper-based systems to the adoption of PDAs as mobile health information management systems for ambulatory care.

References

Answers.com. (n.d.). *Personal digital assistant.* Retrieved July 26, 2005, from http://www.answers.com/topic/personal-digital-assistant

Australian Research Council. (n.d.). *Linkage-programs.* Retrieved August 19, 2005, from http://www.arc.gov.au/grant_programs/linkage_projects.htm

Aydin, C. E., & Forsythe, D. E. (1997). *Implementing computers in ambulatory care: Implications of physician practice patterns for system design.* Retrieved September 2, 2005, from http://www.amia.org/pubs/symposia/D004119.PDF

Azjen, I. (1985). From intentions to actions: A theory of planned behaviour. In J. Kuhl & J. Beckmann (Eds.), *Action control: From cognition to behaviour* (pp. 11-39). Heidelberg, Germany: Springer.

Azjen, I., & Fishbein, M. (1980). *Understanding attitudes and predicting social behaviour.* Englewood Cliffs, NJ: Prentice Hall.

Bremner, D. (n.d.). *Generations ahead* [Powerpoint presentation to Scottish enterprise CommTech]. Retrieved September 15, 2005, from http://www.scottish-enterprise.com/publications/generations_ahead_hr.zip

Brown, S. A., Massey, A. P., Montoya-Weiss, M. M., & Burkman, J. R. (2002). Do I really have to? User acceptance of mandated technology. *European Journal of Information Systems, 11*, 283-295.

Davis, F., & Venkatesh, V. (2004). Toward preprotoype user acceptance testing of new information systems: Implications for software project management. *IEEE Transactions on Engineering Management, 51*(1), 31-46.

Davis, F. D. (1985). *A technology acceptance model for empirically testing new end-user information systems: Theory and results.* Unpublished doctoral dissertation, Massachusetts Institute of Technology, Cambridge: MA.

Dragonsys. (2005a). *Dragon NaturallySpeaking 8 Medical.* Retrieved July 27, 2005, from http://www.dragonsys.com/naturallyspeaking/medical/

Dragonsys. (2005b). *ScanSoft company overview.* Retrieved July 27, 2005, from http://www.dragonsys.com/company/

Goulielmos, M. (2003). Outlining organisational failure in information systems development. *Disaster Prevention and Management, 12*(4), 319-327.

Grayson, L. (1996). *What is hospital in the home?* Proceedings of the Australian Home and Outpatient Intravenous Therapy Association Annual Scientific Meeting, Melbourne, Australia.

Health Level 7 (HL7). (n.d.). *What is HL7?* Retrieved September 21, 2005, from http://www.HL7.org/

Koblentz, E. (n.d.). *The history of PDAs: Evolution of the PDA 1975-1995.* Retrieved July 26, 2005, from http://www.snarc.net/pda/pda-treatise.htm

Krall, M. A. (1998). Achieving clinician use and acceptance of the electronic medical record. *Permanente Journal. 2*(1), 48-53. Retrieved September 20, 2005, from http://xnet.kp.org/permanentejournal/winter98pj/emr.html

Krestoffersen, S., & Ljungberg, F. (1999). Making place to make IT work: Empirical explorations of HCI for Mobile CSCW. *GROUP '99: Proceedings of the International ACM SIGGROUP Conference on Supporting GroupWork,* 276-285.

Liang, H., Xue, Y., & Byrd, T. (2003). PDA usage in healthcare professionals: Testing an extended technology acceptance model. *International Journal of Mobile Communications, 1*(4), 372-389.

McCord, L. (2003). Using a personal digital assistant to streamline the OR workload. *Journal of Association of Perioperative Registered Nurses, 78*(6)996-1001. Retrieved July 28, 2005, from http://www.findarticles.com/p/articles/mi_ m0FSL/is_6_78/ai_111895687

McCreadie, S. R., Stevenson, J. G., Sweet, B. V., & Kramer, M. (2002). Using personal digital assistants to access drug information. *American Journal of Health-System Pharmacy, 59*(15), 1340-1343.

McLeod T.G., Ebbert, J. O., Lymp, J. F. (2003). Survey assessment of personal digitial assistant use among trainees and attending physicans. *Journal of the American Medical Informatics Association, 10*(6), 605-607.

McNurlin, B. C., & Sprague, R. H. (2005). *Information systems management in practice* (7th ed.). NJ: Pearson Prentice Hall.

Monahan, T. (2002). Hot & cold on PDAs. *Health Informatics Online.* Retrieved July 25, 2005, from http://www.healthcare-informatics.com/issues/2002/05_02/ cover.htm

Montalto, M. (2002). *Hospital in the home: Principles & practice.* Ivanhoe, Victoria: Artwords.

Moore, S. K. (2001). Unhooking medicine [wireless networking]. *Spectrum, 38*(1), 107-108, 110.

NSW Health. (n.d.). *CHIME overview.* Retrieved September 20, 2005, from http:// www.chime.gov.au/iasd/chime/overview/index.html

NSW Health. (2001). *Information management division: Information management and technology, strategic plan.*

Perry, M., O'Hara, K., Sellen, A., Brown, B., & Harper, R. (2001). Dealing with mobility: Understanding access anytime, anywhere. *ACM Transactions on Computer-Human Interaction, 8*(4), 323-347.

Raatikainen, K., Christensen, B. H., & Nakajima, T. (2002). Application requirements for middleware for mobile and pervasive systems. *ACM SIGMOBILE Mobile Computing and Communications Review, 6*(4), 16-24.

Sargent, J., Burgess, L., Cooper, J., & Ryan, D. (2005, November-December). *Enhancing user acceptance of mandated technology implementation in a mobile healthcare setting: A case study.* Paper presented at the Australasian Conference on Information Systems (ACIS) 2005: Socialising IT: Thinking About the People, Manly, Australia.

Schou, J. (2001). Information where it's needed. *Health Management Technology, 22*(10), 48.

searchCIO.com. (n.d.). *Definitions: Personal digital assistant.* Retrieved August 7, 2005, from http://searchcio.techtarget.com/sDefinition/0,,sid19_gci214287,00. html

Smordal, O., & Gregory, J. (2003). Personal digital assistants in medical education and practice. *Journal of Computer Assisted Learning, 19*, 320-329.

Venkatesh, V., & Davis, F. D. (2000). A theoretical extension of the technology acceptance model: Four longitudinal field studies. *Management Science, 46*(2), 186-204.

Versel, N. (2002). Wave of the (not-so-distant) future: Annual healthcare IT survey shows rise in technology adoption. *Modern Physician, 6*(11), 12.

Victorian Government Health Information. (n.d.). *Ambulatory care Australia: Innovation in healthcare.* Retrieved July 25, 2005, from http://www.health.vic. gov.au/aca/

VoiceRecognition.com. (2005). *Industry trends.* Retrieved July 27, 2005, from http://www.voicerecognition.com/trends/

Walsh, D. (2003). *Discovering the requirements for a personal digital assistant based electronic medical records system for ambulatory care services.* Unpublished honour's thesis, University of Wollongong, Wollongong, New South Wales, Australia.

Walsh, D., Alcock, C., Burgess, L., & Cooper, J. (2004). PDAs and effective community healthcare delivery: A mobile technology solution to point-of-care health services delivery for ambulatory care. *Proceedings of CollECTeR LatAm 2004.* Retrieved September 29, 2005, from http:// www.collecter.org/coll2004latam/pdf/MobileTech/1_Walsh.pdf

Wikipedia. (n.d.). *Personal digital assistants.* Retrieved July 20 from http:// en.wikipedia.org/wiki/Personal_digital_assistant

Section V

Industrial Applications

Chapter XIII

3G Mobile Medical Image Viewing

Eric T. T. Wong, The Hong Kong Polytechnic University, Hong Kong

Carrison K. S. Tong, Pamela Youde Nethersole Eastern Hospital, Hong Kong

Abstract

Tele-radiology is the technology of remote medical consultation using X-ray, computed tomographic, or magnetic resonance images. It was commonly accepted by clinicians for its effectiveness in making diagnoses for patients in critical situations. Because of the huge size of data volume involved in tele-radiology (American College of Radiology [ACR], 2003), clinicians are not satisfied with the relatively slow data-transfer rate. It limits the technology to fixed-line communication between the doctor's home and his or her office. In this project, a mobile high-speed wireless medical image viewing system using a 3G (third-generation) wireless network (Collins & Smith, 2001), virtual private network, and one-time two-factor authentication (OTTFA) technologies is presented. Using this system, tele-radiology can be achieved by using a 3G PDA (personal digital assistant) phone to query, retrieve, and review the patient's record at anytime and anywhere in a secure environment. Using this technology, the patient-data availability can be improved significantly, which is crucial to timely diagnoses of patients in critical situations.

Introduction

Tele-radiology is the technology of remote medical consultation using X-ray, computed tomographic (CT), or magnetic resonance (MR) images. This technique is commonly accepted by clinicians for its effectiveness in making diagnoses for patients in critical situations. For effective implementation of tele-radiology, many technical problems including data integrity, accessibility, size of data volume, compression method, and bandwidth of linkage should be considered. Hitherto, due to the huge size of data volume involved, clinicians are not satisfied with the slow data-transfer rate. It limits the use of the technology to fixed-line communication between a doctor's office and his or her home. In this project, a mobile high-speed wireless medical image viewing system using third-generation (3G) mobile technology, a virtual private network (VPN), common gateway interfacing (CGI), dynamic JPEG (Joint Photographic Experts Group) compression, the Web, the structural query language (SQL; DuBois, 2002), digital imaging and communication in medicine (DICOM; National Electrical Manufacturers Association [NEMA], 2004), and one-time two-factor authentication (OTTFA) technologies were developed. Using this system, tele-radiology has been enhanced to a large extent; image-data query and retrieval can be transferred from a hospital data centre to any notebook personal computer (PC), or to any 3G personal digital assistant (PDA) phone at anytime and anywhere in a secure environment. Hence, the patient-data availability can be improved significantly, which is quite important for patients in critical situations.

Background

Tele-radiology involves the process of sending radiographic images from one point to another through digital standard telephone lines, wide area networks (WANs), or over a local area network (LAN). The radiographic images can be acquired either by a video-capture board such as a frame grabber or the console of a medical imaging modality. After acquisition, the images were digitally stored in a tele-radiology workstation in which the images were ready to be sent to a remote site over a network such as an Ethernet.

In the field of medical imaging, most of the images were stored in DICOM formation. DICOM is a standard that is a framework for medical imaging communication. It was developed by the American College of Radiology (ACR) and the National Electrical Manufacturers Association (Bidgood & Horii, 1992) with input from various vendors, academia, and industry groups. It is referred to as Version 3.0 because it replaces Versions 1.0 and 2.0 of the standard previously issued by ACR and NEMA, which was called the ACR-NEMA standard. It provides standardized formats for images, a common information model, application service definitions, and protocols for communication. Based upon the open-system interconnect (OSI) reference model, which defines a seven-layer protocol, DICOM is an application-level standard, which means it exists inside Layer 7 (the uppermost layer).

Today, most tele-radiology systems run over standard telephone lines (Oguchi, Murase, Kaneko, Takizawa, & Kadoya, 2001) and ISDN (integrated services digital network), which are available in most parts of the world. Other high-speed lines, including T1 lines and SMDS (shared multi-megabit data services), will also become more popular as their prices continue to drop. Over the next couple of years, we should see a substantial migration to wireless networks such as IEEE (Institute of Electrical and Electronics Engineers) 802.11x wireless and 3G (Collins & Smith, 2001) networks, which offer higher flexibility than fixed-line networks.

Digital images, whether viewed on a computer monitor, transmitted over a phone line (Reponen et al., 2000), or stored on a hard disk or archival medium, are pictures that have a certain spatial resolution. The spatial resolution, or size, of a digital image is defined as a matrix with a certain number of pixels (information dots) across the width of the image and down the length of the image. The more the number of pixels, the better the image resolution. This matrix also has depth. This depth is usually measured in bits and is commonly known as shades of grey: A 6-bit image contains 64 shades of grey, 7-bit 128 shades, 8-bit 256 shades, and 12-bit 4,096 shades. The size of a particular image is referenced by the number of horizontal pixels by (or times) the number of vertical pixels, and then by indicating the number of bits in the shades of grey as the depth. For example, an image might have a resolution of 640x480 and contain 256 shades of grey (8 bits deep). The number of bits in the data set can be calculated by the product of 640x480 and 8, which equals 2,457,600 bits. Since there are 8 bits in a byte, the 640x480 image with 256 shades of grey is 307,200 bytes or 0.3072 megabytes of information.

Although images should be permanently archived as raw data or with only lossless data compression (i.e., no data is destroyed), hardware and software technology exists that allows tele-radiology systems to compress digital images into smaller file sizes so that the images can be transmitted faster. Compression (Nelson & Gailly, 1995) is usually expressed as a ratio: 3:1, 10:1, or 15:1. A 10:1 compression factor means that for each piece of information in the original image's matrix, 10 are compressed. Certain images can withstand a substantial amount of compression without a visual difference: CT and MR images have large areas of black surrounding the actual patient image information in virtually every slice. The loss of some of those pixels has no impact on the perceived quality of the image nor does it significantly change reader-interpretive performance.

Transmission time has to follow the laws of size. The only way to reduce the transmission time is either to increase the speed of the modem or reduce the number of bits (compress the image) being sent.

The following formula is used to calculate the time to transmit an image:

```
(Matrix Size) x (Matrix Depth + X bits) x (Percentage of Compression) /
          (Network Speed) = Transmission Time (Seconds),
```

where matrix depth is expressed as shades of grey: 256 shades of grey equal 8 bits, 128 shades of grey equal 7 bits, and 64 shades of grey equal 6 bits. For transmission control, all devices add X bits as overhead when transmitting data.

Client-server computing developed from the need to move systems used for application development and operations from expensive mainframes to more efficient, less expensive, yet just as powerful, workstations. A client-server architecture involves the use of two types of computers: a client computer, which runs applications and makes requests for data and other resources, and a server, which processes the client's requests by distributing the requested resources. Client-server computing is known as a cooperative distribution system because both the client and the server cooperate in performing a task. For example, if a client requests a record from the server, the server uses its resources to process the entire file while the client computer uses its resources to run an application that reads and writes individual records on the file. The server does not need to send the entire file to the client, thus diminishing network traffic or traffic over communication lines.

Client-server computing has numerous advantages for the medical world and for the field of radiology in particular. It permits workstations to achieve computing power previously only available from mainframes at a fraction of the mainframe costs. By efficiently dividing resources, client-server computing reduces network traffic and improves response time. This efficiency offers a significant advantage to physicians who need to receive images quickly and who require real-time image navigation and manipulation to perform diagnostic tasks effectively. Client-server computing facilitates the use of graphical user interfaces (GUIs), making tele-radiology and picture archiving and communication system (PACS) applications (Tong & Wong, 2005) easier to use and more responsive. Additionally, clients and servers can be run on different platforms, allowing end users to free themselves from particular proprietary architectures. Software applications designed for client-server computing can interface seamlessly with most hospital information systems (HISs; Royal College of Radiologists [RCR], 1999) or radiology information systems (RISs) while providing rapid soft-copy image distribution.

While most tele-radiology systems used over the last decade were intended for on-call purposes, the past 2 years have seen a rapid increase in the use of tele-radiology to link hospitals and affiliated satellite facilities, other primary hospitals, and imaging centers. As tele-radiology allows radiologists to use their time more efficiently, the volume of images transmitted increases without substantially increasing the costs. Although a number of enabling technologies have been developed for effective

Figure 1. Schematic diagram of traditional tele-radiology

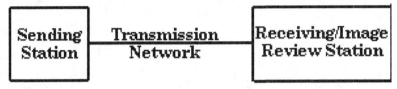

overread networks, such as the more affordable high-speed telecommunications networks and improved data-compression techniques (Nelson & Gailly, 1995) in recent years, the traditional tele-radiology technology is only available within fixed-line communication systems between the doctor's home and his or her office, and this requires manual image selection from the hospital side. Both factors have limited the flexibility of the tele-radiology technology.

Solution

3G (Collins & Smith, 2001) is a generic name for a set of mobile technologies launched at the end of 2001 using a host of high-tech infrastructure networks, handsets, base stations, switches, and other equipment to allow mobiles to offer high-speed Internet access, data, video, and CD-quality music services. Data speeds in 3G networks, being up to 2 Mbps, show an improvement in the current technology. Today, various standards of 3G techniques are found in the commercial market including WCDMA (wideband code division multiple access), CDMA (code division multiple access) 2000, UMTS (universal mobile telecommunications system), and EDGE (enhanced data rates for global evolution) technologies.

Types of 3G

WCDMA

WCDMA is a technology for wideband digital radio communications for the Internet, multimedia, video, and other capacity-demanding applications. WCDMA has been selected for the third generation of mobile telephone systems in Europe, Japan, and the United States. Voice, images, data, and video are first converted to a narrowband digital radio signal. The signal is assigned a marker (spreading code) to distinguish it from the signal of other users. WCDMA uses variable-rate techniques in digital processing and it can achieve multi-rate transmissions. WCDMA has been adopted as a standard by the International Telecommunication Union (ITU) under the name International Mobile Telecommunications-2000 (IMT-2000).

CDMA 2000

Commercially introduced in 1995, CDMA quickly became one of the world's fastest growing wireless technologies. In 1999, the ITU selected CDMA as the industry standard for new 3G wireless systems. Many leading wireless carriers are now building or upgrading to 3G CDMA networks in order to provide more capacity for voice traffic, along with high-speed data capabilities. Today, over 100 million consumers worldwide rely on CDMA for clear, reliable voice communications and leading-edge data services.

UMTS

UMTS is the third-generation mobile telephone standard in Europe, standardized by the European Telecommunications Standards Institute (ETSI). It uses WCDMA as the underlying standard. To differentiate UMTS from competing network technologies, UMTS is sometimes marketed as 3GSM, emphasizing the combination of the 3G nature of the technology and the GSM (Global System for Mobile Communications) standard that it was designed to succeed. At the air-interface level, UMTS itself is incompatible with GSM. UMTS phones sold in Europe (as of 2004) are UMTS/GSM dual-mode phones; hence, they can also make and receive calls on regular GSM networks. If a UMTS customer travels to an area without UMTS coverage, a UMTS phone will automatically switch to GSM (roaming charges may apply). If the customer travels outside of UMTS coverage during a call, the call will be transparently handed off to available GSM coverage. However, regular GSM phones cannot be used on the UMTS networks.

EDGE

EDGE is a technology that gives GSM the capacity to handle services for the third generation of mobile telephony. EDGE was developed to enable the transmission of large amounts of data at a high speed: 384 kbps. EDGE uses the same TDMA (time division multiple access) frame structure, logic channel, and 200 kHz carrier bandwidth as today's GSM networks, which allows existing cell plans to remain intact.

Implementation

The proposed system has been designed for image transfer using a UMTS type of 3G wireless network (Figure 2), which could provide a data-transfer rate of at least 384 kbps, that is, about 10 times the speed of GSM (Tong, Chan, & Wong, 2003). The security risk for this kind of connection to the hospital enterprise network is also 10 times greater. One way to enhance the security is to use one-time two-factor authentication technologies for effective access control as shown in Figures 4, 5, and 6. The access-right control and audit trail are based on the VPN technology. For the management of terabytes of image data, an SQL-based database could be used. Users can search the database for any interested studies with the related images (Figures 7, 8, 9, 10, and 11). This design allows the clinician to search and retrieve the required studies without any help from the hospital side.

Web technology, today, is the most robust client-server computing technology. Web technology has been applied in many areas including desktop and mainframe computers. A Web browser is also available in some mobile phones such as the PDA phone used in this project. In the proposed system, all CT and MR images were stored in a DICOM image-archiving server. Apache-based Web server software (Kabir, 1999) was installed in an existing DICOM image server connected with an SQL server for image-data management. The image data stored in the DICOM server would allow queries from the Web browser through the 3G mobile PDA phone (Figure 3). The retrieved images were first converted from 16 bits into 8 bits, and then JPEG compressed before being sent to the remote client as a Web page with configurable parameters such as image size, window level and width, and rotation and flip using the CGI method (Tong et al., 2003). CGI is one of the common techniques for the manipulation of data in Web technology. The installed CGI software is an interface between the DICOM server and Web client sides. It talks to the DICOM server using SQL and DICOM commands and communicates with the Web clients using the hypertext markup language (HTML). The clinical user can change the display parameters for image processing of the data dynamically in order to improve the view of the images.

Figure 2. Schematic diagram of 3G medical image viewing system

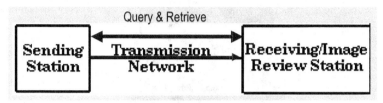

Result

In the initial test of the system, the transfer of 20 slices of CT images took about 45 seconds, which is much shorter than the traveling time required by the doctor driving from home to the hospital. During a remote consultation, the clinician can also work independently to search for the target studies without any help from the hospital. Hence, unnecessary traveling of doctors to hospitals can be minimized for non-urgent cases. For remote consultation using medical images, two major limitations have been identified: the processing power of the portable image receivers and the speed of the wireless network.

A typical 3G PDA phone is shown in Figure 3. On this phone, the display size is 2.9 inches with a resolution of 208×320. Comparing the resolutions of CT and MR images of resolutions of 256×256 and 512×512 pixels, the PDA phone was sufficient or nearly sufficient to display those images. The one-time two-factor authentication token in which the displayed password existed kept changing with an interval of 1 minute as shown in Figure 4. The log-in page of the system is shown in Figure 5. The data entry was done using a virtual keyboard, which was a basic function of the PDA phone as shown in Figure 6. After log-in, the user might require the search of related studies. He or she could use the data-management system for the search of interesting examinations using the patient identity number or study date as shown in Figure 7. For the patients with multiple studies or series of images, the thumbnail image of the first image of each series was displayed as in Figure 8. After the display, the user could select the study by clicking the thumbnail image. A few minutes later, the user could see multiple CT or MR images as shown in Figures 9, 10, and 11.

Figure 3. A 3G PDA phone

Figure 4. One-time two-factor authentication token

Figure 5. Log-in page of image-viewing system

Figure 6. Log-in of the image-viewing system using a virtual keyboard

Figure 7. Patient search in image-viewing system

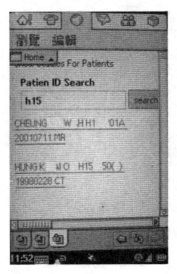

Figure 8. Series selection in image-viewing system

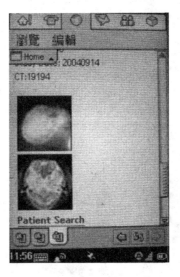

Figure 9. Multiple-image display

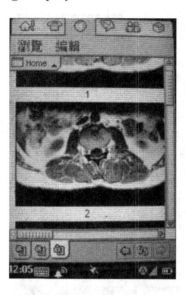

Figure 10. Display of CT image

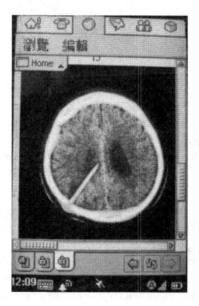

Figure 11. Display of MR image

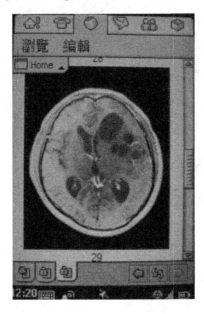

Future Trends

Today, tele-radiology is mainly limited by the speed of the wireless network. With the next generation of wireless networks such as IEEE 802.11 with a speed up to 108 Mbps, it is anticipated that tele-radiology can be performed more efficiently.

The fourth generation (4G) is the next generation of wireless networks that will replace 3G networks sometime in future. In another context, 4G is simply an initiative by academic research and development labs to move beyond the limitations and problems of 3G, which is having trouble getting deployed and meeting its promised performance and throughput. In reality, as of the first half of 2002, 4G is a conceptual framework for or a discussion point to address future needs of a universal high-speed wireless network that will interface with wireline backbone networks seamlessly. 4G also represents the hope and ideas of a group of researchers in Motorola, Qualcomm, Nokia, Ericsson, Sun, HP, NTT DoCoMo, and other infrastructure vendors who must respond to the needs of multimedia messaging service (MMS), multimedia, and video applications if 3G never materializes in its full glory.

Conclusion

It can be seen that with the proposed scheme involving the integration of Web, SQL, DICOM, 3G, and JPEG technologies, a practical solution has been developed to tackle the commonly noted tele-radiology implementation problems using high-speed wireless networking technologies. In addition, data integrity and accessibility have been enhanced significantly. In conclusion, the proposed 3G mobile medical image viewing system can provide an efficient tool for clinicians' remote consultations, thus allowing timely diagnosis for patients with critical conditions at anytime and anywhere.

References

American College of Radiology (ACR). (2003). *ACR technical standard for teleradiology.* Annapolis Junction, Maryland: American College of Radiology.

Bidgood, W. D., & Horii, S. C. (1992). Introduction to the ACRNEMA DICOM standard. *Radiographics, 12,* 34-35.

Collins, D., & Smith, C. (2001). *3G wireless networks.* New York: McGraw-Hill Professional.

DuBois, P. (2002). *MySQL cookbook.* California: O'Reilly & Associates.

Kabir, M. J. (1999). *Apache Server administrator's handbook.* New York: Hungary: Minds Inc.

Marcus, E., & Stern, H. (2003). *Blueprints for high availability* (2nd ed.). New York: John Wiley & Sons.

National Electrical Manufacturers Association (NEMA). (2004). *The DICOM standard.* Rosslyn, Virginia: National Electrical Manufacturers Association.

Nelson, M., & Gailly, J. L. (1995). *The data compression book* (2nd ed.). New York: Hungary: Minds Inc.

Oguchi, K., Murase, S., Kaneko, T., Takizawa, M., & Kadoya, M. (2001). Preliminary experience of wireless teleradiology system using Personal Handyphone System. *Nippon Igaku Hoshasen Gakkai Zasshi, 61*(12), 686-687.

Reponen, J., Ilkko, E., Jyrkinen, L., Tervonen, O., Niinimaki, J., Karhula, V., et al. (2000). Initial experience with a wireless personal digital assistant as a teleradiology terminal for reporting emergency computerized tomography scans. *J Telemed Telecare, 6*(1), 45-49.

Royal College of Radiologists (RCR). (1999). *Guide to information technology in radiology: Teleradiology and PACS.* Board of Faculty of Clinical Radiology, RCR.

Tong, C. K. S., Chan, K. K., & Wong, C. K. (2003). Common gateway interfacing and dynamic JPEG techniques for remote handheld medical image viewing. In *Proceedings of International Conference on Computer Assisted Radiology and Surgery, 1256*(6), 815-820. London.

Tong, C. K. S., & Wong, E. T. T. (2005). Picture archiving and communication system in healthcare. In M. Pagani (Ed.), *Encyclopedia of multimedia technology and networking* (pp. 821-828). Hershey, PA: Idea Group Reference.

Appendix: Terms and Definitions

3G Technology: 3G (or 3-G) is an abbreviation for third-generation technology. It is usually used in the context of mobile phones. The services associated with 3G provide the ability to transfer both voice data (a telephone call) and non-voice data (such as downloading information from the Internet, exchanging e-mail, and multimedia messaging).

Common Gateway Interface (CGI): This is a standard protocol for interfacing external application software with an information server, commonly a Web server. This allows the server to pass requests from a client Web browser to the external application. The Web server can then return the output from the application to the Web browser.

Digital Imaging and Communications in Medicine (DICOM): This is the industry standard for the transferal of radiologic images and other medical information between computers. Patterned after the open-system interconnection of the International Standards Organization, DICOM enables digital communication between diagnostic and therapeutic equipment and systems from various manufacturers.

Enhanced Data Rates for Global Evolution (EDGE): This is an enhancement of the GSM and TDMA digital cellular phone systems that provides data transmission up to 384 Kbps. Like cell phones in general, the EDGE service is more ubiquitous for mobile users than searching for Wi-Fi (wireless fidelity) hotspots; however, data rates are much lower than Wi-Fi.

Global System for Mobile Communications (GSM): This is a digital cellular phone technology based on TDMA (time division multiple access), which is a satellite and cellular phone technology that interleaves multiple digital signals onto a single high-speed channel. Operating in the 900MHz and 1.8GHz bands in Europe and the 1.9GHz Personal Computer System (PCS) band in the United States, GSM defines the entire cellular system, not just the air interface (TDMA, CDMA, etc.).

One-Time Two-Factor Authentication (OTTFA): This is any authentication protocol that requires two independent ways to establish identity and privileges. This contrasts with traditional password authentication, which requires only one factor (knowledge of a password) in order to gain access to a system.

Tele-radiology: This is a means of electronically transmitting radiographic patient images and consultative text from one location to another.

Universal Mobile Telecommunications System (UMTS): The European implementation of the 3G wireless phone system. UMTS, which is part of IMT-2000, provides service in the 2GHz band and offers global roaming and personalized features.

Wideband Code Division Multiple Access (WCDMA): This is also known as Universal Mobile Telecommunications System (UMTS) in Europe, and is a 3G standard for GSM in Europe, Japan, and the United States. It supports very high-speed multimedia services such as full-motion video, Internet access, and video-conferencing. It uses one 5 MHz channel for both voice and data, offering data speeds up to 2 Mbps.

Chapter XIV

Intelligent Agents Framework for RFID Hospitals

Masoud Mohammadian, University of Canberra, Australia

Ric Jentzsch, Compucat Research Pty Ltd., Canberra, Australia

Abstract

When dealing with human lives, the need to utilize and apply the latest technology to help in saving and maintaining patients' lives is quite important and requires accurate, near-real-time data acquisition and evaluation. At the same time, the delivery of a patient's medical data needs to be as fast and as secure as possible. One possible way to achieve this is to use a wireless framework based on radio-frequency identification (RFID). This framework can integrate wireless networks for fast data acquisition and transmission while maintaining the privacy issue. This chapter discusses the development of an agent framework in which RFID can be used for patient data collection. The chapter presents a framework for the knowledge acquisition of patient and doctor profiling in a hospital. The acquisition of profile data is assisted by a profiling agent that is responsible for processing the raw data obtained through RFID and a database of doctors and patients.

Introduction

The use and deployment of radio-frequency identification (RFID) is a relatively new area and it has been shown to be a promising technology (Glover & Bhatt, 2006; Lahiri, 2005; Shepard, 2004). This technology has the capability to penetrate and add value to nearly every field, lowering costs while improving service to individuals and businesses. Although many organizations are developing and testing the deployment of RFIDs, the real value of RFID implementation is achieved in conjunction with the use of intelligent systems and intelligent agents. The real issue is how intelligent-agent technologies can be integrated with RFID to be used to achieve the best outcome in business and services areas.

In this research, a new method for integrating intelligent-agent technologies with RFIDs in managing patients' healthcare data in a hospital environment is given. Knowledge acquisition and profiling of patients and doctors in a hospital are assisted by a profiling agent that is responsible for processing the raw data obtained through RFID data that are stored in a hospital database. There are several perspectives for profiling that could be used in a healthcare and hospital environment.

An intelligent agent can assist in profiling patients based on their illness and ongoing diagnostics as reported by the RFIDs. There are certain data and knowledge about each patient in the hospital. This knowledge could be the description of what the patient's symptoms are, monitoring status, and why the patient was admitted to the hospital. Using this information, an evolving profile of each patient can be built.

This data and knowledge can assist in deciding what kind of care he or she requires, the effects of ongoing care, and how to best care for this patient using available resources (doctors, nurses, beds, etc.). The intelligent agent will build a profile of each patient. Along with a profile of each patient, a profile for each doctor can also be developed. Then the patient and doctor profiles can be correlated to find the best doctor to suit the patient.

The patient-doctor profiling can be useful in several situations:

- Providing personalized services to a particular patient, for example, by identifying the services that a patient requires and hence speeding his or her recovery progress in or even out of the hospital.

- Disambiguating a patient's diagnostic based on the patient profile and matching this profile to the available doctor's profile. This may help in matching doctors with the suitable specialization to a patient.

- Providing speedy, reliable reentry of patients into the hospital by having the patients allocated to visit the relevant doctors.

- Presenting information in a way suitable to the doctor's needs, for example, presenting the information about the patients on a continuous basis for the doctors.
- Providing tailored and appropriate care to assist in cost reduction.

Personalization, user modeling, and profiling have been used for many e-commerce applications by IBM, ATG Dynamo, BroadVision, Amazon, and Garden. However, the use of such systems in hospital and personal care and profiling has not been reported. It should be noted that the definitions of personalization, user modeling, and profiling that these companies discuss are not quite the same as our intended meanings.

Many user models try to predict the user's preference in a narrow and specific domain. This works well as long as that domain remains relatively static and as such the results may be limited.

One of the main aims of profiling and user modeling is to provide users with correct and timely responses for their needs. This entails an evolving profile to ensure that as the dynamics of the user and real world change, the profile and user model reflects these changes.

A patient's visit to the hospital can simply be classified as a regular visit, an emergency visit, or an ad hoc appointment (on a need basis). In each of these situations, the needs of the patients are different. During a regular visit, the patient visits the hospital at a regular interval and usually a doctor is assigned to that patient. The patient's profile can assist the patient in a situation where the assigned doctor suddenly becomes unavailable. In this situation, the profile of the patient can be matched with the available doctors with suitable specializations for the needs of the patient. The patient-doctor assignment here is a kind of timetabling problem as we know the profile of the patient and doctors as well as the available doctors. Timetabling of doctors is out of the scope of this research study.

However, in an emergency visit, there is no assigned doctor for such a patient. The doctor in the emergency section of the hospital will provide information about a patient after examination, and a patient profile then can be created. In this case, the intelligent agent can assist the patient by matching the profile of the patient with the doctors suitable for the needs of the patient. Also, the doctors can be contacted in a speedier manner as they are identified and their availability is known.

An appointment visit is very similar to a regular visit, but it may happen only once and therefore the advantages mentioned for regular patients apply here.

We will endeavor to describe several of these, but will expand on one particular potential use of RFIDs in managing patient health data. First some background on RFIDs will be presented in this chapter including some definitions. We will discuss the environment that RFIDs operate in and their relationship to other available wireless technologies such as the IEEE (Institute of Electrical and Electronics Engineers) 802.11b, IEEE 802.11g, and so forth in order to fulfill their requirements effectively and efficiently.

This research is divided into five main sections. The following is based on the patient-doctor profiling and intelligent agents. Then the chapter gives an RFID background that will provide a good description of RFIDs and their components. This section discusses several practical cases of RFID technology in and around hospitals. It lists three possible applicable cases assisting in managing patients' medical data. Next we discuss the important issue of maintaining patients' data security and integrity, and relate that to RFIDs. Finally, the conclusion and further research directions are given.

Patient-Doctor Profiling

A profile represents the extent to which something exhibits various characteristics. These characteristics are used to develop a linear model based on the consensus of multiple sets of data, generally over some period of time. A patient or doctor profile is a collection of information about a person based on the characteristics of that person. This information can be used in a decision situation between the doctor, domain environment, and patient. The model can be used to provide meaningful information for useful and strategic actions. The profile can be static or dynamic. The static profile is kept in prefixed data fields where the period between data-field updates is long, such as months or years. The dynamic profile is constantly updated as per evaluation of the situation. The updates may be performed manually or automated. The automated user-profile building is especially important in real-time decision-making systems. Real-time systems are dynamic. These systems often contain data that are critical to the user's decision-making process. Manually updated profiles are at the need and discursions of the relevant decision maker.

The profiling of patients and doctors is based on the patient-doctor information:

- The categories and subcategories of doctor specialization and categorization. These categories will assist in information processing and patient-doctor matching.
- Part of the patient's profile is based on symptoms (past history problems, dietary restrictions, etc.) and can assist in the prediction of the patient's needs specifically.
- The patient's profile can be matched with the available doctor profiles to provide doctors with information about the arrival of patients as well as presentation of the patient's profile to a suitable available doctor.

A value denoting the degree of association can be created from the above evaluation of the doctor-patient profile. The intelligent-agent based on the denoting degrees can identify (and allocate) an appropriate available doctor to the patient.

In the patient-doctor profiling, the intelligent agent will make distinctions in attribute values of the profiles and match the profiles with the highest value. It should be noted that the intelligent agent creates the patient and doctor profiles based on data obtained from the doctors and patient using the following:

- Explicit profiling based on the data entered by hospital staff about a patient.

- Implicit profiling by filling that gap for the missing data by acquiring knowledge about the patient from his or her past visit or other relevant databases if any, and then combining all these data to fill in the missing data. Using legacy data for complementing and updating the user profile seems to be a better choice than implicit profiling. This approach capitalizes on the user's personal history (previous data from previous visits to the doctor or hospital).

The proposed intelligent-agent system architecture allows user profiling and matching in such a time-intensive and important application. The architecture of the intelligent-agent profiling system using RFID is given in Figure 1.

Figure 1. Intelligent-agents profiling model using RFID

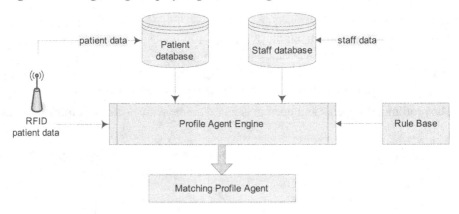

Integration of Intelligent Agents, RFID Technologies, and Profile Matching

Intelligent agents have been used in order to provide the needed transformation of RIFD passive data collection into an active organizational knowledge assistant. An intelligent agent should be able to act on new data and already stored profile knowledge and thereafter examine its current actions based on certain assumptions. It then inferentially plans its activities. Furthermore, intelligent agents must be able to interact with other agents if required (Bigus & Bigus, 1998; Watson, 1997; Wooldridge & Jennings, 1995) and be able to substitute for a range of human activities in a situated context. In our case, the activities are medical patient assignment and the context is a hospital environment. The integration of RFID capabilities and intelligent-agent techniques provides promising development in the areas of performance improvements in RFID data collection, inference and knowledge acquisition, and profiling operations.

Profile matching is based on a vector of weighted attributes. To get this vector, a rule-based system can be used to match the patient's attributes (stored in the patient's profile) against a doctor's attributes (stored in the doctor's profile). If there is a partial or full match between them, then the doctor will be informed based on his or her availability from the hospital doctor database. The matching is done through the rule-based system by examining the attributes of the patient profile and matching them using the rules already created to the doctor's profile based on the availability of doctors. Figure 1 displays the integration of intelligent agents, RFID technologies, and the profile-matching module.

RFID Description

RFID is a progressive technology that has been said to be easy to use and is well suited for collaboration with intelligent agents. Basically, an RFID can be read only, volatile read and write, or write once and read many (WORM) times. RFIDs are non-contact and non-line-of-sight operations. Being non-contact and non-line-of-sight will make RFIDs able to function under a variety of environmental conditions while still providing a high level of data integrity (Glover & Bhatt, 2006; Lahiri, 2005; Shepard, 2004). A basic RFID system consists of four components. These are the RFID tag (sometimes referred to as the transponder), a coiled antenna, a radio-frequency transceiver, and some type of reader for the data collection.

The RFID tag (transponder) emits radio waves in ranges of anywhere from 2.54 centimeters to 33 meters. Depending upon the reader's power output and the radio frequency used, and if a booster is added, that distance can be somewhat increased. When RFID tags pass through a specifically created electromagnetic zone, they detect the reader's activation signal. Transponders can be online or off line and electronically programmed with unique information for a specific application or purpose. A reader decodes the data encoded on the tag's integrated circuit and passes the data to a server for data storage or further processing.

RFID tags can be categorized as active, semi-active, or passive. Each has and is being used in a variety of inventory-management and data-collection applications today. The condition of the application, place, and use determines the required tag type.

Active RFID tags are powered by an internal battery and are typically read-write. Tag data can be rewritten and/or modified as the need dictates. An active tag's memory size varies according to manufacturing specifications and application requirements; some tags operate with up to 5 megabytes of memory. For a typical read-write RFID work-in-process system, a tag might give a machine a set of instructions, and the machine would then report its performance to the tag. This encoded data would then become part of the tagged part's history. The battery-supplied power of an active tag generally gives it a longer read range. The trade-off is greater size, greater cost, and a limited operational life that has been estimated to be a maximum of 10 years, depending upon operating temperatures and battery type.

The semi-active tag comes with a battery. The battery is used to power the tag's circuitry and not to communicate with the reader. This makes the semi-active tag more independent than the passive tag, and it can operate in more adverse conditions. The semi-active tag also has a longer range and more capabilities than a passive tag. Linear bar codes that reference a database to get product specifications and pricing are also data devices that act in a very similar way. Semi-passive tags are preprogrammed, but can allow for slight modifications of their instructions via the reader or interrogator. However, they are bigger, weigh more, and are more complete than passive tags. A reader is still needed for data collection.

Passive RFID tags operate without a separate external power source and obtain operating power generated from the reader. Passive tags have no power source embedded in them and are consequently much lighter than active tags, less expensive,

Figure 2. Semi-passive tag

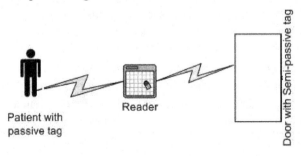

Patient with
passive tag

Reader

Door with Semi-passive tag

and offer a virtually unlimited operational lifetime. However, the trade-off is that they have shorter read ranges than active tags, and require a higher powered reader. Read-only tags are typically passive and are programmed with a unique set of data (usually 32 to 128 bits) that cannot be modified.

RFID systems can be distinguished by their deployment and frequency range. RFID tags generally operate in two different types of frequencies that make them adaptable for nearly any application. These frequency ranges are as follows.

Low-frequency (30 KHz to 500 KHz) systems have short reading ranges and lower system costs. They are most commonly used in security access, asset tracking, and animal-identification applications.

High-frequency (850 MHz to 950 MHz, or in industry, science, and medical applications, 2.4 GHz to 2.5 GHz) systems offer longer reading ranges (greater than 33 meters) and high reading speeds. These systems are generally used for such applications as railroad-car tracking, container dock and transport management, and automated toll collection. However, the higher performance of high-frequency RFID systems incurs higher system-operating costs.

The coiled antenna is used to emit radio signals to activate the tag and read or write data to it. Antennas are the conduits between the tag and the transceiver that controls the system's data acquisition and communication. RFID antennas are available in many shapes and sizes. They can be built into a door frame, mounted on a tollbooth, or embedded into a manufactured item such as a shaver or software case so that the receiver tags the data from things passing through its zone. Often, the antenna is packaged with the transceiver and decoder to become a reader. The decoder device can be configured either as a handheld or a fixed-mounted device. In large, complex, often chaotic environments, portable or handheld transceivers would prove valuable.

RFID for Hospital Environment

In hospitals, systems need to use rules and domain knowledge that is appropriate to the situation. One of the more promising capabilities of intelligent agents is their ability to coordinate information between the various resources.

In a hospital environment, in order to manage patient medical data, there is a need for both types: fixed and handheld transceivers. Transceivers can be assembled in ceilings, walls, or door frames to collect and disseminate data. Hospitals have become large complex environments.

Nurses and physicians can retrieve the patient's medical data stored in transponders (RFID tags) before they enter a ward or patient's room.

Given the descriptions of the two types of RFID tags and their potential use in hospital patient data management, we suggest the following:

- It would be most useful to embed a passive RFID transponder into a patient's hospital wristband.

- It would be most useful to embed a passive RFID transponder into a patient's medical file.

- Doctors should have PDAs (personal digital assistants) equipped with RFID or some type of personal area network device. Either would enable them to retrieve some patient information whenever they are near the patient instead of waiting until the medical data is pushed to them through the hospital server.

- Active RFID tags are more appropriate for the continuous collection of patient medical data since the patient's medical data need to be continuously recorded to an active RFID tag and an associated reader needs to be employed. Using an active RFID means that the tag will be a bit bulky because of the needed battery for the write process, and there is a concern about radio-frequency emissions. Thus, it is felt that an active tag would not be a good candidate for the patient wristband. However, if the patient's condition is to be continuously monitored, the collection of the data at the source is essential. The inclusion of the tag in the wristband is the only way to record the medical data on a real-time basis using the RFID technology. As more organizations get into the business of manufacturing RFIDs and the life and size of batteries decrease, the tag size will decrease and this may be a real possible use.

- Passive RFID tags can be used as well. These passive tags can be embedded in the doctors PDA, which is needed for determining their locations whenever the medical staff requires them. Also, passive tags can be used in patients' wristbands for storage of limited amount of data on an off-line basis, for example, information such as the date of hospital admission, medical record number (MRN), and so forth.

Low-frequency-range tags are suitable for the patients' wristband RFID tags since it is expected that the patients' bed will not be too far from an RFID reader. The reader might be fixed over the patient's bed, in the bed itself, or over the door frame. The doctor using his or her PDA would be aiming to read the patient's data directly and within a relatively short distance. High-frequency-range tags are suitable for the physician's tag implanted in the PDA. As physicians move from one location to another in the hospital, data on their patients could be continuously being updated.

Finally, the transceivers and interrogators can differ quite considerably in complexity, depending upon the type of tags being supported and the application. The overall function of the application is to provide the means of communicating with the tags and facilitating data transfer. Functions performed by the reader may include quite sophisticated signal conditioning and parity error checking and correction. Once the signal from a transponder has been correctly received and decoded, algorithms may be applied to decide whether the signal is a repeat transmission, and may then instruct the transponder to cease transmitting or temporarily cease asking for data from the transponder. This is known as the command-response protocol and is used to circumvent the problem of reading multiple tags over a short time frame. Using interrogators in this way is sometimes referred to as hands-down polling. An alternative, more secure but slower tag polling technique is called hands-up polling. This involves the transceiver looking for tags with specific identities, and interrogating them in turn. A further approach may use multiple transceivers multiplexed into one interrogator.

Hospital patient data management deals with sensitive and critical information (patients' medical data). Hands-down polling techniques in conjunction with multiple transceivers that are multiplexed with each other form a wireless network. The reason behind this choice is that there is a need for high-speed transfer of medical data from medical equipment to or from the RFID wristband tag to the nearest RFID reader and then through a wireless network or a network of RFID transceivers or LANs (local area networks) to the hospital server. From there it is a short distance to be transmitted to the doctor's PDA, a laptop, or a desktop through a WLAN (wireless LAN) IEEE 802.11b or 803.11g, or a wired LAN, which operates at the 5.2 GHz band with a maximum data transfer rate of 54 Mbps.

The hand-down polling techniques, as previously described, provide the ability to detect all detectable RFID tags at once (i.e., in parallel), preventing any unwanted delay in transmitting medical data corresponding to each RF tagged patient.

Transponder programmers are the means by which data are delivered to WORM and read-write tags. Programming can be carried out off line or online. For some systems, re-programming may be carried out online, particularly if it is being used as an interactive portable data file within a production environment, for example. Data may need to be recorded during each process. Removing the transponder at the end of each process to read the previous process data and to program the new data would naturally increase process time and would detract substantially from the intended flexibility of the application. By combining the functions of a transceiver

Figure 3. Patient and outpatient

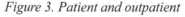

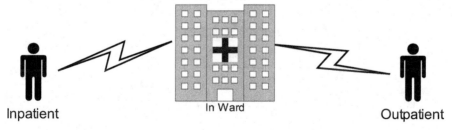

Inpatient   In Ward   Outpatient

and a programmer, data may be appended or altered in the transponder as required without compromising the production line.

Practical Cases using RFID Technology

This section explains in detail three possible applications of the RFID technology in three applicable cases. Each case is discussed step by step and then represented by a flowchart. These cases cover issues of the acquisition of a patient's medical data, locating the nearest available doctor to the patient's location, and how doctors stimulate the patient's active RFID tag using their PDAs in order to acquire the medical data stored in it.

Case 1: Acquisition of Patient's Medical Data

Case 1 will represent the method of acquisition and transmission of medical data. This process can be described in the following points as follows:

- A biomedical device equipped with an embedded RFID transceiver and programmer will detect and measure the biological state of a patient. This medical data can be an ECG (electrocardiogram), EEG (electroencephalogram), BP (blood pressure), sugar level, temperature, or any other biomedical reading.

- After the acquisition of the required medical data, the biomedical device will write this data to the RFID transceiver's EEPROM (electronically erasable programmable read-only memory) using the built-in RFID programmer. Then the RFID transceiver with its antenna will be used to transmit the stored medical data in the EEPROM to the EEPROM in the patient's transponder (tag)

Figure 4. Acquisition of patient data

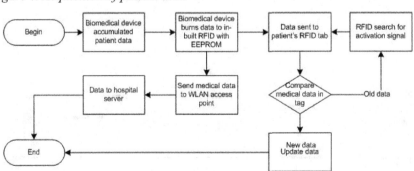

that is around his or her wrist. The data received will be updated periodically once new fresh readings are available by the biomedical device. Hence, the newly sent data by the RFID transceiver will be accumulated with the old data in the tag. The purpose of the data stored in the patient's tag is to make it easy for the doctor to obtain medical information regarding the patient directly via the doctor's PDA, tablet PC (personal computer), or laptop.

- Similarly, the biomedical device will also transfer the measured medical data wirelessly to the nearest WLAN access point. Since a high data-transfer rate is crucial in transferring medical data, IEEE 802.11b is recommended for the transmission purpose.

- The wirelessly sent data will then be routed to the hospital's main server to be sent (pushed) to the following:

  o   Other doctors available throughout the hospital so they can be notified of any newly received medical data.

  o   An online patient-monitoring unit or a nurse's workstation within the hospital.

  o   An expert (intelligent) software system running on the hospital server to be then compared with other previously stored abnormal patterns of medical data and to raise an alarm if any abnormality is discovered.

- Another option could be using the embedded RFID transceiver in the biomedical device to send the acquired medical data wirelessly to the nearest RFID transceiver in the room. Then the data will travel simultaneously in a network of RFID transceivers until reaching the hospital server.

Case 2: Locating the Nearest Available Doctor to the Patient's Location

This case will explain how to locate the nearest doctor, who is needed urgently, to attend an emergency medical situation. This case can be explained as follows:

- If a specific surgeon or physician is needed in a specific hospital department, the medical staff in the monitoring unit (e.g., nurses) can query the hospital server for the nearest available doctor to the patient's location. In our framework, an intelligent agent can perform this task.

- The hospital server traces all doctors' locations in the hospital through detecting the presence of their wireless mobile device, for example, PDA, tablet PC, or laptop in the WLAN range.

- Another method that the hospital's server can use to locate the physicians is making use of the RFID transceivers built into the doctor's wireless mobile device. Similar to the access points used in WLAN, RFID transceivers can

Figure 5. Locating nearest doctor

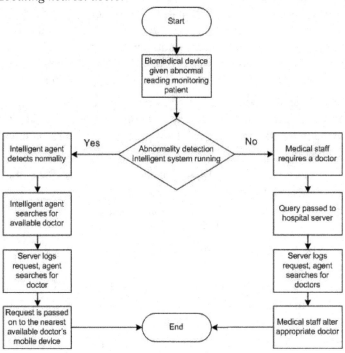

assist in serving a similar role of locating a doctor's location. This can be described in three steps.

o   The fixed RFID transceivers throughout the hospital will send a stimulation signal to detect other free RFID transceivers, which are in the doctors' PDAs, tablets, laptops, and so forth.

o   All free RFID transceivers will receive the stimulation signal and respond back with an acknowledgement signal to the nearest fixed RFID transceiver.

o   Finally, each free RFID transceiver cell position would be determined by locating to which fixed RFID transceiver range it belongs to or currently is operating with.

•   After the hospital server locates the positions of all available doctors, the RFID determines the nearest requested (condition evaluation) physican (pediatries, neuologist, etc...) to the patient's location.

•   Once the required physician is located, an alert message will be sent to his or her PDA, tablet PC, or laptop indicating the location to be reached immediately.

This alert message would show the following.

o   Case profile over application period

o   The building, floor, and room of the patient (e.g., 3C109).

o   The patient's case (e.g., heart stroke, arrhythmia, etc.).

o   A brief summary description of the patient's case.

- If the hospital is running an intelligent agent on its server as described in our proposed framework, the process of locating and sending an alert message can be automated. This is done through comparing the collected medical data with previously stored abnormal patterns of medical data, then sending an automated message describing the situation. This system could be used in the patient-monitoring unit or the nurse's workstation, who observe and then send an alert message manually.

Case 3: Doctors Stimulate a Patient's Active RFID Tag using Their PDAs in Order to Acquire the Medical Data Stored in It

This method can be used in order to get rid of medical files and records placed in front of the patient's bed. Additionally, it could help in preventing medical errors (reading the wrong file for the wrong patient) and could be considered an important step toward a paperless hospital.

This case can be described in the following steps:

- The doctor enters the patient's room or ward. The doctor wants to check the medical status of a certain patient. Instead of picking up the hard-copy paper medical file, the doctor interrogates the patient's RFID wrist tag with his or her RFID transceiver equipped in a PDA, tablet PC, laptop, or so forth.

- The patient's RFID wrist tag detects the signal of the doctor's RFID transceiver coming from his or her wireless mobile device and replies back with the patient's information and medical data.

- If there was more than one patient in the ward possessing RFID wrist tags, all tags can respond in parallel using hands-down polling techniques back to the doctor's wireless mobile device.

- Another option could be that the doctor retrieves only the patient's number from the passive RFID wrist tag. Then, through the WLAN, the doctor could access the patient's medical record from the hospital's main server.

RFID technology has many potential important applications in hospitals, and the three cases discussed are real practical examples. Two important issues can be concluded from this section. WLAN is preferred for data transfer given that IEEE's wireless networks have much faster speed and greater coverage area as compared to RFID transceiver and transponder technology. Yet, RFID technology is the best for data storage and locating positions of medical staff and patients as well.

Here there is a need for the RFID transceiver and programmer to be embedded in biomedical devices for data acquisition and dissemination. A RFID transceiver embedded in a doctor's wireless mobile device enables the doctor to obtain medical data. With the progress of the RFID technology, it could become a standard as other wireless technologies (Bluetooth, for example) and eventually manufacturers will be building them in electronic devices, or biomedical devices, in our case.

Maintaining Patients' Data Security and Integrity

Once data are transmitted wirelessly, security becomes a crucial issue. Unlike in wired transmission, wirelessly transmitted data can be easily sniffed out, leaving the transmitted data vulnerable to many types of attacks. For example, wireless data could be easily eavesdropped on using any mobile device equipped with a wireless card. In worst cases, wirelessly transmitted data could be intercepted and then possibly tampered with, or in best cases, the patient's security and privacy F be compromised. Hence emerges the need for data to be initially encrypted from the source.

This section of the chapter discusses how we could apply encryption to the designed wireless framework that was explained in the previous section, suggesting exactly where data need to be encrypted and\or decrypted depending on the case that is being examined.

Next, a definition of the type of encryption that would be used in the design of the security (encryption or decryption) framework will be discussed, and this will be followed by a flowchart demonstrating the framework in a step-by-step process. There are two layers of encryption that are recommended to be used.

- **Physical:** (hardware) layer encryption means encrypting all collected medical data at the source or hardware level before transmitting them. This ensures that the patient's medical data would not be compromised once exposed to the outer world on its way to its destination. Even if a person with a malicious intent and also possessing a wireless mobile device steps into the coverage range of the hospitals' WLAN, this intruder will gain actually nothing since all medical data are encrypted, making all intercepted data worthless.

- **Application:** (software) layer encryption means encrypting all collected medical data at the destination or application level once receiving it. Application-level encryption runs on the doctor's wireless mobile device (e.g., PDA, tablet PC, or laptop) and on the hospital server. Once the medical data are received, they will be protected by a secret password (encryption or decryption key) created by the doctor who possesses this device. This type of encryption would prevent any person from accessing the patient's medical data if the doctor's wireless mobile device gets lost, or even if a hacker hacks into the hospital server via the Internet, intranet, or some other means.

Figure 6. Functional flow

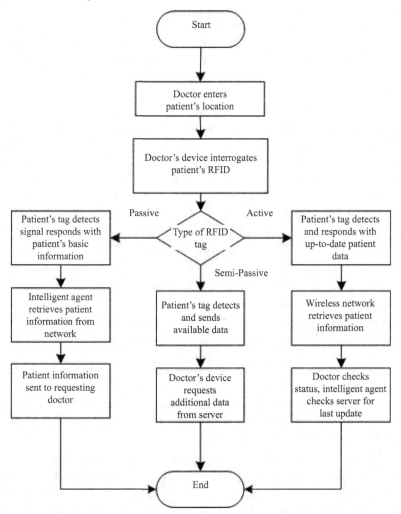

Framework of Encrypting a Patient's Medical Data

The previous section (practical cases using RFID technology) focused on how to design a wireless framework to reflect how a patient's medical data can be managed efficiently and effectively leading to the elimination of errors, delays, and even paperwork. Similarly, this section will focus on the previously discussed framework from a security perspective, attempting to increase security and data integrity during the acquisition of a patient's medical data and when doctors stimulate the patient's active RFID tag using their wireless mobile devices in order to acquire the medical data stored in it. The third case on locating the nearest available doctor to the patient's location is more concerned about locating doctors than transferring a patient's data and it is not discussed here.

The lower part of Figure 6 represents the physical (hardware) encryption layer. This part is divided into two sides. The left side demonstrates the case of a doctor acquiring a patient's medical data via a passive RFID tag located in a band around the patient's wrist. The passive RFID tag contains only a very limited amount of information such as the patient's name, date of admission to the hospital, and above all his or her MRN, which will grant access to the medical record containing the acquired medical data and other information regarding the patient's medical condition. This process is implemented in six steps, and involves two pairs of encryption and decryption. The first encryption occurs after the doctor stimulates the RFID passive tag to acquire the patient's MRN, so the tag will encrypt and reply back with the MRN to the doctors PDA, for example. Then the doctor will decrypt the MRN and use it to access the patient's medical record from the hospital's server. Finally, the hospital server will encrypt and reply back with the medical record, which will be decrypted once received by the doctor's PDA.

The right side of Figure 6 represents a similar case, but this time using an active RFID tag. This process involves only one encryption and decryption. The encryption happens after the doctor stimulates the active RFID tag using a PDA that has an equipped RFID transceiver, so the tag replies with the medical data encrypted. Then the received data is decrypted through the doctor's PDA. The upper part of Figure 6 represents the application encryption layer requiring the doctor to enter a password to decrypt and then access the stored medical data. So, whenever the doctor wants to access a patient's medical data, he or she simply enters a certain password to get access to either the wireless mobile device or a hospital server, depending on where the medical data actually resides.

Securing medial data seems to be uncomplicated, yet the main danger of compromising such data comes from the people managing it, for example, doctors, nurses, and other medical staff. We have seen that even though the transmitted medical data are initially encrypted from the source, doctors have to run application-level encryption on their wireless mobile devices in order to protect the important data if the device gets lost, left behind, stolen, and so forth. Nevertheless, there is a compromise. Increasing security through using multiple layers and increasing the length of encryption keys decreases the encryption-decryption speed and causes unwanted time delays, whether we are using application- or hardware-level encryption. As a result, this could delay medical data being sent to doctors or online monitoring units. Figure 6 represents the case of high and low levels of security in a flowchart applied to the previously discussed two cases.

We conclude that there are two possible levels of encryption: the software level (application layer) or hardware level (physical layer) depending on the level of security required. Both physical-layer and application-layer encryption are needed in maintaining collected medical data on hospital servers and doctors' wireless mobile devices. Encrypting medical data makes the process of data transmission slower, while sending data unencrypted is faster. Here there is a need to compromise between speed and security. For medical data, it has to be sent as fast as possible to medical staff, yet the security issue has the priority.

Conclusion

Managing patients' data wirelessly (paperless) can prevent errors, enforce standards, make staff more efficient, simplify record keeping, and improve patient care. In this chapter, both passive and active RFID tags were used in acquiring and storing medical data, and in linking to the hospital's server via a wireless network. Moreover, three practical applicable RFID cases were discussed and it was explained how the RFID technology can be put in use in hospitals while at the same time maintaining the acquired patient data's security and integrity.

This research in the wireless medical environment introduces some new ideas in conjunction with what is already available in the RFID technology and wireless networks. The aim here is to link both technologies with each other to achieve the research's main goal, which is delivering patient medical data as fast and secure as possible to pave the way for future paperless hospitals.

References

Bigus, J. P., & Bigus, J. (1998). *Constructing intelligent agents with Java: A programmers guide to smarter applications.* New York: Wiley. (ISBN:0-471-19135-3)

Glover, B., & Bhatt, H. (2006). *RFID essentials.* O'Railly Media Inc. (ISBN:0-596-0094-4)

Kaptelinin, V. (1992). Human computer interaction in context: The activity theory perspective. In *Proceedings of the East-West Human Computer Interaction Conference*, St. Petersburg, Russia.

Lahiri, H. (2005). *RFID sourcebook.* New York: IBM Press. (ISBN:0-131-8513-73)

Odell, J. (Ed.). (2000). *Agent technology* (OMG Document 00-09-01). OMG Agents interest Group.

Shepard, S. (2004). *RFID.* McGraw-Hill. (ISBN: 0-071-4429-95)

Watson, M. (1997). *Intelligent Java applications for the Internet and intranets.* Morgan Kaufmann. (ISBN: 1-55860-420-0)

Wooldridge, M., & Jennings, N. (1995). Intelligent agents: Theory and practice. *The Knowledge Engineering Review, 10*(2), 115-152. (ISBN: 0-269-888-9)

Chapter XV

Rescheduling Dental Care with GSM-Based Text Messages

Reima Suomi, Turku School of Economics and Business Administration, Finland

Ari Serkkola, Helsinki University of Technology Lahti Center, Finland

Abstract

The Internet has opened new avenues for customer communication, even for public services. In this chapter, we propose a framework for an integrated electronic health platform. Most of the platform is still at the planning stage, but the first applications are already up and running, among them, dental-service appointment rescheduling. In this application, new patients to fill canceled dental-service appointments are searched from an existing waiting list using GSM SMS messages. The first few months of operation have already shown that the new application, in conjunction with other methods in use, could limit the share of time slots that dentist completely lose through cancellations to under 10% percent of all canceled times.

We present and analyze the function of the SMS-message-based dental-service appointment-reservation system, which is being implemented in Lahti, Finland. This analysis contains a description of the system functions, as well as some assessment of the success from a service-provider and customer point of view.

Introduction

Internet technology is penetrating every aspect of modern life. We speak of e-commerce, e-learning, e-health, and of e-everything. Healthcare is one of the industries in modern-day society where the adoption of information technology is now happening very fast. We consider e-health as a subset of e-government as far as public health services are concerned. The industry was, however, late in starting the adoption process. Currently, the development of information systems in healthcare is several years behind the general development in most other industries (Ragupathi, 1997). We have found the following reasons for the late adoption of modern ICT in the healthcare sector (Suomi, 2000):

- Fragmented industry structure.
- Considerable national differences in processes.
- Strong professional culture of medical-care personnel.
- One-sided education.
- Traditions of manual work.
- Weak customers.
- Hierarchical organization structures.

Healthcare institutions, especially hospitals, must emphasize professional information management more strongly in their organizations (Haux, Ammenwerth, Herzog, & Knaup, 2002). A part of professional management is the application of different available infrastructure and architecture solutions on the market to avoid solutions that would emerge as technology

In our chapter, we propose an integrated architecture for electronic mobile health applications. This architecture is defined as a typical care-taking chain for a patient, and also as a series of interactions. A typical interaction chain begins from the emergence of a need for healthcare, and ends with curing of the sickness, with the development of the sickness to a new qualitative phase, or, in the worst case, death. Another key part of the architecture is the definition of the actors: the patient, the healthcare staff, and the system. Especially important is to see how the system can add value through performing different functions automatically, or supporting the staff or patient very efficiently.

The proposed architecture has already been successfully implemented in two areas in the city of Lahti, Finland, and further implementations are under way.

Background

The healthcare service consists of many different elements. Traditionally, human contact has been seen as an integrated and obligatory element in healthcare services, but current practice in and research into e-health has made it clear that many elements of healthcare provisioning can be performed through electronic means. Health is a very complicated concept. Ewles and Simnett (1992) define the following areas of health:

- **Physical health:** Concerned with the mechanistic functioning of the body.
- Mental health: The ability to think clearly and coherently, distinguished from emotional and social health, although there is a close association between the three.
- **Emotional (affective) health:** The ability to recognize emotions such as fear, joy, grief, and anger and to express such emotions appropriately; also the means of coping with stress, tension, depression, and anxiety.
- **Social health:** The ability to make and maintain relationships with other people.
- **Spiritual health:** For some people, they are connected to religious beliefs and practices; for others, it has to do with personal creeds, principles of behavior, and ways of achieving peace of mind.
- **Societal health:** Personal health is inextricably related to everything surrounding that person and it is impossible to be healthy in a sick society.

As we take such a broad definition of health, it is clear that almost all activities of humans are connected with health maintenance. Within such a broad definition, e-health becomes very meaningful: As the Internet has penetrated all parts of human life, it cannot be excluded from health issues.

Despite its clear and documented benefits, e-health has suffered from weak development and a reluctance to adopt. As Morrissey (2003) puts it, "Nobody wants to be on the front end of this wave (IT Investments in healthcare), especially when it comes to issues related to physicians" (p. 7).

Architecture for an Integrated E-Health Platform

Some Basic Taxonomies of Patients and Healthcare

Electronic or mobile health service delivery is a cooperative effort by many different stakeholders. The service providers organize themselves around complicated value chains, or care-taking chains as they are often termed in the healthcare industry. The customers also form a multifaceted group of stakeholders. Luck (2000) defines the following roles of health and social care customers:

- **Customer:** Someone who has financial resources to make his or her own decision whether or not to purchase goods or services.
- **Client:** A person for whom a service is being provided.
- **Patient:** A person in receipt of medical or nursing care.
- **Consumer:** A person who receives a service.
- **Citizen:** A person with the democratic right to vote and who might have a constitutional right to receive some health or social care services.

All in all, Luck's categorization reflects the change that has taken place in the role of the patient. A person may have the role of a traditional patient, but may also feel that he or she is a customer and a consumer of health services. Similarly, the doctor's role varies, from healer to expert and health consultant. In the latter case, the doctor is needed to reinforce patients' own understanding of their state of health. Customers need information channels in order to identify their own feelings of illness.

Some basic taxonomies can be helpful when discussing e-health. To begin with, the customers or patients can be divided into three different classes (Utbult, 2000). First, we have well-off customers who are not suffering from any diagnosed and acute sickness or disease. They use healthcare services to maintain their good health and to prevent diseases from emerging. Second, there are those patients who have an acute sickness diagnosed. They usually have a very intensive demand for service and information. Third, we have the chronically ill people, who are not usually in a hurry with the service and have varying needs as regards care intensity and timing. These basic groups can be further divided into special areas according to models of health promotion and medical treatment, and into the illness types that represent these. A challenging aspect of the e-health perspective is how to plan for the different customer groups an electronic communication and consultation service that operates on a uniformly consistent information-technology platform.

Proposed Architecture

Basic Premises of Healthcare Architecture

According to Smolander, Rossi, and Purao (2005), the concept of architecture can be understood in four dimensions or metaphors: blueprint, decision, language, and literature. The blueprint emphasizes that an architecture is an overall plan for the planned object. The decision metaphor stresses that the structure is born under a flow of socially constructed decisions. The language metaphor stresses the importance of communication between the different stakeholders of the object for which the architecture is made. Finally, the literature metaphor emphasizes written documentation of the architecture as a means of communication. Here in our chapter, we understand the proposed architecture for an integrated e-health platform as a blueprint that must be communicated and decided upon.

The procedures at the customer consultation office and the events of the interaction create the basis for the electronic architecture. The architecture should take into account the functions that Tang and McDonald (2001) use to define the criteria for the patient report:

- Integrated view of patient data
- Clinical decision support
- Clinician order entry
- Access to knowledge resources
- Integrated communication tool (for medical consultation)

The functions concerned are part of the care-taking process. Therefore, the architecture is built around the idea of a care-taking chain. This chain begins with the first contact with the customer and ends when the care-taking activity is finished, normally once the patient is cured. Other outcomes might be the transition of the patient to a fundamentally new state (for example, acute or chronic, or curable vs. incurable), or, in the worst case, death.

The starting point is that the architecture is based on user trust. Luo and Najdawi (2004) define the following features that encourage customer trust in e-health portals:

- Security.
- Privacy.

- Information content.
- Technical functionality.

Even though our architecture is not seen as just a portal, we see that the environment should cater to the features listed above.

The idea of the architecture is to turn activities to electronic channels when appropriate. Reasons for electronic communication might be those defined by Malone, Yates, and Benjamin (1987).

1. Electronic Communication Effect
   - More information can be communicated in the same amount of time or the same amount of information in less time.
2. Electronic Brokerage Effect
   - Data can be collected from several places.
   - The cost of the entire data-collection process decreases.
3. Electronic Integration Effect
   - There is a tighter coupling of processes that create and use information.

In our architecture, there are three types of activities:

1. Activities initiated and performed by the customer or patient.
2. Activities initiated and performed by the healthcare staff.
3. Activities initiated and performed by the automatic system.

The Proposed Electronic and Mobile Healthcare Architecture

In the electronic or mobile healthcare architecture, the service interaction takes place between the patient, the staff, and the automatic functions. Within the architecture, certain functions or tasks are specified for each actor and data system. The architecture is further based on a typical scenario of care taking. The stages of the scenario are as follows:

1. **The customer makes the first contact.**
   Customers contact the contact centre or the duty office. They receive information about self-treatment and/or they can seek information themselves from the health portal.

2.   **The (first) clinical service appointment is organized.**

Customers receive electronic authorization to reserve a consultation appointment (for example, an identification code for making an appointment), and authorization is given for restricted user rights to their patient records (for the treatment of a coded illness, for instance).

3.   **The first clinical appointment is made.**

Customers are given an appointment time, or they can reserve a time for themselves.

4.   **A need for further consultations or operations emerges.**

The customer can reserve a time for further examination (laboratory or X-ray), for example.

a.   Stages 2 and 3 are repeated, maybe several times.

b.   The results from the consultations are received.

Figure 1. The care-taking scenario used in our architecture

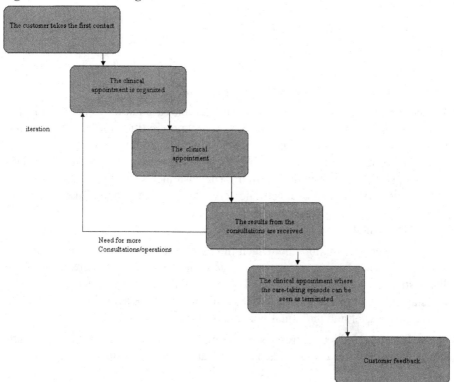

5. **The clinical appointment during which the care-taking episode takes place is terminated.**

6. **Customer feedback is given.**

    The customer gives electronic feedback on the treatment and/or recovery.

The care-taking scenario is depicted graphically in Figure 1.

By service channels, we mean the forms of information management and the technology by which the patient relationship is handled. The service channels are where the communications technology and the forms of interaction with patients meet. Well-known service channels are health advice, duty consultation, reception desks, telephone enquiries, letters of invitation, and control questionnaires.

As possible electronic communication channels, we see the following:

1. SMS (short-message service) messages on the GSM (Global System for Mobile Communication) network (or other equivalent mobile networks).

2. WAP- (wireless application protocol) based services on the GSM network (or other equivalent mobile networks).

3. Secure e-mail.

4. Web-based interactive services.

5. Digital television.

Our architecture components converge as depicted in Table 1.

A combination of consultation office procedures, the care-taking scenario, and electronic communication channels will offer many possibilities for the organization of appointment reservations, data transmission, and data processing in situations of healthcare interaction. The handling of transactions can be differentiated into customer, staff, and automatic functions. The customer's functions include giving authorization for the use of electronic services; receiving appointment times by text message; reserving appointments by text message, e-mail, and/or Internet; and giving electronic feedback.

The staff gives information about such things as free appointment times and canceled times. They also send electronic instructions concerning preparation, send diagnostic results by text message and e-mail, and monitor and direct waiting lists. The starting point is that the customer should feel that the availability of services is becoming easier and their quality is improving.

The architecture and its electronic services make the customer responsible for handling their use of the services independently and spontaneously. The staff, in turn,

Table 1. Main service components of an architecture for an integrated e-health platform

Phase	Activities by the customer supported by electronic communication	Activities by the healthcare staff supported by electronic communication	Possibilities for automatic electronic communication
The customer makes the first contact	The customer studies self-care guidance and other material on the Net. First contact can happen by electronic means	Staff can maintain different queues with the help of information technology	The system selects a patient automatically and offers a consultation time (for routine treatment or checkups)
The (first) clinical appointment is organized	The customer fills in a patient data form beforehand.	Staff gives the customer authorization to reserve a consultation time. The staff send preparation instructions	Free and canceled consultation times can be offered automatically to the patient. The system sends preparation instructions
The clinical encounter or appointment	The customer supplements the electronic patient data form beforehand if necessary.	Re-staffing of consultation hours when needed	The system sends the patient an automatic reminder. Free and canceled consultation times can be offered automatically to the patient
A need for further consultations or operations emerges	The customer can reserve a laboratory, X-ray, or other appointment electronically	Staff can invite the patient for further examination	The system transmits the results automatically to the doctor responsible
The results from the consultations are received	The customer receives the electronic results	Medical staff send the results to the patients	The results are sent automatically to the patient
The clinical appointment where the care-taking episode can be seen as terminated	The customer receives instructions on self-care	Staff send instructions on self-care	The system informs the patient that the care-taking episode has terminated and sends instructions on self-care
Customer feedback	The customer gives feedback electronically	Staff receive electronic feedback	The system transmits and stores the electronic feedback automatically

feels that their work is made easier, and the automatic functions take care of many routine appointment-reservation and data-transmission operations.

Each organization has its own way of working in consultation and customer work. The architecture should be flexible enough so that the electronic services support the organization, and its clientele can be arranged within its framework. The focus will now shift from utilization of the architecture to the healthcare processes and their medical treatment models. In the final analysis, it is a question of which forms and channels of electronic communication are suited to which health service and its consultation procedures. We shall examine this question with reference to the Lahti City oral health services in Finland, where the architecture and its SMS functions have been applied in practice.

The Lahti Pilot Project with SMS-Based Reservations

As a concrete example of a component of the architecture above that has already been implemented, we discuss here the SMS-based reservation system for canceled dental care appointments in the city of Lahti.

Introduction to the Environment

The City of Lahti in Finland, with 98,500 residents, offers its inhabitants the possibility to obtain oral healthcare services at public dental clinics. The oral healthcare is provided by 28 dentists, 11 oral-hygiene specialists, and 46 dental assistants and receptionists. Specialist dental care is offered in clinical dental care, the correction of irregularities of the teeth, and oral surgery. In 2003, visits by clients numbered around 84,000, and half of these were by clients under the age of 18. In addition, about 60 full- or part-time private dentists work in the city. There are 21 dental clinics for the use of clients in different parts of the city. Of these, 3 are larger in size, and the 15 operate near schools for the most part. In addition, mobile staff makes visits to the wards of care institutions and hospitals, and to homes.

A centralized contact centre has been set up for oral healthcare in the City of Lahti. A customer who calls the contact centre will access one of the following: an oral health advisor, an emergency duty office, or an appointment with an oral hygienist to assess the need for treatment. Another possible outcome is that the caller is put onto a waiting list, where (after the introduction of the treatment-guarantee law in March 2005 in Finland) a person can remain for a maximum of 6 months. Urgent cases are directed from the contact centre to a duty office, where they have access to treatment the same day. Electronic communication services support the activities of the contact centre.

The Finnish dental-care system was reformed in 2002 so that all citizens, irrespective of age, became entitled to public oral healthcare. Citizens could thus seek either municipal oral healthcare or private care subsidized by health insurance. Dental care has become a market where customers have a choice. There is evidence that the availability of different communication channels with healthcare professionals and efficient scheduling solutions are key selection factors as customers select their healthcare providers (Gopalakrishna & Mummaleni, 1993).

Previously, subsidized care was restricted to the younger age groups, those born in 1956 or later. The reform resulted in increased demand for services. In those towns and cities that were not prepared for the increase in demand, waiting lists for municipal dental care lengthened. In the City of Lahti, the waiting time lengthened from about 12 months to 18 months. The waiting list in question consists only of persons aged 18 and over. Oral healthcare for children and youth is based on an individually defined care program. For specialist care, the waiting times are usually less than 3 months.

Previously, before GSM appointment reservation, clients were, as now, put on a waiting list at a dental clinic, and appointments were distributed from there to the other dental clinics. The waiting list was maintained manually, however, and was rather laborious to handle. Clients on the waiting list were invited to the clinics either by telephone or by letter. A particular problem was posed by appointments that were canceled the same day. These often remained unfilled. The Lahti City oral healthcare services use a computerized client information system (Winhit), which operates online at all the branches. This system has no facility for the management of waiting lists. GSM appointment reservation is an example of how social and health-service information systems can be supplemented with mobile technology in serving clients.

Operational Model for GSM-Based Appointment Reservation

In our case study relating to oral healthcare, we categorize customers into groups based on dental science. In oral health services, the treatment models are divided as follows: care of those under 18, care of adults (over 18 years old), and care of those who need special disease treatment. In addition, patients are divided into urgent and non-urgent groups. The definition of the patient segments has also meant the implementation of different consultation procedures. The consultation types are as follows:

- Duty consultation for urgent care.

- Appointment reservation consultation for non-urgent care.

- Recall consultation for those under 18.

- Special disease-treatment consultation.

The clinics are situated within the same regional organizational structure. This comprises 21 dental clinics, of which 2 have duty consultation and 1 has special disease-treatment consultation. People under 18 and non-urgent adults are catered to at all the dental clinics.

Next, we turn to an examination of two stages within the framework of the electronic communication architecture: the first contact of the customer and the first clinical appointment. We discuss the making of appointment reservations by GSM mobile phones in order to arrange emergency dental appointments and to reallocate canceled dental appointments. The studied activity covers GSM-based appointment reservation and waiting-list management in oral healthcare services in the City of Lahti, Finland.

Finland has been one of the pioneers in the development of mobile communication solutions (Aarnio, Enkenberg, Heikkilä, & Hirvola, 2002). The environment in Finland is that of GSM. Webopedia (http://www.webopedia.com) defines SMS, which is a core technology in our chapter, as follows: "Abbreviated as SMS, the transmission of short text messages to and from a mobile phone, fax machine and/ or IP address. Messages must be no longer than 160 alpha-numeric characters and contain no images or graphics." Usually, the user interface of an SMS is that of a mobile phone, but other solutions may exist. SMS is a central application platform in the GSM system.

Forming the basis for our architecture is the customer database, which is in use at all dental clinics in the City of Lahti. The customer database contains the lists of urgent emergency patients as well as non-urgent appointment-reservation patients. The patients in the database are in chronological order, so that first on the list are those that contacted the duty office earliest or those that have been on the non-urgent waiting list for the longest time. The customer database is common to all the dental clinics, and the clinics can make additions to it and deletions from it online. The database thus commits the clinics to handling and assigning appointments from the same waiting list (Serkkola & Suokas, 2004)

GSM Emergency Duty Service

Emergency duty means urgent outpatient treatment given within 24 hours of the treatment-seeking contact or the receipt of a referral. In the emergency duty service,

Box 1. Emergency duty message

```
Your  emergency  dental  appointment  is
approaching.
Your  estimated  appointment  time  is
13:15.
Welcome to the City Hospital emergency
dental clinic!
```

the customer contacts the receptionist or dental assistant, who records the customer's name on a waiting list. The customer gives his or her mobile-phone number, to which the customer will receive a message concerning the time when he or she has to be at the clinic. The customer receives an invitation 1 hour before the intended arrival time. The patient who is currently sixth on the list receives an invitation requesting him or her to come to a clinic. A message is sent automatically straightaway when a customer comes out from a treatment session and the session is recorded as finished. The average length of a treatment session is reckoned to be about 20 minutes. If a customer does not want this service, his or her name is visible on the waiting list in order of initial contact, and so he or she can be invited for treatment from the clinic premises. In this way, the emergency client can move around freely and need not wait in the waiting room. The customer will receive a message on his or her mobile phone as seen in Box 1.

GSM Reallocation Service for Canceled Appointments

When managing the appointments and the reservations by GSM, the dental clinic informs clients on the waiting list by mobile phone about the canceled appointment time. The GSM-based reservation in this application concerns canceled appointment times that are reallocated within the same day (2-8 hours) or days before. In the database, the clients may be grouped into either one waiting list or several, depending on the grouping principle. The basis for the grouping may be, for example, the

Box 2. Cancelled appointment invitation

```
Cancelled  appointment  time  at  Laune
Dental Clinic,
Launeenkatu 74 B, 31 March at 12.00.
Please reply IMMEDIATELY: write HA and
send to the number 18444.
Wait for confirmation!
```

Box 3. Reminder to the school student

```
Reminder: An appointment has been re-
served for you at
Kivimaa Dental Clinic on 8 April 2006.
If you wish, you can cancel the ap-
pointment
by replying to this message: MU KI
```

Box. 4. Reminder to the parent or guardian

```
Reminder: An appointment has been re-
served for Mikko Mallikas at
Kivimaa Dental Clinic on 8 April 2006.
```

treatment unit, the medical priority, the waiting time, or the local area. Clients can be added or removed from the waiting list according to need. The opportunity for GSM reservation is presented to clients who wish to accept canceled appointment times by mobile phone.

A receptionist or dental assistant sends the canceled available appointment time by one press of a button to five clients simultaneously. On the computer screen, he or she enters only the free appointment time. The first of the five clients to reply to the text message can reserve the free appointment time. The other four clients are informed that the appointment has been filled, and they return to the top of the list to await the next time that becomes available. This electronic invitation, which may be either free to the client or subject to a charge, costs the service provider the price of one text message. The content of the text-message invitation can be seen in Box 2.

GSM Reminder Service

In the first phase of the experiment, the reminder service has been intended for school students under the age of 18. It has been designed for recall consultation, in which the person under 18 is invited for a regular checkup every 2 years. The young person is given an appointment time and is informed about it at least a month beforehand, or he or she is given a time at the clinic in accordance with the treatment plan. The young person and his or her parent or guardian receives an automatic SMS reminder about the appointment time 48 hours before the appointed time. The service is intended to reduce students' absence from dental hygienists' and dentists' appointments. According to our research, absences from young people's recall appointments constitute 78% of all unused appointments (Serkkola & Suokas 2004). The SMS messages are seen in Box 3 and Box 4.

In the architecture, reminder and cancellation time services have been integrated with each other, and they operate automatically in customer-interaction situations. Reminders are sent automatically to a specified GSM number 48 hours before the appointment time. Similarly, if a customer cancels a time given, the system searches automatically for a new patient to take his or her place. Information about the free time may be sent either automatically to the next person on the list, or manually by authorization of the person reserving the appointment. The same message is also sent automatically to a customer's e-mail address. The service operates by a multi-channel system so that messages of identical content are sent to both the mobile phone and the e-mail address. In addition, preparation instructions can be sent to customers by e-mail. Wireless appointment-reservation services have thus become part of electronic communications in oral healthcare.

Evaluation of the GSM Services

A regional database common to all the dental clinics forms the basis for using the new electronic services. For a population of around 100,000 inhabitants, centralized customer service and waiting-list management are the most practical and efficient ways to organize emergency treatment and appointment reservation (Serkkola, Suomi, & Mikkonen, 2005). A centralized customer database commits all the city's dental clinics to treating a customer group in the same chronological order. The waiting list can be monitored online. What is more, a centralized contact centre brings cost savings in consultation and appointment reservation.

According to our research, the establishment of a customer database, that is, a waiting-list database, has resulted in the following benefits for dental care (Serkkola & Suomi, 2004). First, the waiting list is visible in real time at each dental clinic, and each receptionist or dental assistant can make additions to it or deletions from it. It has thus been possible to monitor customers' positions on the list in real time. Second, according to staff, the waiting-list database has made it easier to monitor the waiting list and assign appointments from it. One person handling the list is able to direct customers on the list to the different dental clinics more efficiently than reception offices operating in many different dental clinics. Third, besides other organizational measures, the waiting list has been reduced from 4,000 to 2,000 customers in 2 years. Similarly, waiting times have shortened from 16 months to less than a year. The target is to achieve a waiting time of less than 6 months by the end of 2006.

GSM appointment reservation has had an effect on the reallocation of canceled appointments and on the costs arising from these. In oral healthcare in Finland, canceled times constitute on average 6% of all appointments. By means of mobile appointment reservation, it is possible to reallocate nearly all appointment times canceled the same day (within 2-8 hours) and canceled earlier (within 1-7 days). On the other hand, the service can do nothing about appointments that remain unused. The GSM service results in calculated savings to Lahti if it sells at least 50 canceled appointments per month. Mobile appointment reservation is only suitable for municipalities and municipal federations with at least 100,000 inhabitants (Serkkola et al., 2005).

GSM appointment reservation has given rise to certain procedural changes in customer reception work. Besides the person's name, the GSM number has become a key personal data item. Observations show that the method of contact has shifted from phone invitations to text messages. Text messages are easier to send in between other work tasks. GSM appointment reservation saves about 15 to 20 minutes of receptionists' or dental assistants' working time per appointment sold. In this way, GSM reservation does, to some extent, increase receptionists' and dental assistants' control over their own work. Dental assistants can issue GSM invitations alongside their other work tasks. Earlier, a basic premise in customer service was that canceling an appointment was not a responsible use of the services. In the new system, appointment cancellation has been made easier. Through GSM reservation, the aim is to keep unused times to a minimum.

Finally and most importantly, customers have accepted the GSM service with satisfaction. In a research study to obtain feedback on a GSM emergency duty service, the persons interviewed included not only ones who had used the mobile phone service, but also ones who had not used it (total 153). Those that had used the service were unanimous about the fact that the text-message invitations were a necessary part of client service in present-day society (100%). Nonusers, too, considered the service necessary (73%), but among them were also those who had more reservations regarding it (14%), or who did not express any opinion (13%; Serkkola & Mikkonen, 2005). In a study to evaluate a GSM reallocation service for canceled appointments, young and middle-aged people regarded the service as trouble free to use (89% of 59; Serkkola & Mikkonen, 2004). A GSM reminder service was intended to remind school students about approaching appointment times. Amongst a small survey sample (83), the service appeared to have had hardly any effect on unused appointments. According to the results, the GSM emergency duty service and reallocation service are necessary in the clients' view, while the reminder service is less necessary.

Future Trends

Our research shows that electronic communication solutions require the redesigning of the entire customer strategy for the service under study, in our case, for regional oral healthcare. They involve integrating the consultation procedures and customer services with each other, both internally within the dental clinics and also between the different clinics.

Capability and willingness to undertake such redesigning initiatives will be a key success factor for healthcare organizations in the future. In addition to process changes, changes in governance structures enabled by and for information technology are needed.

Electronic and mobile health applications are penetrating the health industry with speed and force. It is of key importance to design them well from the very beginning so that the users get a positive image of them: Mistakes made now will maybe take decades to repair.

Based on the good experiences in Lahti, we feel that e-health solutions are going to spread in healthcare, even to medically more complex environments than basic dental care.

Conclusion

We presented in this chapter an integrated architecture for e-health applications. The architecture defines the phases of a healthcare episode and the different roles different stakeholders have in this process.

The architecture has been a basis for the implementation of two pilot applications in the city of Lahti, Finland. Both pilot applications have already turned out to be successes after a very short application period. In addition, there are other applications under development and other pilot projects not reported in this chapter.

The architecture proposed here helps developers of health applications to take into account all the functions the systems could have. The activities of the patient or customer, the medical staff, and the system itself must all be coordinated.

GSM appointment reservation has proved easy to use, and people of all ages succeed in reserving a time with no trouble. In the opinion of all age groups, the quality of customer service is improving through GSM reservation. In the beginning, customers need to receive written instructions on GSM emergency duty service and reallocation service.

Our chapter shows that well-designed systems can find an audience, even in the healthcare sector, which is documented as being an environment that is rather hesitant to use IT and even hostile toward it.

References

Aarnio, A., Enkenberg, A., Heikkilä, J., & Hirvola, S. (2002). *Adoption and use of mobile services: Empirical evidence from a Finnish survey.* Paper presented at the 35th Hawaii International Conference on System Sciences, Waikoloa, HI.

Ewles, L., & Simnett, I. (1992). *Promoting health: A practical guide.* London: Scutari.

Gopalakrishna, P., & Mummaleni, V. (1993). Influencing satisfaction for dental services. *Journal of Healthcare Marketing, 13*(1), 16-22.

Haux, R., Ammenwerth, E., Herzog, W., & Knaup, P. (2002). Healthcare in the information society: A prognosis for the year 2013. *International Journal of Medical Informatics, 66,* 3-21.

Luck, M. (2000). The relevance of market research for health and social care. In M. Luck, R. Pocock, & M. Tricker (Eds.), *Market research in health and social care* (pp. 3-9). London: McGraw-Hill.

Luo, W., & Najdawi, M. (2004). Trust-building measures: A review of consumer health portals. *Communications of the ACM, 47*(1), 109-113.

Malone, T. W., Yates, J., & Benjamin, R. I. (1987). Electronic markets and electronic hierarchies: Effects of information technology on market structure and corporate Strategies. *Communications of the ACM, 30*(6), 484-497.

Morrissey, J. (2003). An info-tech disconnect. *Modern Healthcare, 6-7,* 36-48.

Ragupathi, W. (1997). Healthcare information systems. *Communications of the ACM, 40*(8), 81-82.

Serkkola, A., & Mikkonen, M. (2004). Matkapuhelin hammashoidon peruutusaikojen täytössä. *Suomen Hammaslääkärilehti, 13,* 736-738.

Serkkola, A., & Mikkonen, M. (2005). Mobiilikutsu päivystyspoliklinikalle: Asiakaspalautetta kokeilun tuloksista. *Suomen Hammaslääkärilehti, 8,* 520-522.

Serkkola, A., & Suokas, L. (2004). *GSM-ajanvaraus suun terveydenhuollossa: Hammasaikapilotin arviointi lahden kaupungissa.* Retrieved March 30, 2004, from http://www.aluenet.com

Serkkola, A., Suomi, R., & Mikkonen, M. (2005). *GSM-based reservation system for dental care.* Manuscript submitted for publication.

Smolander, K., Rossi, M., & Purao, S. (2005). *Going beyond the blueprint: Unraveling the complex reality of software architectures.* Paper presented at the European Conference on Information Systems 2005 (ECIS 2005), Regensburg, Germany.

Suomi, R. (2000). *Leapfrogging for modern ICT usage in the healthcare sector.* Paper presented at the Proceedings of the Eighth European Conference on Information Systems, Vienna.

Tang, P. C., & McDonald, C. J. (2001). Computer-based patient record systems. In E. H. Shortliffe (Ed.), *Medical informatics: Computer applications in healthcare and biomedicine.* New York: Springer.

Utbult, M. (2000). *Näthälsa internetpatiente möter surfande doktore: Uppstår konfrontation eller samarbete?* Stockholm: Reldok och KFB-Kommunikationsforskninsgberedningen.

World Health Organisation (WHO). (1985). *Targets for health for all.* Copenhagen, Denmark: Author.

Section VI

Conceptual Frameworks

Chapter XVI

Conceptual Framework for Mobile-Based Application in Healthcare

Matthew W. Guah, School of Business Economics,
Erasmus University Rotterdam, The Netherlands

Abstract

The significance of aligning IT with corporate strategy is widely recognised, but the lack of an appropriate framework often prevents practitioners from integrating emerging Internet technologies (like Web services and mobile technologies) within organisations' strategies effectively. This chapter introduces a framework that addresses the issue of deploying Web services strategically within a mobile-based healthcare setting. A framework is developed to match potential benefits of Web services with corporate strategy in four business dimensions: innovation, internal healthcare process, patients' pathway, and management of the healthcare institution. The author argues that the strategic benefits of implementing Web services in a healthcare organisation can only be realized if the Web-services initiatives are planned and implemented within the framework of an IT strategy that is designed to support the business strategy of that healthcare organisation.The chapter will use case studies to answer several questions relating to wireless and mobile technologies and how they offer vast opportunity to enhance Web services. It also investigates what challenges are faced if this solution is to be delivered successfully in healthcare. The healthcare industry globally, with specific emphasis on the USA and United Kingdom, has been extremely slow in adopting emerging technologies that focus on better practice management and administrative needs. The chapter elaborates on certain emerging information technologies that are currently available to aid the smooth process of implementing mobile-based technologies into healthcare industry.

Introduction

This chapter is based on research—using a longitudinal case study—into the National Programme for Information Technology (NPfIT). NPfIT is an initiative that has been budgeted to cost the UK government £6.3 billion for the purpose of improving the information systems in the National Health Service (NHS), with emphasis on IT infrastructure and the creation of a nationwide patient database.

The significance of aligning IT with corporate strategy in healthcare organisations is widely recognised, but the lack of an appropriate framework often prevents medical practitioners from integrating emerging Internet technologies (like Web services and mobile technologies) within healthcare organisations' strategies effectively. This chapter introduces a framework that addresses the issue of deploying Web services strategically within a mobile-based healthcare setting. A framework is developed to match potential benefits of Web services with corporate strategy in four business dimensions: innovation, internal healthcare process, patients' pathway, and the management of healthcare institution. The author argues that the strategic benefits of implementing Web services in a healthcare organisation can only be realized if the Web-services initiatives are planned and implemented within the framework of an IT strategy that is designed to support the business strategy of that healthcare organisation.

The chapter will also consider certain essential issues regarding the deployment of any mobile data solution (i.e., reliability, efficiency, and security) in the healthcare industry and how such deployment can support healthcare professionals in saving patients' lives. Using case studies, the chapter will answer the following questions:

- Wireless and mobile technologies offer vast opportunities to enhance services, but what challenges are faced if this solution is to be delivered successfully in healthcare?

- Why has the global healthcare industry, with specific emphasis on the USA and United Kingdom, been extremely slow in adopting technologies that focus on better practice management and administrative needs?

- How complacent can IS strategists be to the productivity paradox in the wake of HIPAA (Health Insurance Portability and Accountability Act in USA) and NPfIT (in UK)?

- What emerging information technologies are there to aid the smooth process of implementing mobile-based technologies into the healthcare industry?

The existing economics and IS literature on information-technology adoption often considers network externalities as one of the main factors that affect adoption decisions (Brown & Venkatesh, 2003). It is generally assumed that potential adopters achieve a certain level of expectations about network externalities when they have to decide whether to adopt a particular technology. However, there has been little discussion on how the potential adopters reach their expectations. This chapter attempts to fill a gap in the literature on the adoption of mobile healthcare technology by offering an optimal control perspective motivated by the rational expectations hypothesis and exploring the process dynamics associated with the actions of decision makers in the healthcare industry. They must adjust their expectations about the benefits of a mobile healthcare technology over time due to bounded rationality. The model posed in this chapter addresses mobile healthcare technologies that exhibit strong network externalities. It stresses adaptive learning to show why different healthcare organisations that initially have heterogeneous expectations about the potential value of a mobile healthcare technology eventually are able to arrive at contemporaneous decisions to adopt the same technology, creating the desired network externalities. This further allows these organisations to become catalysts to facilitate processes that lead to healthcare industry-wide adoption.

Background

The NHS has been responsible for the provision of healthcare and services in the United Kingdom for the past 56 years on the basis of being free for all at the point of delivery. The traditional perception of the NHS is one of a healthcare system organised as a professional guild, with unlimited finance from the government. This type of NHS is experiencing an irrevocable change as taxpayers are no longer complaisant and paternalistic employers are reacting against inflating costs and escalating complaints from the patients. The employer is reacting to the continuous massive flow of subsidies for inefficient physician practices, fragmented delivery systems, and cost-unconscious consumer demand. The patients are increasingly assertive as to their preferences and few have expressed their willingness to make additional contributions for particular health benefits and medical interventions.

Web services are technologies with roots in the Application Service Provision (ASP) business model that are used mostly to automate linkages among applications (Hagel, 2002). They are generally anticipated to make critical system connections not only possible but also easy and cheap (Kreger, 2003; Sleeper & Robins, 2001).

One of the perceived benefits of Web services is that organisations would be able to concentrate on their core competencies (Perseid Software Limited, 2003). Service providers argued that the remote delivery of software applications would release managers from the perennial problems of running in-house IT departments, allowing more time to develop IT and e-business strategy rather than the day-to-day operations (Currie, Desai, & Kahn, 2004). This justification has been used in traditional forms of outsourcing over many years (Willcocks & Lacity, 1998).

The NHS is experiencing massive changes in the structure of information systems provision markets and organisations. The local service provision (LSP) and national service provision (NSP) models in use by the NPfIT are in a state of ferment. The payment methods borrow from both capitation and fee for service, and methods of utilisation management are compromised between arm's-length review and full delegation (Guah & Currie, 2006). LSP and NSP consist of large and more complex entities. These are the result of merger, acquisition, and product diversification. The service providers involved have had to take on a visible feature of ceaseless acquisition and divestiture, integration and outsourcing, and combination and recombination. Providers of medical systems, hospital administration systems, and health plans are coming together and then coming apart. They are substituting contracts for joint ownership, creating diversified conglomerates and refocused facilities, and experimenting with ever-new structures of ownership, finance, governance, and management (Robinson, 2000). These would give the NHSIA (National Health Service Information Authority) the benefits not only of a middle ground between the extremes of vertical integration and spot contracting, but also a balance of co-ordinated and autonomous adaptation in the face of its ever-new challenges.

The general assumption is that expenditures in the nation's health will outpace the overall growth in the economy (Collins, 2003; Pencheon, 1998). This is reflected in the percentage of the GDP (gross domestic product) of USA (13%), Germany (10.7%), France (9.6%), and the United Kingdom (7.6%) being devoted to the total cost of healthcare resources (Brown, 2002). Unlike the United Kingdom, however, some of these countries are faced with limitations in social willingness to pay. It has been documented that millions of U.S. residents currently lack the most basic insurance coverage (Institute of Medicine, 2002).

Response to Emerging Technologies in the NHS

Over the years, nontechnologists in the NHS have managed to muddle through one powerful new system after another. Generational strategy is one continuously being used to deal with some of the pressures induced by IS. Adopting such innovations as PCs (personal computers) and the Internet requires the personal and organisational costs of unfreezing deeply ingrained old habits. Many workforces ignore, deny, or deal awkwardly with such technologies.

Srinivasan, Lilien, and Rangaswamy (2004) found several reasons why an organisation should respond to new technology development. Two major reasons are listed below:

*   Technological change is a principal driver of competition. This is principally because it destroys monopolies, creates new industries, and renders products and markets obsolete.
*   Additional sources—both within and outside the organisation or industry—are increasingly complementing in-house technology development efforts.

A common response to new systems is the "not invented here" (NIH) syndrome (Collins, 2003; Guah & Currie, 2004; Haines, 2002). This often leads to certain organisations rejecting a perfectly useful system based on an implicit assumption that the system does not fully recognise or accommodate their own needs and idiosyncrasies (Brown & Venkatesh, 2003; Davis, 1989). Davis sees this as a likely result of a decline in communication with external sources. NIH syndrome could also result from competences that can be proven to be outdated and inefficient in comparison to an existing technology. One Trust, which places a central role in the direction of regional IS strategy, had to reject a system promoted by the Department of Health because the system was not as familiar as another bespoke system (Haines).

The common characteristics of new systems in the NHS are uniformity in products and prices in the face of great variability in consumer preferences and the actual costs of providing service (Collins, 2003). This one-size-fits-all approach usually leads to services that are of excessive costs for some users and insufficient quality for others, impeding the use of price flexibility to enhance capacity utilisation (Robinson, 2000). Also of concern is a combination of overcapacity and low load factors in some regional trusts with undercapacity and shortages elsewhere. Concerns are growing in the NHS that this may generate cross-subsidies from trusts for which the cost of service will be low to trusts for which the cost of service will be high (McGauran, 2002). Additionally, deregulation of healthcare costs has spurred an outpouring of new services. Consequently, several of these services are the following (Collins, Pencheon, 1998):

*   A different cost structure.
*   An impact on IS budgets.
*   A better match between supply and demand.

Incomplete information has been a fascinating attribute of the NHS's unusual system's organisational and normative characteristics. The asymmetry of NHS information between patients and medical practitioners has changed in an exogenous fashion over its 56 years. The amount of healthcare information available to patients is usually the result rather than the cause of changes in the economic and political environment (Robinson, 2000).

Project Description: National Programme for Information Technology

The NPfIT is an initiative by the National Health Service Information Authority, born as a result of several plans to devise a workable IS strategy for the NHS (NHSIA, 2003; Wanless, 2002). The NPfIT was designed to connect the capabilities of modern IT to the delivery of the NHS plan devised in 1998. The core of this strategy is to take greater control of the specification, procurement, resource management, performance management, and delivery of the information and IT agenda (NHSIA).

The NPfIT is an essential element in delivering the NHS plan. It has created £6 billion information infrastructure, which could improve patient care by increasing the efficiency and effectiveness of clinicians and other NHS staff. The intention of the plan is to address the following (http://www.npfit.nhs.uk):

- Create an NHS Care Records Service to improve the sharing of consenting patients' records across the NHS.
- Make it easier and faster for GPs (general practitioners) and other primary care staff to book hospital appointments for patients.
- Provide a system for electronic transmission of prescriptions.
- Ensure that the IT infrastructure can meet NHS needs now and in the future.

The decision to implement a national programme for IT into the NHS system complexity is only the first step in the IS modernisation journey for a multifaceted organisation. There are many examples of new technologies disrupting organisational routines and relationships in the NHS (Atkinson & Peel, 1998; Majeed, 2003; Metters, Abrams, Greenfield, Parmar, & Venn, 1997). These usually require both medical professionals and NHS regional trusts managers to relearn how to work together. Orlikowski (1993) and Edmondson (2003) suggest that one technology can be seen differently by two groups of people in an organisation. Findings from Barney and Griffin (1992) and Orlikowski have showed how this could result in the elicitation of different responses for members of that organisation.

Scope of Project Work

The chapter takes a more in-depth look at the role of the NHSIA (seen as the project leader for the NPfIT), currently the most visible spokesperson and translator for the potential implications of the resulting new technologies. Research has shown NHS IS staff to pay particular attention to what the NHSIA says and does in regard to information systems (Collins, 2003; Ferlie & Shortell, 2001). This research builds on a framework that identifies the key dimensions of the NHSIA tactic that is situation specific for NPfIT assumptions. The work looks at the NHSIA goal and roles for the NPfIT, as well as the role of the private-sector service providers in the implementation of NPfIT.

Here are a few objectives the NPfIT hopes to accomplish:

- To have a series of tightly specified and priced framework contracts on a short list (of about five) primary service providers (PSPs) who can work at the regional and local Strategic Health Authority (StHA) level. This should enforce the integration and implementation partnership—at a national level—during all aspects of the NPfIT project. Each PSP will have an aligned consortium of service providers and vendors for the integrated care resource service element of the NPfIT, and will be mandated to work with the domain PSP for electronic booking, the infrastructure providers, and healthcare providers. StHA PSPs may not make their products exclusive or mandatory to their StHA.

- To create priced packages of national services and applications that the PSPs and StHAs can together implement locally. This activity will include managing the creation of a single Human Resource Information Systems (HRIS) and other national services to access and move health-record information as required.

- To create service-level agreements for the national services and other services out of the scope of the PSP consortium that the PSPs can work toward in providing an integrated service to the StHA

- To develop and maintain the national standards and specifications that all vendors must use. It is also anticipated to create the national business cases required for the Department of Health governance (required by the National Treasury), and to support the local decision-making business cases required at the StHA level.

- To procure, under national contracts, a backbone network infrastructure

Such an arrangement provides the greatest clarity in respect to the appropriate allocation of responsibilities and should be well understood in the public and private sectors (see Table 1). Services will be procured on a long-term basis so the com-

Table 1. PSP implementation timetable (as of July 2002)

Activity/Output	Target Date
Agree on procurement strategy (Department of Health (DoH) & local health authorities)	End Jul 2002
Service requirement finalized and approved	End Sep 2002
Outline business case developed and approved	End Sep 2002
Official Journal of European Communities (OJEC) advert	Oct 2002
Procurement of systems and implementation services for electronic booking begins	Oct 2002
National long list of PSPs created	Dec 2002
Invitation to negotiate issued	Jan 2003
National short list of PSPs created	Apr 2003
First local health authorities begin detailed planning with PSPs	Aug 2003
PSP framework contract finalized	Oct 2003
Infrastructure provider(s) contract agreed	Oct 2003
First local health authorities begin implementation	Nov 2003
Infrastructure migration begins	Mar 2004

bination of local and central funding will be required for at least 5, and preferably 10, years at guaranteed levels.

The Research Study

This study intended to address the gap in the existing literature with regard to the complex issues surrounding the adoption of mobile technology in healthcare. I define mobile healthcare as the use of all kinds of wireless devices (cell phones, personal digital assistants, mobile e-mail devices, handheld computers, etc.) to provide health information and patient-care records to healthcare practitioners, patients, and their caregivers, employers, and employees of health service providers and public regulars of healthcare and services.

The findings reported in this chapter are part of a larger 5-year research study that was developed to investigate the deployment, hosting, and integration of the ASP and Web-services technologies from both a supply-side and demand-side perspective. The overall research was in two phases. The first phase, comprising of a pilot study, was conducted in the USA and United Kingdom (Currie et al., 2004). An exploratory-descriptive case-study methodology (Yin, 1994) was used to investigate 28 ASP vendors and seven customer sites in the United Kingdom. The dual focus upon supply side and demand side was critical for obtaining a balanced view between vendor aspirations about the value of their business models and customer experiences, which may suggest a less optimistic picture. The unit of analysis was the business model (Amit & Zott, 2001), not the firm or industry level, so a case-study methodology was anticipated to provide a rich data set for analysing firm activities and behaviour (Currie et al.).

The result from the pilot study led to the funding of two additional research studies by the Engineering and Physical Sciences Research Council (EPSRC) and Economic and Social Research Council (ESRC) respectively. Industrial collaborators were selected for the roles of technology partners, service providers, and potential or existing customers. These studies were concerned with identifying sources of value creation from the ASP business model and Web-service technologies in different vertical sectors (including health).

Research Methodology

The research followed a number of stages involving the use of both qualitative and quantitative data-collection techniques and approaches (Walsham, 1993). A questionnaire survey was distributed by e-mail to businesses and healthcare organisations all over the United Kingdom. These organisations were listed on a national database maintained by the NHSIA, plus those maintained by the university. To ensure relevant managers and practitioners responded, the covering note clearly stated the purpose of the questionnaire and requested that it be passed on to the person(s) with responsibility for managing healthcare e-business strategy. Scales to address the research questions were not available from the literature, so the questionnaire was developed based on the theory of strategic value (Banker & Kauffman, 1988). It included a checklist, open-ended questions, and a section seeking organisational data. Research questions under Part 1 required respondents to answer yes or no if the application of Internet technologies in healthcare were bringing value to patients. Data in Part 2 of the questionnaire were collected by open-ended questions seeking respondents' views on the best approach to healthcare performance improvement and Web-service value creation. This line of questioning was used to increase the reliability of data since all respondents were asked the same questions, but some added additional information. The purpose was to impose uniformity across the sample of representation rather than to replicate the data obtained from each participant (Yin, 1994).

Patient-Information Management

Healthcare organisations are showing a clear interest in accelerating the transformation of clinical care through the routine use of appropriate emerging technologies by clinicians when diagnosing problems and subsequently planning and administering patient care. To support such noble efforts toward delivering better healthcare to the public, President George Bush and other national leaders have publicly called for the development of a national health information infrastructure. The U.S. government (in 2005) has over the past 2 years published plans for all Americans to have an electronic health record by 2014. Similar to the NPfIT in the United Kingdom, the plans call for a hierarchical set of local, regional, state, and national networks that facilitate peer-to-peer sharing of patient records.

When considering the deployment of any mobile data solution, reliability, efficiency, and security are essential, none more so than in the emergency services if lives are to be saved. To support such communication of critical and personal information, there has been an increased demand for the creation of electronic methods for storing and tracking clinical information (see Figure 1). This requires the solution for some fundamental architectural problems within the healthcare environment: scalability, reliability, recoverability, interchangeable vocabularies, and integration:

- Most service providers can support several thousands of simultaneous log-ons. Many are finding it difficult to demonstrate scalability in thin-client, rules-based order entry or structured clinical information.

- Service providers need to show evidence that they have appropriate schedules for downtime because healthcare organisations require reliability of 99.999% due to the critical nature of the information in the healthcare industry.

- There has to be a recognisable solution for a very quick data-recoverable plan in the event that downtime or a system failure occurs. Healthcare organisations have to ensure there are fault-proof backup plans to provide medical practitioners with information that is only available in an electronic form in the event that systems become suddenly unavailable. Certain work processes in the healthcare industry (including emergency care, scheduling, registration, order entry, and clinical procedure recording) would need to continue seam-

Figure 1. Flexible and independent patient care (http://www.healthpia.us/services.asp)

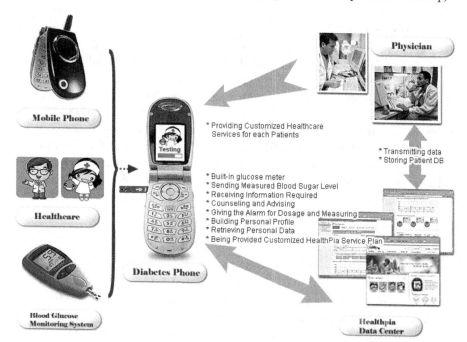

lessly, even with a primary-system interruption (see Figure 1). After there has been an interruption, recovery must be complete with no loss of information. Backup, therefore, must prevent any IT failure from making care to the patient impossible. Where mobile healthcare technology is involved, adequate hardware, infrastructure, and tested processes should exist as part of a complete implementation to guarantee this recoverability.

Due to a previous lack of harmonized acquisition, healthcare organisations in both the United Kingdom and USA are frequently challenged by a variety of code sets and files that have proliferated across various healthcare institutions. HIPAA attachment transactions (in the United States) and the NPfIT (in the United Kingdom) are beginning to dictate that the future exchange of patient information be carried out electronically between healthcare organisations. To facilitate this portable and interoperable mobile healthcare technology, certain local vocabularies need to be replaced and the use of government-specified code sets should be synchronized. The way forward is to maintain current systems and historic data through mapping infrastructures that manage the correct translation, giving semantic meaning to patient data with the hope that one day soon there will be a complete migration to common vocabularies.

Many applications currently in the NHS have been designed with the assumption that that the approach and architecture does not need to coexist and interoperate. While some of these may support integration with other applications that also have significantly different philosophical stances, they do not fully recognise that the need for a healthcare organisation to implement a total solution involves the practicalities of many different dimensions of time, scope, economics, and service providers' organisational politics. Toward this goal, all interoperability for a mobile healthcare technology should require that all features and functions work across all applications. The NPfIT project is proving that all service providers have to significantly alter their current approach to internal and external integration, security, and nomenclatures during the life of the project.

Case 1: MotoHealth

Motorola, along with its partners, initiated a telemedicine service at Harvard Teaching Hospital called MotoHealth (http://www.motorola.com/mediacenter/news/detial) late 2004. The Motorola solution uses mobile phones to help healthcare professionals to monitor chronically ill patients during their normal daily routines. This product was designed to meet the customer's convenience, and as a discreet way of monitoring patients in the mobile environment, can replace in-home hospital and home monitoring devices. As a result, it gives the patient more independence to continue daily activities virtually anywhere they like. This method to providing healthcare pushes healthcare and services out of high-cost health institutions. It enables the patient's body to become the point of care and the mobile healthcare technology becomes the bridge to the patient's body, thus, enabling the delivery of care, educational advice, and support remotely and transparently.

This case has proven that when a mobile healthcare technology is implemented as part of a comprehensive healthcare program, it can give healthcare providers useful daily updates on a patient's physiological levels such as blood pressure, glucose level, and weight control. Such a method of healthcare facilitates proactive treatment action, resulting in fewer hospitalizations and visits to emergency rooms, potentially lowering the increasing demand on the costs of providing healthcare and services to the public.

Policy Issues

Arguably, the most viable techniques for successful mobile healthcare technology implementation are practical guidelines and good management practices. Policies established by a healthcare organisation are the first steps toward this goal. There are, however, few steps for establishing a policy for mobile healthcare technology:

Guidelines must be developed for the acquisition of mobile healthcare technologies. This would balance the need to encourage innovative applications against wasteful spending, which can be seen by certain members of the staff as duplication of effort. This in turn makes it the responsibility of every medical practitioner who may need a particular type of application to strictly adhere to this policy.

There must be regular inventory taken. This helps to identify and evaluate all installed hardware and software before setting or changing the standards. These would certainly affect policy decisions and future acquisitions. While standards can sometimes be looked upon as restrictive in the IT industry, medical practitioners actually see these to offer benefits for the care providers and the patients alike. Nearly all healthcare organisations have standards that cover many aspects of their operations within the healthcare process. Generally, standards in the NHS are recorded in formal standards and procedures manuals, but in certain cases, we came across informal handwritten notes (i.e., "this is the way we do things here") that are also considered to be standards.

Ruyter, Wetzels, and Kleijnen (2001) show how organisations implementing e-commerce first learn to exploit the Internet for information transfer before supporting transactions, and then finally use it for commercial trading and collaboration among various actors. Considering mobile healthcare is still in its infancy, borrowing from

Figure 2. Home visits and general-practice consultation (http://www.bmj.com, 2004)

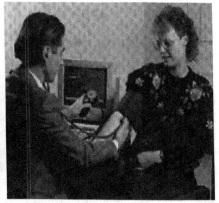

the e-commerce experience will mean that healthcare organisations will probably adopt wireless e-business methodologies first to support their existing healthcare processes and improve efficiency before they come up with new business models to transform the competitive landscape in the healthcare industry.

In the case of the NHS, wireless enterprise implementation issues frequently extend well beyond the technology domain and into areas of medical practices and organisational culture (see Figure 2). Nearly all the regional healthcare trusts that are actively pursuing wireless enterprise strategies at the moment are handcrafting solutions around their own local IT infrastructures and their own homegrown healthcare processes partly because there are very few packaged mobile healthcare solutions on the market.

The focus, currently, is on accessing information via wireless mobile healthcare messaging. However, the future should hold more applications like mobile access, telemedicine, and alerts for facilitating better disease management and controls. Given the emerging state of mobile technology and its potential impact on healthcare, mobile healthcare can be seen as truly radical. Mobile healthcare has the potential to remake this entire industry and obsolete established strategies. Most healthcare organisations in Western Europe and North America feel they must participate in this emerging healthcare technology in order to survive the increasing demand to service a continuously evolving patient environment.

The research found two reasons why healthcare organisations are beginning to pay keen attention to mobile healthcare:

- These organisations are being defensive. There is a general belief that newcomers in the healthcare-provision market may be plotting to use new functionalities available through the use of mobile healthcare to attack the incumbents' core providers.
- Converse to the previous reason is the understanding that mobile healthcare could realize its potential. If this happens successfully, mobile healthcare would be too attractive a proposition to ignore, and joining in at a later date may prove too expensive.

The author poses a conceptual framework for the research upon which this chapter is based, showing the different stages of the technology-adoption process for healthcare organisations as well as the main factors operating at each stage (see Figure 3). In the preadoption stage, healthcare organisations take an internal perspective and analyse the fitness of the mobile healthcare technology for the contemplated task as well as the value of this technology. These should be the drivers of the adoption decision. The phase after that shows the healthcare organisation analysing whether organisational and environmental factors are favourable for continuing with this

Figure 3. Conceptual framework of mobile healthcare technology adoption

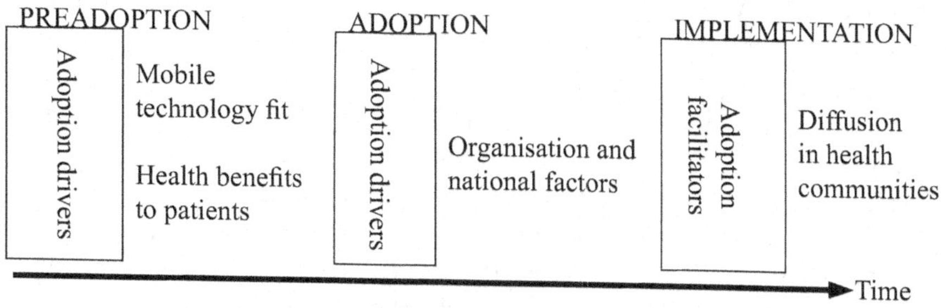

PREADOPTION	ADOPTION	IMPLEMENTATION
Adoption drivers — Mobile technology fit — Health benefits to patients	Adoption drivers — Organisation and national factors	Adoption facilitators — Diffusion in health communities

→ Time

novel technology. This may uncover inhibitors that could slow down the adoption process. In situations where the healthcare organisation decides to implement the mobile healthcare technology in the next stage, it should find adoption facilitators that can help in the diffusion of mobile technology within the healthcare environment. Readers should note that implementation is beyond the scope of this framework.

This framework borrows from the technology-organisation-environment framework put forward by Tornatzky and Fleischer (1990) as well as the deinstitutionalisation framework by Tolbert and Zucker (1994). Tornatzky and Fleischer's framework identifies three aspects of an organisation's context that influence the process by which it adopts and implements innovation: technological context, organisational context, and environmental context. The omission of environmental context here is due to the fact that this chapter is about a single industry in which the environment is held constant.

The author considers innovation in healthcare as an idea, medical practice, or any material artefact in the healthcare process that is perceived to be new by the relevant unit of adoption in medical treatment. The relative advantage, compatibility, complexity, trial, and observation of such innovation can usually be used to determine the tendency for its adoption in the healthcare environment.

By relative advantage, the author means the degree to which an innovation within the health industry is perceived as better than the idea it supersedes. Compatibility, on the other hand, is seen as the degree to which an innovation within the health industry is perceived as being consistent with the existing values given by a particular healthcare community, or with past experiences and needs of potential adopters within the healthcare process. Compatibility of mobile healthcare technology can also be explained in terms of a combination of what healthcare practitioners feel or think about a particular innovation. This would also involve a critical look at the practical and operational compatibility with what healthcare practitioners are doing in the ongoing healthcare process. This interpretation of compatible innovation is in conjunction with perceived usefulness, the degree to which an end user believes a certain system can help perform a certain medical task. Complexity is the degree to which an innovation within the health industry is seen to be difficult to understand and use within the healthcare industry.

The trial of an innovation within the healthcare industry is an important part of evaluating new technologies within this critical industry. It is the degree to which an innovation within the health industry is experimented with on a limited basis. Given an opportunity to experiment with a new mobile healthcare technology before decisions are made about the adoption is an important benefit, especially for emerging technologies. This is an industry where practitioners take very highly the availability of information, while learning from experiences with previously disappointing IT projects.

Observation is a reliable means by which the healthcare industry evaluates innovations. This process identifies the degree to which the performance of a mobile healthcare technology and related benefits to the patients are visible to the medical practitioners and not only the service providers.

The determinants of mobile healthcare technology adoption are the benefits to the patients vs. the cost of such adoption. Most often, the NHS measures this in terms of the difference in costs for the shift from a previous technology to a mobile healthcare technology. Also worth considering are several factors that are important to the health service, such as the improvements made to the healthcare process as a result of a mobile healthcare technology after its introduction. There might even be a discovery of new uses for the mobile healthcare technology and the development of certain complementary inputs.

Hartmann and Sifonis (2000) relates to some of these features of mobile healthcare technology application by the identification of four dimensions within an organisation: leadership, governance, technology, and operational competencies. By leadership, they referred to the process of managing the initiatives and how the host organisation should stay motivated throughout the adoption process. By governance, they referred to the process of organising the innovation as regards the structure and operating procedures. Technology is where Hartmann and Sifonis looked at the organisation's ability to rapidly develop and implement new applications. They finally explained operational competencies as the way the host organisation manages the coordination of the relationship between leadership, governance, and technology as well as exploiting the available resources.

Levy and Powell (2005), on the other hand, presented evidence—from their study of small and medium-sized businesses in the United Kingdom—that the adoption of emerging technologies posits a similar framework as adoption related to the readiness of organisations taking into consideration the perceived ease of use and perceived usefulness. The readiness of the NHS to adopt mobile healthcare technology can be determined by reviewing the financial and technological resources available as well as various other factors dealing with the compatibility and consistency of emerging technologies with organisational culture and values.

Case 2: Pervasive Monitoring System

Oracle, along with its London partners, piloted a wireless sensor interface technology platform in mid-2005 (http://www.toumaz.com/news.php?act). It used advanced transactional database capabilities and offered the potential to transform the treatment and management of chronic diseases for millions of patients. This project was meant to bring the economies of scale of semiconductors into the healthcare industry with its advantages of real-time, personalised care and the delivery of some form of breakthrough. The system was based on a low-cost, disposable, integrated sensor interface chip.

Due to the chip's ultra low power and very small battery size, it could be worn on the body with complete freedom of movement, or it could be implanted. Such a mobile healthcare technology is compatible with a wide range of sensors (see Figure 4) and can therefore be configured to detect vital signs such as ECG (electrocardiogram), blood oxygen and glucose, body temperature, and mobility. The device can also dynamically process and filter event data (including irregularities in heartbeat or blood pressure) and send the details to a mobile phone or PC via an ultra low-power, short-range radio telemetry link.

Figure 4. Mobile healthcare technology with built-in sensor (http://www.healthpia.us)

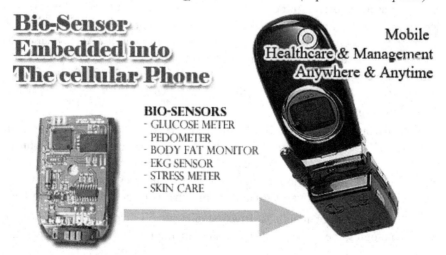

Further improvements to this kind of mobile healthcare technology could enhance the quality and efficiency of the healthcare patients of the NHS receive in the future. It could permit the following in future healthcare:

- Provide more timely and personalised care.

- Deliver unprecedented freedom, flexibility, and control for patients.

- Include exciting possibilities for medical practitioners to ultimately offer consumer items for which selection is based on quality and efficiency.

Conclusion

This chapter has provided examples of mobile healthcare technologies (case studies) for which the successful delivery of mobile solutions can help with certain kinds of emergency services. When considering the deployment of any mobile data solution, reliability, efficiency, and security are essential, none more so than in the emergency services if lives are to be saved. Wireless and mobile technology offers the opportunity to vastly enhance services, but there are still challenges to be faced if a cost-efficient solution is to be delivered.

Although some of these initiatives described as e-health can deliver certain benefits (including increased productivity and effectiveness of healthcare personnel and improved delivery of information and services), they will be faced with a number of challenges. Service providers in the private sector are looking to government for more leadership on identification issues and, as such, mobile healthcare technology should be a welcomed measure.

IT is seen as a key driver in the delivery of an efficient public sector, but how can departments justify further expenditure and eventually provide a clear road map to return on investment whilst delivering what has been promised? Also, what are the key short-term issues and, more importantly, the solutions that government departments can focus on? This panel will examine the savings that ICT investment is expected to deliver in the public sector, and how to serve more people by making things more efficient.

The United Kingdom has certainly increased its uptake on open-source software since the Office of Government Commerce's (OGC's) announcement that open source is a viable desktop alternative for the majority of users. However, many government organisations have chosen to remain with their existing proprietary software. This panel will examine the advantages and the drawbacks of both software solutions.

While this chapter is not intended to give specific guidelines for using mobile healthcare technologies, the author finds it useful to mention the following two points:

- Have clear objectives. Mobile healthcare technology is only a means to an end. It is advisable for managers of healthcare organisations to not be dazzled by the technology.

- Mobile healthcare technology is a unique medium, requiring management to capitalize on its uniqueness. The information being transmitted by mobile healthcare technology is the same, but is just delivered in a different way.

A conceptual framework has been posed from the research upon which this chapter is based, showing the different stages of the mobile healthcare technology-adoption process for healthcare organisations as well as the main factors operating at each stage.

In conclusion, the author has argued that Web services can aid the strategic planning of a healthcare organisation and can be used for competitive advantage. Web services can also contribute to improving our understanding and management of the critical issues surrounding mobile-based healthcare. Such understanding not only avoids disastrous consequences during the adoption of information systems, but also proves essential in supporting healthcare professionals to effectively manage the current trend of rapid increase in healthcare costs.

References

Amit, R., & Zott, C. (2001). Value creation in e-business. *Strategic Management Journal, 22*,493-520.

Atkinson, C. J., & Peel, V. J. (1998). Transforming a hospital through growing, not building, an electronic patient record system. *Methods of Information in Medicine, 37*, 285-293.

Banker, R., & Kauffman, R. (1988). Strategic contributions of information technology: An empirical study of ATM networks. In *Proceedings of the Nineth International Conference on Information Systems*, Minneapolis.

Barney, J. B., & Griffin, R. W. (1992). *The management of organizations: Strategy, structure, behavior.* Boston: Houghton Mifflin.

Brown, S. (2001). NHS finance: The issue explained. *The Guardian*, 30 May (pp. 21).

Brown, S. A., & Venkatesh, V. (2003). Bringing non-adopter along: The challenge facing the PC industry. *Communications of the ACM, 46*(4), 76-80.

Collins, T. (2003). Doctors attack health IT codes. *ComputerWeekly,* February 6, (pp. 21).

Currie, W., Desai, B., & Khan, N. (2004). Customer evaluation of application services provisioning in five vertical sectors. *Journal of Information Technology, 19*, 39-58.

Davis, F. D. (1989). Perceived usefulness, perceived ease of use and user acceptance of information technology. *Management Information Systems Quarterly, 13*(4), 982-1003.

Edmondson, A. C. (2003). Framing for learning: Lessons in successful technology implementation. *California Management Review, 45*(2), 34-54.

Ferlie, E. B., & Shortell, S. M. (2001). Improving the quality of healthcare in the United Kingdom and the United States: A framework for change. *Milbank Quarterly, 79*, 281-315.

Guah, M. W., & Currie, W. L. (2004). Application service provision: A technology and working tool for healthcare organisation in the knowledge age. *International Journal of Healthcare Technology and Management, 6*(1/2), 84-98.

Guah, M. W., & Currie, W. L. (2006). *Internet strategy: The road to Web services solutions.* PA: IRM Press.

Hagel, J. (2002). *Out of the box: Strategies for achieving profits today and growth tomorrow through Web services.* Boston: Harvard Business School Press.

Haines, M. (2002). *Knowledge management in the NHS: Platform for change.* Department of Health. Retrieved November 2002 from http://www.health-knowledge.org.uk

Hartmann, A., & Sifonis, J. (2000). *NetReady-strategies for the success in the e-economy.* New York: McGraw-Hill.

Institute of Medicine. (2002). *Crossing the quality chasm: A new health system for the 21st century.* Washington, DC: Committee on Quality Healthcare in America, National Academy Press.

Kreger, H. (2003). Fulfilling the Web services promise. *Communications of the ACM, 46*(6), 29-34.

Levy, M., & Powell, P. (2005). *Strategies for growth in SMEs: The role of information and information systems.* Oxford: Butterworth Heinemann.

Majeed, A. (2003). Ten ways to improve information technology in the NHS. *British Medical Journal, 326*, 202-206.

McGauran, A. (2002). Foundation hospitals: Freeing the best or dividing the NHS? *British Medical Journal, 324*(1), 1298.

Metters, J., Abrams, M., Greenfield, P. R., Parmar, J. M., & Venn, C. E. (1997). *Report to the Secretary of State for Health of the professional committee on the appeal of Mr. D. R. Walker under paragraph 190 of the terms and conditions of service of hospital medical and dental staff (England and Wales).* London: Department of Health.

National Health Service Information Authority (NHSIA). (2003). *Annual operating plan: To be the national provider of information and infrastructure services.* London: UK, Department of Health.

Orlikowski, W.J. (1993). CASE tools as organizational change: Investigating incremental and radical changes in systems development. *MIS Quarterly, 17*(3).

Orlikowski, W. J., & Tyre, M. J. (1994). Windows of opportunity: Temporal patterns of technological adaptation in organisations. *Organisation Science*, May, 98-118.

Pencheon, D. (1998). Matching demand and supply fairly and efficiently. *British Medical Journal, 316*, 1665-1667.

Perseid Software Limited. (2003). *The strategic value of Web services for healthcare and the life sciences.* Retrieved August 2003 from http://www.perseudsiftware.com

Robinson, J. C. (2000). Deregulation and regulatory backlash in healthcare. *California Management Review, 43*(1), 13-33.

Ruyter, K. D., Wetzels, M., & Kleijnen, M. (2001). Customer adoption of e-service: An experimental study. *International Journal of Service Industry Management, 2*, 184-206.

Sleeper, B., & Robins, B. (2001). *Defining Web services.* Retrieved April 2002 from http://www.stencilgroup.com

Srinivasan, R., Lilien, G. L., & Rangaswamy, A. (2004). Technological opportunism and radical technology adoption: An application to e-business. *Journal of Marketing, 66*(3), 47-60.

Tolbert & Zucker. (1994). *Institutional Analysis of Organizations: Legitmate but not Institutionalized.* Institute for Social Science Research working paper, University of California, Los Angeles, Vol. 6, No. 5.

Tornatzky, L. G., & Fleischer, M. (1990). *The process of technology innovation.* Lexington: Lexington Books.

Walsham, G. (1993). *Interpreting information systems in organisations.* Chichester, United Kingdom: Wiley.

Wanless, D. (2002). *Securing our future health: Taking a long-term view. Final report of an independent review of the long-term resource requirement for the NHS.* London: Department of Health.

Willcocks, L., & Lacity, M. (1998). *Strategic sourcing of information systems.* John Wiley & Sons, New York.

Yin, R. K. (1994). *Case study research: Design and methods.* CA: Sage Publications.

Chapter XVII

IDEF3-Based Framework for Web-Based Hospital Information System

Latif Al-Hakim, University of Southern Queensland, Australia

Abstract

This chapter presents a framework for a Web-based hospital information system to manage the surgery-management process (SMP). The framework can be used to manage any other hospital information system processes. The developed framework challenges the traditional hospital Web strategies with a dual aim: first, to improve customer satisfaction in an environment that often imposes unexplained deviation from planned activities, and second, to create a system that is an effective decision-support system for SMP. The chapter identifies factors affecting SMP decisions and employs a descriptive modeling technique known as IDEF3 to map the information flow within and between elements of SMP. The IDEF3 process mapping becomes part of an integrated Web-based system of multiple stages. Each stage has three levels of accessibility. The first level of the Web system is accessible to the public, the second level is accessible to patients and their designated representatives, and the third level is accessible only to hospital professionals.

Introduction

One of the challenges before healthcare executives is to design systems that improve customer satisfaction when admitting and waiting for consultation or surgery. The term customer in this research refers to patients, their designated relatives, and the hospital's personnel. Healthcare executives are increasingly looking to information technology as an opportunity to develop such systems. One of these technologies is Web-based technology. A large number of hospital Web sites allow customers to download and lodge forms, provide necessary information and feedback to patients, and facilitate the arrangement of appointments. Estimates in 2002 suggest that over 500 million people have access to the World Wide Web (WWW), with 50% to 75% of the users having used the World Wide Web to look for health information (Powell & Clarke, 2002). In December 2001, the NHS (National Health Service) Direct Online Web site dealt with 5.2 million hits from 171,900 visitors to their Web site. The figures for the same period in 2000 were 2.8 million hits and 24,830 visitors (KabelNet.Com, 2002). A survey conducted in Canada to explore patients' attitudes toward health services suggests that Internet users expressed interest in using the Web for several reasons, including to learn about their health condition through patient educational materials (84%), to obtain information about the status of their clinical appointments (83%), to renew prescriptions (75%), to consult with their health professional about non-urgent matters (75%), and to access laboratory test results (75%; Rizo, Lupea, Baybourdy, Anderson, Closson, & Jadad, 2005).

Web-based technology, however, does not routinely achieve customer satisfaction unless, from the customer perspective, a value-added service is created (Al-Hakim, 2006). Web-site design needs to be appropriate to the needs of the hospitals and their patients, and should focus on supporting the hospital's business goals (Thelwall, 2000). The development of Web-site information usability, information flow, and security have been sporadic and uneven across the healthcare industry (Mercer, 2001; Smith & Correa, 2005). This chapter deals with a Web-based system for the surgery-management process (SMP) in hospitals taking into consideration SMP objectives and factors affecting customer satisfaction.

The SMP is a complicated healthcare-delivery process starting from the referral of a patient to a hospital and ending with the discharge of the patient from the hospital after surgery has been performed. One key function of SMP is to manage operating-theatre waiting lists (OTWLs). OTWL refers to the operating-theatre schedule. Patients on this list are awaiting a surgical intervention to be conducted in an operating theatre of the hospital. Operating-theatre lists are of great concern to healthcare executives because of their societal and political priority, their link to potential quality of an individual patient's life, their relation to the economic management of operating theatres and the management of patient flow through the hospital, and the distribution of scarce medical resources, such as specialist surgeons and anaesthetists, and technology and instrumentation (Gonzalez-Busto & Garcia, 1999; McAleer, Turner, Lismore, & Naqvi, 1995). OTWL, however, cannot be managed successfully in separation from other SMP functions, including the preadmission appointment list, recovery-unit schedule, capacity management, and discharge plan. The OTWL schedule relies on information from SMP functions or activities that will be implemented at a later stage, such as information from recovery units (Al-Hakim, 2006). Feedback from these functions and activities may affect the OTWL schedule at a later stage. This dynamic nature of OTWL requires that researchers look to the whole SMP as an integrated process in order to redesign the process. Mapping the information forms a platform for redesigning the SMP process. The chapter employs a technique referred to as IDEF3 for mapping information. IDEF3 is a product of the Integrated Computer-Manufacturing (ICAM) initiatives of the United States Air Force. "IDEF" stands for **ICAM Def**inition Language (Mayer et al., 1992; KBSI, 2002). The subsequent sections of this chapter deal with factors affecting SMP decisions, process mapping and information mapping using IDEF3, and the integration of IDEF3 mapping within the Web-based system. The chapter uses an Australian regional hospital as a case study to illustrate the proposed system.

Factors Affecting SMP Decisions

Consulting surgeons historically determine the urgency of patients' conditions, the nature of the required surgery, and the timing of the operation. They make clinical decisions based on their opinion of the physical status and medical needs of their patients (McAleer et al., 1995). Decisions regarding the physical status of a patient include assigning the patient to a certain category of emergency and surgical requirements. Surgeons form their opinion based on the patient's health history, information received from various medical reports, results of various clinical tests, and so forth. Without proper updated and complete information related to a patient's health status, a surgeon may make a wrong assessment. Medical errors resulting from the surgery at a later stage may not merely be attributed to an error that occurred during the surgery itself.

Key targets of SMP are to improve patient flow throughout the surgery process and to eliminate medical errors. To achieve these targets, surgery-management decisions should be based on relevant, timely, and updated information about a patient's health status. In addition, information related to queues in the waiting lists, available resources, available beds, and process variations are critical factors affecting the SMP decision related to patient flow in a hospital. Factors affecting SMP decisions can be classified into four aspects; process complexity, medical errors, OTWL, and capacity management.

Process Complexity

Giddens (as cited in Biazzo, 2002) describes a process as a series of interdependent actions or activities embedded in a structure. Giddens views a structure as a set of rules and resources that can both constrain and enable social actions. The SMP process is a constructive example that fits Giddens' concept and, accordingly, the SMP can be considered as not just a group of activities. Rather, it is a socio-technical system (Keating, Fernandez, Jacobs, & Kauffmann, 2001) in which each operation or activity has additional elements other than that activity's input and output. These additional elements are the resources including the information necessary to perform the activities and the rules that govern the activity's implementation. The complexity of SMP arises from two aspects:

a. The process-mapping aspect that relates to the interdependencies between the elements of various activities of the process.

b. The socio-technical aspect that requires designing work such that a balance can be struck between the technology and the people using the technology. This is known as joint optimisation.

There are four variables that make both defining interdependencies between the elements of the various activities of the operating-theatre management process and achieving joint optimisation extremely important.

- **Object behaviour:** The behaviour of the object (patient) is not predictable and could vary considerably. Significant disruption could result from patient behaviour, for example, the cancellation of surgery by a patient. This makes every patient a unique object in the system.

- **Surgeon effectiveness:** Surgeons differ in skill and expertise. It is hard to measure the effectiveness of a surgeon in dealing with various complexities during surgery.

- **Surgery success:** Because of the level of complexity and variability, it is hard to predict the degree of the success of a surgery.

- **Surgical time:** Though the time required for surgery can be partly predicted, complexities during surgery may considerably affect the surgery time. The high probability of the unpredictability of time required for surgery renders difficult any attempt to precisely schedule operating-theatre waiting lists and the performance of SMP.

While some of the above variables are uncontrollable, their impact on SMP can be considerably reduced and effectively managed with (a) the correct and real-time flow of information, and (b) the coordination of the various interdependencies between elements of the activities involved in the SMP process. Realising the role of information flow and interdependencies as the main drivers of SMP highlights the need for effective information mapping.

SMP is a multistage process with a hierarchical structure. Each stage includes several operations, and each operation has several activities. Brief descriptions of the main stages of SMP are as follows:

- **Preadmission:** This includes receiving the referral form from the general practitioner or specialist by the hospital admission registry and booking an appointment with a surgeon or specialist. The patient at this stage is provided with consent and admission forms.

- **Booked admission:** This stage includes an appointment with a specialist to tentatively evaluate the patient's needs for surgery. The specialist determines the required clinical tests. This stage includes determining the date for the pre-assessment stage and entering the patient into a bed-allocation list. Patients, at this stage, may be told the date they will be admitted to the hospital, possibly months in advance (Gallivan, Utley, Treasure, & Valencia, 2002).

- **Preassessment:** At this stage, all the required clinical tests are performed. Based on the results, the surgeon and anaesthetists assess the surgery requirements.

- **Perioperative:** This stage may include all operations before admitting a patient to an operating theatre. For the purpose of this chapter, the perioperative stage starts after the admission of the patient to the hospital and includes all the activities required to be performed prior to surgery for preparation for surgery.

- **Procedure:** This includes activities to perform the surgery inside the operating theatre.

- **Postoperative:** This stage starts immediately after the surgery. The postoperative stage starts in the recovery room or post-anaesthesia care unit (PACU). This unit is dedicated to minimising postoperative complications. In the PACU, the patient may wear certain devices to automatically monitor vital signs (Rodts & Spinasanta, 2004). The overall condition of the patient is assessed at this stage.

- **Discharge:** This stage includes activities related to the transition of patients from the hospital to their place of residence.

Medical Errors

A simple Internet search reveals an astonishing number of medical errors as a result of judgment based on receiving incomplete or incorrect information. While medical errors are usually attributed to mistakes made by individuals, errors resulting from poor information flow (incomplete data and information, lack of communication, lack of timely data, lack of information accessibility, etc.) play a major role in contributing to this error rate. Unfortunately, cause and effect resulting from poor information flow go unmonitored. Examples of medical errors that can result from the lack of adequate information mapping and information flow are listed below:

- A family received an undisclosed settlement after a 5-year-old girl died following a medical error. She had gone into the hospital to have surgery to correct a hole in the heart and leaky valves. A solution used to prepare the heart for surgery was administered, but it was missing two ingredients. This caused her heart to stop beating irreversibly (WrongDiagnosis, 2004).

- A 67-year-old man went into surgery for bladder cancer, but came out of surgery missing more than he ever expected: his penis. In an interview, the man said, "It was never even discussed. And I felt like he [the surgeon] ought to have at least told us that this might be a possibility so that we could have talked it over even before I was admitted to the hospital" (Urban Scrawl, 2003).

- Another cancer patient was actually admitted to the operating theatre and was about to have his prostate removed, when a call came from the pathology department indicating there had been a mistake.

- In 1998, a lady had sought laser eye surgery to correct her astigmatism. However, after surgery, her vision failed to improve. It was found that wrong data were entered into the LASIK machine, causing the laser to operate on the wrong part of the lady's eye, which made the problem worse rather than better.

- Eighteen months after having a common bowel operation, an X-ray revealed a pair of 15-centimeter surgical scissors, slightly opened, lodged between a patient's lower bowel and spine. The hospital explained it did not count scissors after the surgery because they were considered too large to lose (Pirani, 2004).

Table 1 provides two examples of the same medical error from different countries at different times. Both examples reflect an error in information flow.

The above examples reflect the importance of information-flow mapping within and between various sections of SMP.

Table 1. Two examples of poor information flow

Example from USA	Example from Australia
In May 2002, two doctors switched the pathology slides of an American lady's breast biopsy with another woman's. After the surgery, she was told that she actually had no sign of cancer (McDougal, 2006).	Two women with the same first name attended an Australian hospital on the same day to have breast biopsies. One had breast cancer while one did not. It was discovered that the biopsy results had been mixed up and the patient without breast cancer had endured months of chemotherapy and was minus a breast. The woman with the breast cancer died after 9 months (Pirani, 2004).

Operating-Theatre Waiting List

There are commonly two kinds of waiting lists in hospitals: outpatient and surgery (Gonzalez-Busto & Garcia, 1999). The first type of waiting list includes patients waiting for consultation in certain sections, including the accident and emergency departments. Patients on the second type of waiting list are awaiting a surgical intervention to be conducted in an operating theatre of the hospital. The second type of waiting list is a reference to the operating-theatre schedule. This research deals with the second type, which is referred to as the OTWL.

OTWLs are critical for many reasons: their societal and political impact, their potential link to the patient's life, their relationship with the economic management of operating theatres, and the allocation of scarce resources such as surgeons, specialists, nurses, and equipment (Al-Hakim, 2006).

Once drawn up, the composition of a waiting list is unprotected in that it is subject to multiple alterations and modifications without prior notice or agreement with patients already on the list. Unanticipated changes affect the quality of patient care and may escalate costs considerably. Patients who have their consultation or procedure time altered can suffer prolonged anxiety, pain and discomfort, frustration, anger, and stress (Buchanan & Wilson, 1996). For the organisation, delayed or canceled procedures usually result in activity targets not being achieved.

The OTWL process has a multiple-criteria objective:

1. Compliance with all due dates and times of the scheduled consultations and operations
2. Ensuring a quick and convincing response to unanticipated changes to the waiting list

These criteria may conflict with each other in that the waiting list requires changes in response to unanticipated clinical needs rather than the order of arrivals, as well as logistical factors, such as readily available resources (Al-Hakim, 2006). The difficulty of compromising the two conflicting criteria of the OTWL objective arises mainly from process complexity and the improper flow of information.

Capacity Management

Capacity management is about the admission of the appropriate patients when necessary, providing the appropriate treatment in the correct environment, and discharging patients as the right time (Proudlove & Boaden, 2005). Capacity management includes bed management and resources management.

The ideal situation is where the demand of incoming patients is predictable and matches the capacity of the hospital in terms of available beds and resources. However, literature and common practices indicate that demand considerably exceeds hospital capacity (Buchanan & Wilson, 1996; District Commission, 2002; McAleer et al., 1995). SMP attempts to relieve the common mismatch between new incoming patients and outgoing discharged patients. The variability in the surgeries' times and the lengths of stay of patients in recovery wards considerably disturbs capacity-management decisions and lengthens the OTWL. Time constraints and the accessibility of information may prevent the discovering of medical errors even if the information entered in the system is accurate. There is a clear need to have an effective flow of information that allows revising decisions related to hospital capacity and then better determining the admission date for new patients during the booked admission stage of the OTWL or allocating the scarce resources. One case reported by Davies and Walley (2000) referred to the lack of data from the computer system of a hospital, which either did not keep records of potentially relevant information, or was not kept up to date. As a consequence, many capacity decisions were based on subjective judgment or commonly held beliefs. Hospitals often overlook the opportunity to coordinate short-term demand with capacity.

Specialised beds usually represent the first bottleneck that blocks new patients from being admitted to the hospital (Hill, 2006). Because it is difficult to move patients out from PACU based only on predetermined assessment rather than the patient's actual health status, PACU does not have the capacity to accept new patients, thus creating a bottleneck preventing the admission of patients to the operating theatre.

The effect of a lack of resources on capacity management is readily apparent. The key bottleneck resources are surgery specialists such as surgeons, and specialised surgery equipment. However, cases are reported where delays in surgery have not resulted from a lack of resources, but because of lack of coordination. Buchanan (1998) reports a hospital case in which there were a number of complaints from surgeons and anaesthetists about apparent delays in transporting patients to and from operating theatres.

The unpredictability of surgery cancellation by patients is another factor affecting capacity decisions. Surgery could be canceled by a patient even after admission to the hospital. Cancellation can also occur because of the lack of resources, surgeons, anaesthetists, and so forth. However, Davies and Walley (2000) stress that 2% of operations canceled in one case hospital were due to the hospital's inability to actually find patients who had been admitted to the ward! This is an obvious example of a lack of effective information flow. A study conducted at a major Australian hospital reveals that the percentage of cancellations on the day of intended surgery exceeds 11%. The main reasons recorded include overrun of a previous surgery (18.7%), no postoperative bed (18.1%), changes in patient clinical status (17.1%), procedural reasons (including unavailability of surgeons, list errors, administrative causes, and communication failure; 21%), and finally cancellation by patient (17.5%; Schofield, Rubin, Piza, Sindhusake, Fearnside, & Klineberg, 2005). These data reflect the fact that more than 83% of surgery cancellations on the day of intended surgery are related to the flow of information and could be avoided or minimised with a good information-flow system.

Process Mapping and Information Mapping

Process mapping is a technique used to detail business processes by focusing on the important elements that influence their behaviour (Soliman, 1998). It consists of constructing a model that shows the relationships between the activities, people, data, and objects involved in the production of a specified output (Biazzo, 2002). Lin, Yang, and Pai (2002) argue that process mapping performs two important functions. The first function is to capture existing processes by structurally representing their activities and elements. The second function is to redesign the processes in order to evaluate their performance.

SMP cuts across several stages and functional boundaries that are managed by different staff. At each stage, several activities are performed. The value of each activity is not only derived from its connection with the other SMP activities, but derived also from the consistency of its rules with other activities' rules and the coordination between the resources, which perform the whole SMP. The effectiveness of activity coordination is directly related to process complexity. Browning (1998) stresses that process complexity is a function of at least four factors: (a) the number of activities, (b) the individual complexity of each activity, (c) the number of relationships (or interdependencies) between activities, and (d) the individual complexity of each interdependency. Giddens (1984) seeks additional relationships and interdependencies between resources and rules governing the actions. This research adds the following four additional important factors: (e) the relationships or interdependencies between activities and the associated resources and rules, (f) the interdependencies between the resources, (g) the interdependencies between the resources and rules, and (h) the complexity of these interdependencies. Accordingly, to successfully map SMP, it is necessary to go beyond the detailing of each activity in a chart that only shows the flow of work from one activity to another.

Effective process mapping should therefore consider the interrelationships between activities as well as the objects (patients) involved, the resources needed to perform the process's activities, the rules and policies that govern the implementation of activities, and the flow of information between various elements of the activities.

Traditional process mapping such as flow- and Gantt charts are not robust enough to handle the process complexity. Traditional process-mapping techniques focus only on sequencing the activities or workflow. They do not take into consideration other elements of an activity such as rules and resources, including information flow. Accordingly, these techniques cannot be used to map the interdependencies between the elements of the activities and hence cannot deal with the complexity factors mentioned above. One structured process-mapping technique referred to as IDEF3 can be used to map the various elements of activities.

IDEF3

The IDEF techniques (IDEF0, IDEF2, IDEF3, IDEF1X, etc.) have been developed in projects sponsored by the U.S. Air Force in order to describe, specify, and model manufacturing systems in a structured graphical form (Plaia & Carrie, 1995). IDEF techniques can be classified into two major categories: the modeling and the descriptive techniques. The term description is used as a reserved technical term to mean records of empirical observations; that is, descriptions record knowledge that originates in or is based on observations or experience. The term model is used to mean an idealization of an entity or state of affairs (Mayer, Menzel, Painter, DeWitte, Blinn, & Perakath, 1995). In general, the model is used to predict what the system will do. The description method describes what a system supposes to do (Knowledge Based System, Inc. [KBSI], 2002; Plaia & Carrie, 1995).

IDEF3 is a descriptive method that builds structured descriptions (Mayer, Painter, & deWitte, 1992) and captures precedence and causality relations between the situation and events. It focuses on the abstraction and capture of knowledge about a given real-world system, including the temporal, causal, and logical relations between processes occurring within the system; the objects that participate in those processes; and the state transitions of those objects (KBSI, 2002). IDEF3 is a suitable method for the purpose of mapping SMP as it allows an investigator to capture and describe not what happens, but what fundamentally occurs in an SMP.

IDEF3 uses two knowledge-acquisition strategies: a process-centred strategy and an object-centred strategy. The process-centred strategy organises process knowledge with a focus on temporal, causal, and logical relations within a scenario. The second strategy organises process knowledge with a focus on objects and their state-change behaviour (Mayer et al., 1995). The two strategies are represented by two distinctive graphical diagrams. The process-flow diagram (PFD) portrays the process-centred strategy, and the object-centred strategy yields an object state-transition network (OSTN) diagram. Both strategies are hierarchical in nature. The description at any level can be decomposed into further details. The level of decomposition or mapping varies from an overview map to a very detailed map. Additional levels of mapping provide more information but also create greater cost (Soliman, 1998). However, fewer levels of process mapping could result in a poorly designed process and hence higher skilled operators would be required to understand and operate it. So, the number of process-mapping levels is dependent on how much information is needed (Al-Hakim, 2006). OSTN information flows concentrate on objects involved (patients, surgeons, etc.) and, accordingly, OSTN is more useful for hospital internal purposes. The concentration of this research is on the process-centred strategy, which is more applicable to communicate with customers; provide clear, nonmedical descriptions of the required processes; and to convince customers of any unanticipated changes in SMP.

IDEF3 Process-Centred Strategy

The term process in the process-flow diagram is referred to as a unit of behaviour (UOB). UOB describes general patterns or types of behaviour—situations, functions, operations, decisions, events, and so forth—that occur in a system. A UOB could be similar to an activity or event with a business process. A scenario is a collection of UOBs grouped in a defined sequence to achieve a specific output.

A UOB can have two types of descriptions associated with it (KBSI, 2002):

(a) elaborations that detail the characterizations of objects, facts, and constraints that represent the nature and structure of the corresponding real-world situation, and (b) a decomposition diagram (if any) that provides a more detailed view of the UOB.

UOBs form the building blocks to describe a scenario. In addition, the IDEF3 process-flow diagram comprises three other elements: links, junctions, and referents.

- **Links:** Links are arrows used primarily to define the relationships and the sequence of UOBs in a scenario. There are two types of links: precedence links and relational links. A precedence link demonstrates the temporal precedence relation between the elements of the flow process, while a relational link illustrates the existence of a relationship between the elements, for instance, the transference of one or more objects across the link.

- **Junctions:** A junction determines a mechanism to define the logic of branching in the process-flow diagram. Junctions can be classified into two groups. Fan-in junctions represent a convergence of several parallel flow paths into a single path. In contrast, fan-out junctions represent a divergence of a single path into several parallel paths. A junction from any of the two groups can be one of four forms: asynchronous-and, synchronous-and, asynchronous-or, or synchronous-or. These forms of junctions determine the timing and sequence of branching. For example, a fan-out synchronous-and junction indicates all parallel flow paths branching from the junction should occur simultaneously before the flow continues.

- **Referents:** A referent is a block that is attached to an element of the process flow and can be used to make a reference or to direct the flow to other elements, flow paths, or scenarios. There are four types of referents. An unconditional referent can be used to refer only to other elements, paths, or scenarios that contain useful information relevant to the existing element. An asynchronous or call-and-continue referent is used to activate another element, path, or scenario but, at the same time, allows the flow to continue. In contrast, synchronous or call-and-wait referents activate another element, path, or scenario but prevent the flow from continuing until the completion of the element, path, or scenario referred to. The fourth type of referent is a go-to referent, which simply redirects the process flow to another element, path, or scenario.

The output of IDEF3 can be expressed in two forms of process-flow windows: the PFD and the PFN (process-flow node list). The process-flow diagram shows the elements of the process using IDEF3 standard format. UOBs, junctions, and referents are shown as boxes connected by arrows representing the links. The process-flow node-list window displays the hierarchical sequences of UOBs or the processes of all flow process diagrams for the system under consideration.

Figure 1 shows part of the IDEF3 process PFD window. The process mapping in Figure 1 comprises several UOBs connected by links. The map has two fan-out junctions (J7 and J9) and a referent. The referent attached to UOB number 21.1 via the link L35 is a go-to referent that requires referring to UOB number 18.1 that in turn requires sending a letter to the patient. From the numbering of the UOBs, one can configure that this IDEF3 mapping is a second-level mapping.

Figure 1. PFD window

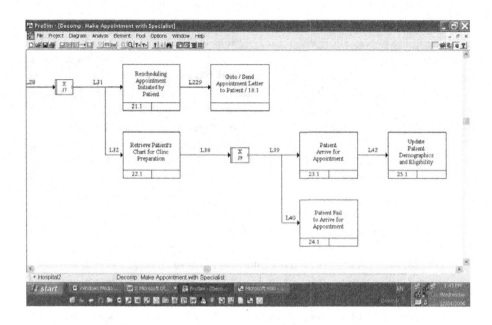

Relationship Between Information Flow, Process Mapping, and Web-Based System

Langefors (as cited in Malmsjo & Ovelius, 2003) formulates the process of obtaining information to be I= i(D, S, t), where I is the information obtained from the interpretation process i of data D, with preknowledge of the user's life experience S at a certain time t. According to Langefors, a certain set of data could be interpreted differently by persons with different experiences or at different times. With the recent explosion of information technology, the absolute user's experience in most business environments is no longer enough for the purpose of the interpretation of data while relying on the developed technologies. Al-Hakim (2006) adds another element to Langefors formula, that is, the technology used. Accordingly, data, experience, technology, and time form prerequisites or inputs for the information process (Figure 2).

Figure 2. Inputs of information process

Traditional Information Process Formula

$$I = i\,(D, S, t\,)$$

Developed Information Process Formula

$$I = i\,(D, S, t, T\,)$$

Where:

i = Reference to a process (i)

I = Information obtained from interpretation of process (i)

D = Data

S = Individual experience of the officer interpreting the data

t = Time

T= Information technology used

The Relationship

The formulation of Langefors (1973) equation and the addition of technology to that equation make information process definition more consistent with the requirements of product-manufacturing systems if we consider the raw material in production to be equivalent to data in information systems. The constraints and environment required by the product system is part of the knowledge. In this regard, the information system requires additional input, that is, time.

An IT system, similar to other systems, has inputs and outputs. The inputs to the IT system are the prerequisites for the information process: data, experience, time, and technology. The output of the system is information with the focus on customer satisfaction. The same formula is applicable to Web-based applications for hospitals. A process is a collection of activities and functions. A process in Langefor's formula can be considered as an activity or a function of the process. An understanding of SMP requires an understanding of detailed activities and interrelationships between them. In other words, understanding SMP necessitates mapping the SMP process. Process mapping, by definition, comprises the resources (people) required to achieve the process. Information obtained from a process is controlled by information flow

Figure 3. Relationships between Web-based application, information flow, and process mapping

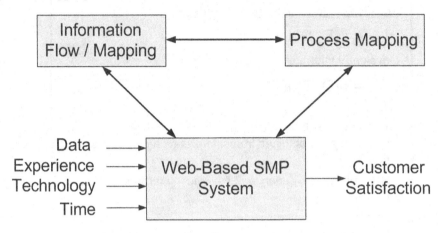

between and within the SMP activities. Accordingly, both process mapping and information flow form a control for achieving the objective of having a Web-based system for SMP (Figure 3).

Case Hospital

This case study is part of a project initiated by the author and has been funded by the Faculty of Business at the University of Southern Queensland, Australia. The project includes two postgraduate students: an honours-degree student and a master's degree student. The selected case is an Australian rural public hospital. The hospital comprises 13 departments including surgery, anaesthetics, orthopaedics, obstetrics and gynaecology, paediatrics, emergency, critical care, medical imaging, medical, renal, public health, oncology, and rehabilitation. It has a 261-bed facility, including 164 acute beds, 57 mental-health beds, and 40 day beds. The hospital employs 2,000 staff in total.

The surgical department of the hospital includes an operating-theatre suite comprising six operating theatres. Four are used for elective lists that are run for two sessions per day. The two other theatres are dedicated to 24-hour emergency surgery services. There are eight recovery wards catering to the theatre patients.

There are 24 surgeons in the hospital; many of them are part-time surgeons. In addition, the hospital has employed nine full-time anaesthetists, 18 consultants, six senior registrars, and a number of registered and clinical nurses.

Every surgeon has his or her own waiting list organised into six types of specialised waiting lists. Each specialty area has at least one clerk to manage the waiting lists and has at least one surgeon. The specialised waiting lists are general surgery; ear, nose, and throat (ENT); gynaecology; ophthalmology; urology; orthopaedic; and fascia maxillary.

Several interviews were conducted with hospital officials including a senior registrar, senior nurse, elective-surgery coordinator, and other related officials and medical professionals. At the start of each interview, the process mapping drafted as a result of information collected from the previous interview was reviewed and revised. After each interview, the IDEF3 process map was updated by combining the changes and new information with the previous process map.

Hospital Smp Process Mapping

The case hospital follows a written framework for elective surgery services. The framework provides general descriptions, policies, and policies for the main functions of SMP. After careful analysis and discussion with the hospital professionals, the SMP was divided into seven main stages: preadmission, scheduling of surgery, day prior to surgery, admission, preoperative, surgery, prooperative, and discharge. Figure 4 and Figure 5 demonstrate the PFD window and PFN window, respectively, for Level 1 of SMP. The shadow around the process elements in the process flow

Figure 4. PFD window for Level 1 of SMP

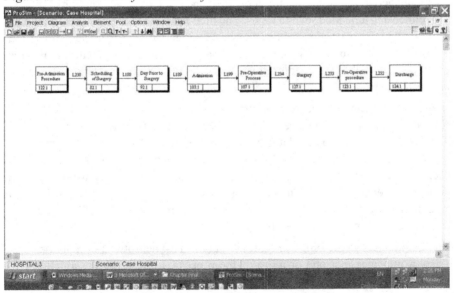

Figure 5. PFN window for the first level of SMP

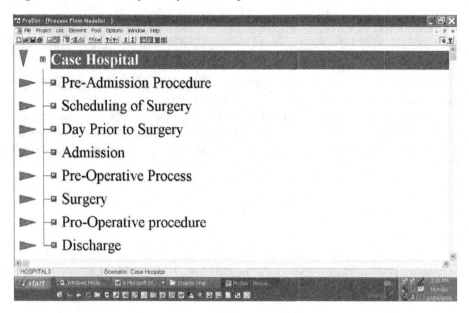

diagram and the symbol ▶ in the PFN indicate that the process has one or more levels of decomposition.

The second stage of interviews with the hospital professionals was devoted to breaking down the main processes into their subprocesses. Initially, the hospital representatives managed a detailed description of SMP and a list of SMP functions. The list was refined and amended several times. The final form of the list includes more than 150 functions. The functions of the list were arranged to reflect the decomposition levels of the main stages. The decomposition was limited to a maximum of three levels. For example, the preadmission procedure was divided into five processes, and each second-level process was divided into further subprocesses (Figure 6).

Figure 6 shows that the receiving-referral process has 10 subprocesses (UOBs). The PFN does not show the sequence or the interrelationship between these subprocesses. However, Figure 6 can be used to list all processes of SMP and their decomposition into subprocesses. In contrast to PFN, PFD displays each decomposition of SMP in a separate window. However, PFD shows the sequence and relationships between various elements of the decomposition. For example, Figure 7 illustrates the PFD for the receiving-referral decomposition that forms part of the Level 3 decomposition of the preadmission procedure.

The first junction in the PFD of Figure 7 is a fan-in asynchronous-or junction. This junction indicates that one or more of the processes preceding the junction (receiving referral from one or more sources) will occur, though not simultaneously,

Figure 6. Subprocesses for preadmission procedure

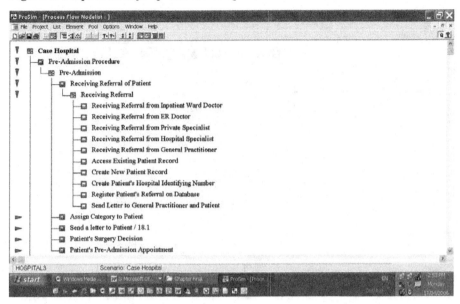

Figure 7. PFD for the receiving-referral process

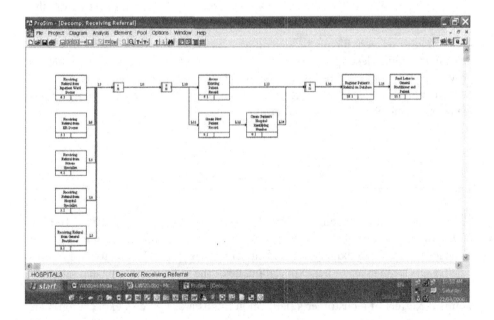

Figure 8. Description attached to the "receiving referral of patient" process

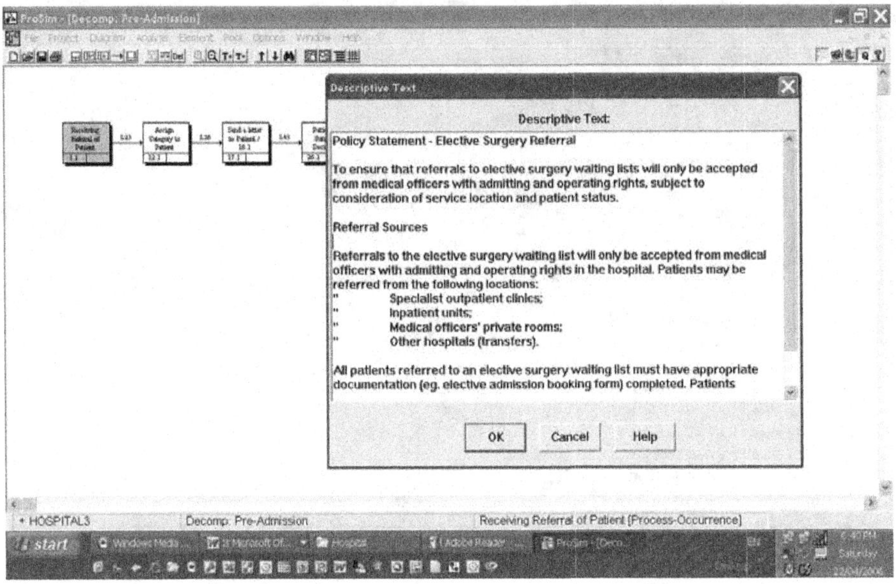

before the flow continues. The junction is a fan-out eXclusive OR(XOR) junction, which indicates that only one of the two processes following the junction ("Access Existing Patient Record" and "Create New Patient Record") will occur before the flow continues.

In addition to referent elements, several attachments can be associated with each process including descriptions, elaborations, and notes to describe the elements, provide instruction, and outline the requirements. Figure 8 shows the description attachments associated with the "Receiving Referral of Patient" process. It illustrates the hospital policy and referral sources.

Web-Based System

Several studies (Pitt, Watson, & Kavan, 1995; Watson & Pitt, 1998; Xu & Al-Hakim, 2005) emphasise that there are two important elements that constitute customer satisfaction: customer expectation and customer perception. If a customer receives a better service than his or her expectation, he or she will leave happy and satisfied. This assertion can be translated to a formula indicating that a customer's satisfaction is the effect of a customer's perception minus a customer's expectation. One factor affecting the customer satisfaction is the waiting time or time of delay. As early as 1976, Kleinrock recognises time as a valuable commodity and stressed that customers are prepared to do whatever necessary to avoid queuing. Patients who have their procedure time altered can suffer prolonged anxiety, pain and discomfort, frustration, anger, and stress (Buchanan & Wilson, 1996). Barlow (2002) concludes that the actual time of delay is not of significant importance, but it is the difference between the customer perceptions and expectations that create the level of (dis)satisfaction. That is, patients will be satisfied only if his or her perceived wait is less than, or equal to, the expected wait. The challenge is to design a system that maintains the waiting time and provides an adequate explanation to customers so as to extend the perceived wait of the customers beyond the initial agreed surgery time. In other words, if customers are informed about their queuing status, and about reasons for changes to that status, customer satisfaction is improved. This challenge necessitates clear understanding of the activities and functions of SMP, the interaction between the SMP process's activities, and the information flow within and between activities and functions. That is, to design a satisfactory Web-based system, it is required first to map both the process and information. In dealing with such a system, it must be kept in mind that not all information and communications at healthcare organisations are accessible to the public, and some are regulated by legislation.

In dealing with the challenge, the IDEF3 process mapping becomes part of an integrated Web-based system of multiple stages and three levels of accessibility. The first stage illustrates the main functions (UOBs) of the SMP as shown in Figure 4 or Figure 5. Other stages represent IDEF3 decompositions into further levels of detail. Each decomposition level represents an SMP stage.

Each stage of the Web-based system has three levels of access. The first level of the Web system is accessible to the public. It provides the following.

1.   Easy-to-follow process mapping that illustrates various units of behaviour (steps, activities, functions, and events) of SMP. This is either in a pictorial flowchart form window (PFD) or sequence form window (NPF).

2.   Clear depiction of the interrelationships and sequence of SMP units of behaviours

3.   Complete and coherent information for each UOB of SMP. This can be performed by either attaching a referent to each UOB or modifying IDEF3 in a way that produces this information from a selected window associated with each UOB

4.   Specific answers to what a patient should do or not to do at each UOB. This can be performed by either attaching a referent to each UOB or modifying IDEF3 in a way that produces this information from a selected window.

5.   Easy to allocate, download, and fill the required forms. Hypertext can be used to access these forms for each UOB.

6.   Easy accessibility and interpretability by patients who have limited skill in navigating the Internet and Web sites. There is no need for patients and other users to understand the technical meaning of IDEF3 elements. The notes and descriptions attached to each IDEF3 element should describe the element and its effect on other elements.

7.   Simple processes to allow users to interact with the hospital, provide feedback, and make appointments. This can be done using software integrated to the Web-based system.

The second level can be accessed by the patient and/or nominated patient representatives using designated identity and password. Unlike the first level, which includes generic IDEF3 process mapping, the second level comprises the specific functions, activities, and events that should be followed for the patient. It includes specific information for each UOB of the system. This level is linked with the hospital database. However, this level is limited to certain information that is not in contradiction with the existing legislation or with the consent of the patient. This level of the Web-based system provides the following.

1.   Information that directs the user to the stage at which the patient is currently on.

2.   The estimated time for the current UOB and successive UOBs.

3.   The specific requirements and duties that the patient should follow or perform at the current time before he/she transfers to the second stage.

4.   Explanation for changes or amendments to the process.

5.   Instructions to allow the patient or designated relative to directly communicate with the related specific professional such as the surgeon.

6.   Instructions to allow the patient to request an urgent appointment or cancel an appointment.

7.   Instructions to allow the patient to explain any unusual event or circumstance that he or she may have faced.

8.   Links to the private e-mail(s) or mobile(s) of the patient and his or her designated relative. The link is used to send messages to the patient reminding or asking the patient to perform certain activities.

9.   Voice recognition can be used to activate this level of the system for certain patients.

The third level of the Web-based system is also linked with the hospital central database and is only accessible to hospital staff. This level allows staff to access, add, and amend data. The degree of accessibility of this level is also controlled. It depends on the actual involvement and needs of the staff in relation to certain information. This level of Web-based system provides real-time communication between hospital staff in order to arrive at various related decisions. The patient-specific clinical and medical data sets can be visible to the consultants or specialists before and during the consultation or during the admission of the patient to the operating theatre. This level also allows the consultant or surgeons to enter the interventions required after the patient leaves the consultation room or the operating theatre (Al-Hakim, 2006). Integrating the Web-based system, especially on the third level, with mobile-based technology will allow surgeons and anesthetists to retrieve information and prepare the relevant documents while on the move. Furthermore, using the dictation facility of the mobile technology will considerably save the time needed to prepare the required forms and documents by hospital professionals.

Conclusion

Hospitals have been experiencing accelerated progress in using and deploying web-based technology to distribute necessary information electronically and to achieve rapid healthcare delivery improvements, which impact both the costs and the quality of healthcare delivery. Web-based technology, however, does not routinely achieve customer (patients, hospitals staff, etc.) satisfaction unless they create, from the customer perspective, a value-added service. This chapter provides a conceptual customer-centric framework of Web-based technology for administrating the SMP in hospitals. The chapter studies the relation between process mapping, information mapping, and the successful application of the Web-based system. It tackles the process complexities from the view of socio-technical theory and emphasises that effective SMP mapping should consider the interrelationships between activities as well as the objects (patients) involved, the resources needed to perform the process's activities, the rules and policies that govern the implementation of activities, and the flow of information between various elements of the activities. The chapter concludes that the traditional process mapping such as flow and Gantt charts are not robust enough to handle the SMP complexity and employs a structured process-mapping technique referred to as IDEF3 to map the information flow between various elements of activities. The chapter uses the SMP of an Australian regional hospital as a case study to illustrate the procedure for information mapping using IDEF3. The IDEF3 process mapping becomes part of an integrated Web-based system of multiple stages. Each stage has three levels of accessibility. The first level of the Web system is accessible to the public. It includes generic mapping of SMP that explains various functions of SMP in a pictorial flow chart. It allows a user to get the necessary information, allocate and download forms, and make appointments.

The second level of the Web-based system is accessible to patients and their designated representatives. Unlike the first level that includes generic process mapping, the second level comprises the specific functions, activities, and events that should be followed for the patient. This level is linked with the hospital database. However, this level is limited to certain information that is not in contradiction with the existing legislation or with the consent of the patient.

The third level of the Web-based system is also linked with the hospital central database and is only accessible to hospital staff. This level allows staff to access, add, and amend data. The degree of accessibility of this level is also controlled. It depends on the actual involvement and needs of the staff in relation to certain information. This level of Web-based system provides real-time communication between hospital staff in order to arrive at various related decisions. It is suggested that the system can be integrated with the mobile-based technology. This integration allows surgeons and other hospital professionals to retrieve information while on the move and to enter information and prepare documents using the dictation facility of the mobile technology.

Acknowledgment

The research was partially sponsored by the University of Southern Queensland. The author would like to acknowledge the thoughtful time of Queensland Health Project Liaison Officers; Ms. Lee Hunter, Mr. Brett Mendezona, and Ms. Sylvaia Johnson. The author would like to acknowledge the participation of two postgraduate students; Ms. Liqin Tan, and Ms. Gerhardine Foo.

References

Al-Hakim, L. (2006). Web-based hospital information system for managing operating theatre waiting list. *International Journal of Healthcare Technology and Management, 7*(3/4), 266-282.

Barlow, G. (2002). Auditing hospital queuing. *Managerial Auditing Journal, 17*(7), 397-403.

Biazzo, S. (2002). Process mapping techniques and organisational analysis: Lessons from sociotechnical system theory. *Business Process Management Journal, 8*(1), 42-52.

Browning, T. R. (1998, July). *Use of dependency structure matrices for product development cycle time reduction.* Proceedings of the Fifth ISPE International Conference on Concurrent Engineering: Research and Applications, Tokyo.

Buchanan, D. (1998). Representing process: The contribution of re-engineering frame. *International Journal of Operations & Production Management, 18*(12), 1163-1188.

Buchanan, D., & Wilson, B. (1996). Re-engineering operating theatres: The perspective assessed. *Journal of Management in Medicine, 10*(4), 57-74.

Davies, C. L., & Walley, P. (2000). Clinical governance and operations mangement methodologies. *International Journal of Health Care Quality Assurance, 13*(1), 21-26.

District Commission. (2002). *Operating theatres: A bulletin for health bodies.* Retrieved January 15, 2005, from http://www.district-audit.gov.uk

Gallivan, S., Utley, M., Treasure, T., & Valencia, O. (2002). Booked inpatient admission and hospital capacity: Mathematical modelling study. *British Medical Journal, 324*(7332), 280-282. Retrieved December 6, 2005, from http://pubmed-central.nih.gov/articlerender.fcgi?artid=65062&rendertype=abstract

Giddens, A. (1984). *The constitution of society.* Cambridge: Polity Press.

Gonzalez-Busto, B., & Garcia, R. (1999). Waiting lists in Spanish public hospitals: A system dynamics approach. *System Dynamics Review, 15*(3), 201-224.

Hill, M. (2006). Medical staff role in hospital capacity. *EMPATH.* Retrieved January 28, 2006, from http://www.empath.md/about_us/MD_Role_in_Capacity.pdf

KableNet.Com. (2002). *The UK's e-health services received a record number of hits over the holiday break.* Retrieved December 6, 2005, from http://www.kablenet.com/kd.nsf/printview/A33471B29784EDD480256B37003DC0EF?OpenDocument

Keating, C. B., Fernandez, A. A., Jacobs, D. A., & Kauffmann, P. (2001). A methodology for analysis of complex sociotechnical processes. *Business Process Management Journal, 7*(1), 33-49.

Kleinrock, L. (1976). Queuing Systems. NY: John Wiley.

Knowledge Based System, Inc. (KBSI). (2002). *ProSim 7.0 user manual.* Retrieved October 10, 2005, from http://www.kbsi.com

Langefors, B. (1973). *Theoretical Analysis of Information Systems*, (4th Eds.) Studentlitterature, Sweden: Lund.

Lin, F., Yang, C., & Pai, Y. (2002). A generic structure for business process modelling. *Business Process Management Journal, 8*(1), 19-41.

Malmsjo, A., & Ovelius, E. (2003). Factors that induce change in information systems. *Systems Research and Behavioral Science, 20*, (pp. 243-253).

Mayer, R. J., Menzel, C. P., Painter, M. K., DeWitte, P. S., Blinn, T., & Perakath, B. (1995). *Information integration for concurrent engineering (IICE) IDEF3 process description capture method report.* Knowledge Based Systems.

Mayer, R. J., Painter, M. K., & deWitte, P. S. (1992). *IDEF family of methods for concurrent engineering and business re-engineering applications.* TX: Knowledge Based Systems, Inc.

McAleer, W. E., Turner, J. A., Lismore, D., & Naqvi, I. A. (1995). Simulation of a hospital theatre suite. *Journal of Management in Medicine, 9*(3), 14-26.

McDougal, L. (2006). *My turn: I trust juries-and American like you.* NSNBC Newsweek. Retrieved December 6, 2005, from http://www.msnbc.msn.com/id/3704876/.

Mercer, K. (2001). Examining the impact of health information networks on health system integration in Canada. *Leadership in Health Services, 14*(3), i-xxx.

Pirani, C. (2004, January 24-25). How safe are our hospitals. *The Weekend Australian,* (p.2).

Pitt, L. F., Watson, R. T., & Kavan, B. (1995). Service quality: A measure of information systems effectiveness. *MIS Quarterly, 19*(2), 173-187.

Plaia, A., & Carrie, A. (1995). Application and assessment of IDEF3: Process flow description capture method. *International Journal of Operations & Production Management, 15*(1), 63-73.

Powell, J., & Clarke, A. (2002). The www of the World Wide Web: Who, what, and why? *Journal of Medical Internet Research, 4*(1), e4.

Proudlove, N. C., & Boaden, R. (2005). Using operational information and information systems to improve in-patient flow in hospitals. *Journal of Health Organisation and Management, 19*(6), 466-477.

Rizo, C. A., Lupea, D., Baybourdy, H., Anderson, M., Closson, T., & Jadad, A. R. (2005). What Internet services would patients like from hospitals during an epidemic? Lessons from the SARS outbreak in Toronto. *Journal of Medical Internet Research, 7*(4), e46.

Rodts, M., & Spinasanta, S. (2004). *What is post-operative care?* Retrieved December 6, 2005, from http://www.spineuniverse.com/displayarticle.php/article613.html

Schofield, W., Rubin, G., Piza, M., Lai, Y., Sindhusake, D., Fearnside, M., & Klineberg, P. (2005). Cancellation of operations on the day of intended surgery at a major Australian referral hospital. *Medical Journal of Australia, 182*(12), 612-615.

Smith, A. D., & Correa, J. (2005). Value-added benefits of technology: E-procurement and e-commerce related to the healthcare industry. *International Journal of Healthcare Quality Assurance, 18*(6), 458-473.

Soliman, F. (1998). Optimum level of process mapping and least cost business process re-engineering. *International Journal of Operations & Production Management, 18*(9/10), 810-816.

Thelwall, M. (2000). Effective Websites for small and medium-sized enterprises. *Journal of Small Business and Enterprise Development, 7*(2), 149-159.

Urban Scrawl. (2003). *Man loses penis.* Retrieved June 10, 2006, from http://demon. twinflame.org/archives/2003/08/man_loses_penis.php

Watson, R. T., & Pitt, L. F. (1998). Measuring information systems service quality: Lessons from two longitudinal case studies. *MIS Quarterly, 22*(1), 61-76.

WrongDiagnosis. (2004). *Medical news summary: Medical error causes girls death following routine surgery.* Retrieved February 10, 2006, from http://www. wrongdiagnosis.com/news/medical_error_causes_girls_death_following_rou-tine_surgery.htm

Xu, H., & Al-Hakim, L. (2005). Criticality of factors affecting data quality of accounting information systems: How perceptions of importance and performance can differ. *AMIS Monograph on Information Quality* (pp. 197-214).

About the Authors

Latif Al-Hakim is a senior lecturer of supply chain management in the Department of Economics and Resources Management of the Faculty of Business at the University of Southern Queensland, Australia. His experience spans 35 years in industry, research, and development organizations and in universities. Dr. Al-Hakim received his first degree in mechanical engineering in 1968. His MS (1977) in industrial engineering and PhD (1983) in management science were awarded from the University of Wales (UK). Dr. Al-Hakim has held various academic appointments and lectured on a wide variety of interdisciplinary management and industrial engineering topics. He has published extensively on facilities planning, information management, and systems modeling. His research papers have appeared in various international journals and have been cited in other research and postgraduate works. Dr. Al-Hakim is editor of the *International Journal of Information Quality* and associate editor of the *International Journal of Networking and Virtual Organisations*.

* * *

Carole Alcock is an associate professor in the School of Computer and Information Science at the University of South Australia where she is associate head of school. Her recent research in the health sphere has included the integrity of electronic health records, recent changes to privacy legislation in relation to health records, and the application of personal digital assistants (PDAs) within the health sector. She has received a number of grants in areas relating to health research, most recently as one of the chief investigators on an ARC (Australian Research Council) Industry Linkage grant on the application of PDAs in ambulatory care with South Eastern Sydney and Illawarra Area Health Services (SESIAHS) and Pen Computing. She is a member of the Health Informatics Group within the Advanced Computing Research Centre at the University of South Australia and also works with health informatics research groups at the University of Wollongong.

Stefano Baraldi is an assistant professor in the Department of Business Administration, Faculty of Economy, at the Catholic University of Milan and Managing Director of the Centre for Research and Studies in Healthcare Management (Ce. Ri.S.Ma.S.).He teaches courses in planning and control, management accounting, financial analysis, performance measurement. He is involved in several research in the area of development of managerial practices in large companies and healthcare organizations.He graduated from Bocconi University in 1988.

Barbara G. Beckerman is the program manager of biomedical engineering at the Oak Ridge National Laboratory, USA. She has developed and managed several new and innovative programs related to health-care technologies, imaging technologies, and advanced networking and communications for both federal and nonfederal customers. Beckerman received her MBA and BA degrees from the Clarks University and University of Miami, respectively. She is a member of the American Telemedicine Association (ATA). She has received numerous awards and recognitions including the 2004 Technology Transfer Award for software developed for the National Digital Mammography Archive.

Paolo Bellavista graduated from the University of Bologna, Italy, where he received a PhD in computer science engineering in 2001. He is now an associate professor of computer engineering at the University of Bologna. His research activities span from mobile agent-based middleware solutions and pervasive wireless computing to location- and context-aware services and adaptive multimedia. He is a member of the IEEE (Institute of Electrical and Electronics Engineers), ACM, and Italian Association for Computing (AICA). He is an associate technical editor of the *IEEE Communication Magazine*.

Elmer V. Bernstam is an associate professor of health informatics and internal medicine at the University of Texas Health Science Center at Houston, USA. Dr. Bernstam heads the clinical-informatics focus at the SHIS. His research currently focuses on consumer informatics and information retrieval. In addition to his MD, Dr. Bernstam holds master's degrees in computer engineering and biomedical informatics. He completed a National Library of Medicine postdoctoral fellowship in biomedical informatics at Stanford Medical Informatics. Dr. Bernstam is board certified in internal medicine and maintains an active clinical practice. He is a fellow of the American College of Physicians.

Dario Bottazzi received the laurea degree in computer engineering from the University of Bologna, Italy. He is currently a PhD candidate in computer engineering at the University of Bologna. His research interests include middleware for mobile environments and pervasive computing technologies. He is a member of the IEEE Computer Society.

Lois Burgess is a lecturer in the School of Information Technology and Computer Science within the Faculty of Informatics at the University of Wollongong, Australia. Burgess has extensive experience in industry-based training, staff development, EEO, HRM, workflow analysis, systems development, planning, policy and strategy development, government accounting, and user-requirements definition in systems development. Burgess is a chief investigator on the collaborative ePOC (electronic point of care) PDA e-health project of ARC with SESIAHS and Pen Computer Systems. Lois is also a member of the editorial board of the *International Journal of Theory and Applied Electronic Commerce Research* (JTAER).

Joan Cooper was appointed deputy vice-chancellor (academic) at Flinders University (Australia) in September 2003. She was previously dean of informatics at the University of Wollongong, where she had also served as chair of the university senate. Professor Cooper has more than 20 years of experience in information technology and was the first female professor of IT appointed in Australia. Her major research areas are in the field of electronic commerce and health informatics, and she was coordinator of the University of Wollongong's Centre for Electronic Business Research. Professor Cooper is one of the three founders of Australia's first interuniversity electronic-commerce research and consulting group CollECTeR (Collaborative Electronic Commerce Technology and Research) and was national deputy chair of Women in Technology, a special-interest group of ACS. She also was a member of the New South Wales (NSW) Privacy Advisory Committee.

Antonio Corradi graduated from the University of Bologna and received his MS in electrical engineering from Cornell University. He is a full professor of computer engineering at the University of Bologna, Italy. His research interests include distributed systems, object systems, mobile-agent platforms, network management, and distributed and parallel architectures. He is a member of ACM and AICA.

Elizabeth Cummings is a PhD candidate in the School of Information Systems at the University of Tasmania, Australia. Cummings commenced her PhD in 2004 and is researching the impact of ICTs on health outcomes. She has over 25 years of experience in health care in Australia through a range of occupations. She is a registered nurse and midwife and has worked in both clinical and administrative positions in nursing. She has also worked as strategic information manager for the state-based organization the Tasmanian General Practice Divisions. She has been involved in developing and implementing statewide and local general-practice ICT initiatives and closely involved with national e-health initiatives. In 2005 she joined the Smart Internet Technology Cooperative Research Centre.

Steve Goldberg founded INET International Inc., a technology-management consulting firm, in 1998. Today, he is the president of INET, overseeing INET online data-collection services for global market research studies and the delivery of INET mobile e-health projects. He also produces an annual INET mini-conference as an executive forum for wireless health care. Mr. Goldberg started his 25-year information-technology career at Systemhouse Ltd., which was at the time Canada's largest systems integrator. He formed an executive collaboration to develop a $30 million IT services business. During his tenure at Cybermation, he transformed the organization from a mainframe environment to a client-server environment. More recently, at Compugen, he successfully built high-performance teams to develop and deploy Internet, intranet, and extranet solutions.

Matthew W. Guah specializes in the organizational issues that surround the use of emerging technologies (i.e., ASP and Web services) in the health-care industry. Dr. Guah came into academia with a wealth of industrial experience spanning over 10 years within Nuffield Hospitals, Merrill Lynch, CITI Bank, HSBC, British Airways, British Standards Institute, and the United Nations. He is the author of *Internet Strategy: The Road to Web Services*. His recent publications include articles in *Information Systems Management*, the *Journal of Information Technology*, the *International Journal of Service Technology and Management*, *Information Technology and Interface*, the *International Journal of Knowledge Management*, the *International Journal of Healthcare Technology and Management*, the *International Journal of Technology and Human Interaction*, and others. He is an associate editor of the *Journal of Management Information Systems* and the *International*

Journal of Electronic Commerce, and is a member of AIS, UKAIS, BMiS, and BCS, among others. He is currently teaching and working on research at Warwick Business School.

Sunil Hazari is an associate professor in the Department of Management and Business Systems, Richards College of Business, University of West Georgia, USA. His teaching and research interests are in the areas of information security, infrastructure design of e-commerce sites, Web usability, and organizational aspects of e-learning. He has authored several peer-reviewed journal publications in information and instructional technology areas, has presented papers at national conferences, and is an editorial board member of information-system journals.

A. R. Hurson is a computer science and engineering faculty member at the Pennsylvania State University, USA. His research has been supported by NSF, NCR Corp., DARPA, IBM, Lockheed Martin, ONR, and Penn State University. He has published over 250 technical papers in areas including database systems, multi-databases, global information-sharing processing, the application of mobile-agent technology, object-oriented databases, mobile computing environments, and the computer architecture of parallel and distributed systems. Dr. Hurson served as the guest editor of several technical journals and has been involved in numerous IEEE/ACM conferences in different positions. Professor Hurson served as a member of the IEEE Computer Society Press editorial board, an IEEE distinguished speaker, editor of *IEEE Transactions on Computers*, and a member of the IEEE/ACM Computer Sciences Accreditation Board. Currently, he is serving as an ACM lecturer and editor of the *Journal of Pervasive and Mobile Computing*.

Ric Jentzsch has over 30 years of experience in the information-technology industry. He has developed and delivered training courses for business and government in Australia, the United States, Canada, and throughout southeast Asia. He has more than 60 published works in journals and international conference proceedings. He has lectured at universities in Australia, the U.S., and Canada. His current research interests include technology diffusion and adoption in SMEs; intelligent-agent frameworks and models, and their incorporation into business; security as a business enhancer; emerging and developing technologies from an SME perspective; and electronic content-management methods, models, and practical applications. He is currently on assignment with Compucat Research Pty Ltd. as the documentation manager and a researcher.

Yu Jiao is a postdoctoral researcher at the Oak Ridge National Laboratory, USA. She received a BS in computer science from the Civil Aviation Institute of China in 1997. She received her MS and PhD degrees from the Pennsylvania State University (2002 and 2005, respectively), both in computer science. Her main research interests include software agents, pervasive computing, and secure global information-system design.

Peter Kastner received an MS in electrical and biomedical engineering from the Graz University of Technology, Austria (1995), and an MBA in biotech and pharmacy management from the Danube University in Krems, Austria (2005). During the last 10 years in business, he gained lots of experience in the field of clinical trials, especially in the integration of patient-reported outcomes via mobile clients such as smart phones. Currently, he is head of the telemonitoring working group at the Department of eHealth Systems of the ARC Seibersdorf Research GmbH, and is responsible for the topic of business development. He is a member of the Austrian Computer Society, the Austrian Society of Biomedical Engineering, the Austrian Scientific Society of Telemedicine, and the IEEE.

Alexander Kollmann received an MS in electrical engineering and biomedical engineering from the Graz University of Technology, Austria, in 2003. He currently works and researches as a PhD student at ARC Seibersdorf Research GmbH in the Department of eHealth Systems in Graz. His research interests include medical informatics in general, focused on telemedicine and home monitoring applications, the development of mobile data-acquisition systems, and human-computer interaction in medical application fields. He is a member of the Austrian Society of Biomedical Engineering and a student member of the IEEE.

Massimo Memmola is a visiting professor in the Department of Business Administration, Faculty of Economy, Catholic University of Milan, and a researcher of the Centre for Research and Studies in Healthcare Management (Ce.Ri.S.Ma.S.). He teaches courses in planning and control, management accounting, financial analysis, performance measurement, and Internet strategy. He is involved in several research projects in the area of the development of managerial practices in large companies and health-care organizations. He graduated from the Catholic University in 1999, and received a PhD in business administration from the Catholic University in 2003.

Funda Meric-Bernstam is an associate professor of surgical oncology at the University of Texas, M. D. Anderson Cancer Center in Houston. Dr. Meric-Bernstam's goal is to improve the care of breast-cancer patients via a translational research program that combines basic molecular biology, clinical research, and consumer informatics. She is board certified in general surgery and has completed a fellowship in surgical oncology at the M. D. Anderson Cancer Center. In addition to her research activities, Dr. Meric-Bernstam maintains an active clinical practice that focuses on breast cancer. Dr. Meric-Bernstam is a fellow of the American College of Surgeons.

Masoud Mohammadian has research interests that lie in adaptive self-learning systems, fuzzy logic, genetic algorithms, and neural networks and their applications in industrial, financial, and business problems. His current research concentrates on the application of computational-intelligence techniques for the learning and adaptation of intelligent agents. He has chaired over nine international conferences on computational intelligence and intelligent agents. He has published over 90 research papers in conference proceedings, journals, and books, as well as editing 15 books. Mohammadian has over 14 years of academic experience and has served as program committee member and/or cochair of a large number of national and international conferences. He was the chair of the IEEE ACT Section and was the recipient of awards from the IEEE in the USA, and the Ministry of Commerce in Austria.

Rebecca Montanari received the laurea degree in electronics engineering and a PhD in computer engineering from the University of Bologna, Italy. She is currently an associate professor at the University of Bologna. Her research interests include policy-based networking and systems-service management, mobile-agent systems, security-management mechanisms, and tools in both traditional and mobile systems. She is a member of the IEEE, the IEEE Computer Society, and the Italian Association for Computing.

Niki Panteli is a senior lecturer in information systems at the University of Bath School of Management, UK. She has a PhD from Warwick Business School. Broadly defined, her research lies in the field of information and communication technologies and emergent organizational arrangements. She is presently engaged in research on virtuality, notably, virtual teams and virtual collaborations.

Andreas Pitsillides, an associate professor of computer science at the University of Cyprus, spent 6 years in industry and 18 years in academia. His research interests include fixed and mobile networks, and mobile e-services (e-health). He participates in European Community (EC) and locally funded projects, and has presented invited lectures and short courses at major research centers and conferences. He also extensively consults those in industry. He regularly serves on international conference executive committees (e.g., INFOCOM 2001-2003) and technical committees, and has served as guest editor and as a reviewer. He serves on the editorial board of the *Computer Networks Journal*.

Barbara Pitsillides has a master's in palliative care (2000) and a graduate certificate in palliative care (1996) from Flinder's University, South Australia. She also received a nursing diploma in general and psychiatric nursing from B. G. Alexander College of Nursing, South Africa (1982). She is a specialist palliative home-care nurse with the Association of Cancer Patients and Friends (PA.SY.KAF), Cyprus. Her research interests include home health care for cancer patients, e-health, and health service evaluation.

Thomas E. Potok is the leader of the Applied Software Engineering Research Group at the Oak Ridge National Laboratory, where he manages a staff of 17 researchers. He is the principal investigator on a number intelligent software agent research projects. Prior to this, he worked for 14 years at IBM's Software Solutions Laboratory in Research Triangle Park, North Carolina. Dr. Potok has a BS, MS, and PhD in computer engineering, all from North Carolina State University. He is an adjunct faculty member at the University of Tennessee, and a member of the ACM and the IEEE Computer Society. He has authored numerous publications, has filed seven software patents, and organized several workshops.

Damian Ryan (MBBS FRACGP) is a medical director at The Ambulatory Care Team (TACT) SESIAHS. He is also a conjoint lecturer of the University of NSW, an honorary senior lecturer at the University of Wollongong, and a medical educator at CCCT. Ryan has experience in ambulatory care, emergency medicine, and general practice. He became a fellow of the Royal Australian College of General Practitioners in 1999. Damian was selected for the position of medical director of ambulatory care for northern Illawarra (TACT) in March 1999. He is also involved in medical education both at an undergraduate and postgraduate level. He was appointed as a conjoint lecturer of the University of NSW in 2001, an honorary senior lecturer at the University of Wollongong in 2005, and a medical educator for CCCT in 2005, and he is an examiner for the Royal Australian College of General Practitioners.

George Samaras is an associate professor of computer science at the University of Cyprus. He worked in an applied-research program at IBM's Communications and Networks Center, Research Triangle Park, North Carolina (1990-1993), and taught at the University of North Carolina at Chapel Hill (visiting assistant professor, 1990-1993). He was a member of IBM's International Standards Committees for issues related to distributed transaction processing. His research interests include relational and object-oriented databases, distributed transaction processing, commit protocols and mobile computing, and mobile e-services (e-health).

Jason Sargent is a research assistant in e-health at the School of Economics & Information Systems (SEIS), University of Wollongong, Australia. He has broad knowledge of technical, theoretical, and social issues as they relate to ICTs and IM. He holds first-class honors for the bachelor of information and communication technology degree and a professional membership with the Health Informatics Society of Australia (HISA). Presently, Sargent performs the dual roles of ePOC research assistant and joint project manager. Jason has presented to CEOs (chief executive officers), CIOs (chief information officers), medical directors, and clinicians at conferences in Australia, and authored or coauthored papers for conferences in America, Germany, Italy, China, and Chile. His work has been published in health-informatics and e-commerce texts and online.

Guenter Schreier received a diploma in English and the doctor-of-technology degree in electrical engineering and biomedical engineering from the Graz University of Technology, Austria, in 1991 and 1996 respectively. He also received an MS degree in communications and management from the Danube University in Krems, Austria, in 2003. Since 2000, following positions in research and industry, he has been the head of the e-health-systems department of the biomedical engineering division of the ARC Seibersdorf Research GmbH. He received a number of research awards and prizes, among those, first place in the "Computers in Cardiology Challenges" in 2001 and 2004. His area of expertise includes biosignal processing, medical informatics, telemedicine, and e-health. Dr. Schreier is a member of the board of the Austrian Society of Biomedical Engineering, a member of the Austrian Scientific Society of Telemedicine, and a member of the IEEE EMBS.

Ari Serkkola has been working at the Helsinki University of Technology Lahti Center since 1999 as research director and more recently as head researcher. His scientific background is multidisciplinary, including social sciences, health sciences, and applied data and mobile-phone technology. He has master's and licentiate degrees in social science from Tampere University, and a doctor's degree in health services from Kuopio University. At the Helsinki University of Technology, he has acquired in-depth knowledge of applied research and development in technology and data technology in the 15 research and development (R&D) projects that he has led there. He has written over 70 articles and four books on information technology in health services, on service innovations, on health anthropology, and on regional development. From 1987 to 1993, Serkkola worked at the Helsinki University Institute of Development Studies as a project researcher and an Academy of Finland research fellow. His tasks during that time included health anthropology research in Somalia. Mr. Serkkola has also gained experience in the operation of health organizations from the Health Promotion Center in Helsinki, where he was project director from

1994 to 1999, as well as project-management consultation, which he has done for companies, public organizations, and nongovernmental organizations (NGOs).

Reima Suomi has been a professor of information-systems science at the Turku School of Economics and Business Administration, Finland, since 1994. He is a docent for the Universities of Turku and Oulu, Finland. From 1992 to 1993, he was a *vollamtlicher dozent* in the University of St. Gallen, Switzerland, where he led a research project on business-process reengineering. Currently, he concentrates on topics on the management of telecommunications, including issues such as the management of networks, electronic and mobile commerce, virtual organizations, telework, and competitive advantage through telecommunication-based information systems. Different governance structures applied to the management of IS and that are enabled by IS also belong to his research agenda, as well as the application of information systems in health care. Suomi has together over 300 publications, and has published in journals such as *Information and Management, Information Services and Use, Technology Analysis and Strategic Management*, the *Journal of Strategic Information Systems*, the *Journal of Management History*, and *Behaviour and Information Technology*. For the academic year of 2001 to 2002, he was a senior researcher *varttunut tutkija* for the academy of Finland. With Paul Jackson, he has published the book *Virtual Organization and Workplace Development* published by Routledge, London.

Carrison K. S. Tong graduated from Southampton University in engineering and got his MS in engineering computation from Queen's University of Belfast. He finished his PhD in medical imaging at the Royal Postgraduate Medical School, Imperial College. Dr. Tong was one of the active researchers in the fields of picture archiving and communications systems, medical image processing, virtual-reality photography, robotics, the Monte Carlo technique, large-scale computation, medical rapid prototyping, and mathematical modeling. He has published a number of papers in the *Society of Computer Assisted Radiology and Surgery* and the *European Journal of Nuclear Medicine*. Recently, Dr. Tong has been working on the information security of PACS (Picture Archiving Communication Systems).

Dawn-Marie Turner is president of DM Turner Informatics Consulting Inc. She has 20 years of experience in health and information technology. Her experience in health has included direct patient care, health administration and management, and health programming. Her experience in information technology has included project management, business-process redesign, executive coaching, and change management and transition planning.

Paul Turner is a senior research fellow in the School of Information Systems at the University of Tasmania. Prior to moving to Australia in 1999, Paul was a research fellow at CRID (Computer, Telecommunications and Law Research Institute) in Belgium where he worked on a variety of European Commission contracts in the technology domain. Paul has also worked as an independent information and tele-communications consultant in a number of countries in Europe and was for 3 years editor of the London-based monthly publication *Telecommunications Regulation Review.* Since moving to Tasmania, Turner has been involved in conducting and coordinating IT research at basic, applied, and strategic levels across a range of in-dustry sectors with a particular focus on Internet technologies, computer forensics, and e-health. At the University of Tasmania, he is responsible for research with the Smart Internet Technology Cooperative Research Centre, which involves collabo-ration with 10 other Australian universities, a number of major corporations, and a large number of SMEs (small and medium-sized enterprises). Recently, Turner's team conducted a trial of a wireless handheld care-management system in an aged care facility in Launceston, Tasmania. Turner has been the vice president of the Tasmanian e-Health Association since 2003. He has produced more than 75 peer-reviewed academic publications for conferences, journals, and books.

Marieke W. Verheijden earned her MS in human nutrition and health from Wa-geningen University in 2001. With her research on nutrition counseling in general practice using the transtheoretical model, she earned a PhD degree in 2004. She was subsequently employed by the Netherlands Organization for Applied Scien-tific Research TNO as a researcher and consultant on workplace health promotion. Verheijden's scientific work has concentrated on face-to-face and Web-based ap-plications to promote healthy dietary patterns and physical activity levels. She has authored several scientific and popular publications on health promotion through behaviour-change programs. Verheijden is a member of several national and inter-national scientific communities.

Nilmini Wickramasinghe, PhD, currently researches and teaches in several areas within information systems including knowledge management, e-commerce and m-commerce, and organizational impacts of technology. Over the last 6 years, she has been instrumental in introducing courses on knowledge management at the graduate and undergraduate level in Australian, European, and U.S. academic institutions. She is well published in all these areas with over 50 referred papers, and she regularly presents her work throughout North America as well as in Europe and Australasia. Dr. Wickramasinghe is the U.S. representative of the Health Care Technology Management Association (HCTM), an international organization that focuses on critical health-care issues and the role of technology within the domain of health care. She is the associate director of the Center for Management Medical

Technologies (CMMT), a unique research-oriented center with key research foci on knowledge management, health care, and the confluence of these domains. She also holds an associate-professor position at the Stuart Graduate School of Business, IIT. In addition, she is a research associate at the University of Coventry's BICORE HCKM group in the United Kingdom. Dr. Wickramasinghe is the new editor in chief of the *International Journal of Networking and Virtual Organisations*, published by InderScience.

Eric T. T. Wong received his MS degree in plant engineering from the Loughborough University (UK) and a PhD from the Leicester University (UK). His research interests in the area of information science and technology include medical imaging, picture archiving and communication systems for the health-care industry, trust in virtual enterprises, quality management, and e-learning. He has publications relating to knowledge and information-technology management, the integration of information and communication technology in smart organizations, the impacts of e-commerce on modern organizations, as well as multimedia technology and networking.

Index

PEF value 248
personal
  computer (PC) 11, 233, 277, 301, 358
  digital
    assistant (PDA)
      55, 154, 176, 228, 274–
      278, 286–289, 294, 301–
      306, 324, 332
    digital communication (PDC)
      communication (PDC) 236
  emergency response system (PERS) 152
Personal Health Information Act (PHIA)
      168
physical
  (hardware) encryption layer 330–332
picture archiving and communication sys-
      tem (PACS) applications 303
Picture Archiving Communication Systems
      (PACS) 285
pioneers 74, 80, 86
plain old telephone systems (POTS) 235
portal 30
post-anaesthesia care unit (PACU) 380
PQL query language 201
preventative control 171
primary service provider (PSP) 361
principal stakeholder 3
process
  -flow
    diagram (PFD) 386–387
    node list (PFN) 387, 392
project-management office (PMO) 52
PSs 88, 92, 93, 97
public-key infrastructure (PKI) 172

Q

query integration system (QIS) 201

R

radio-frequency identification
      (RFID) 150, 252, 316–
      318, 320, 326, 330–331
  passive data 321
  reader 324
  tag 322–323, 332–333

  transceiver 328
radiology information systems (RIS) 41
RC4 symmetric key 171
related term (RT) 211
relational database 201
relation definitions 208

S

scanner 402
SCI 75
Science Panel on Interactive Communica-
      tion and Health (SciPICH) 120
SciPICH 135
SDLC 47–48, 57
search engine 185
second-generation (2G) wireless networks
      236
secure socket layer (SSL) 249
self-management of chronic illness 260
semantic-distance metric (SDM) 202
semi-active
  RFID tag 322
  tag 322
semi-passive RFID tag 322
shared
  -key authentication 171
  multi-megabit data services (SMDS) 302
short-message service (SMS)
      151, 234, 237–239, 243–
      248, 250, 342, 344, 346
single booking centre (SBC) 65
SIRC 87, 92, 97
social cognitive theory 120
software-development kit (SDK) 252
South Eastern Sydney and Illawarra Area
      Health Service (SESIAHS) 285
SSM 200
stages-of-change model 120
stakeholder 2, 103, 113–114, 114
Stanford model 262
Strategic Health Authority (StHA) 361
structural query language (SQL) 301, 306
subject categories (SCs) 211
summary-schemas
  hierarchy 205
  model 202